Lecture Notes in Computer Science 9897

Commenced Publication in 1973
Founding and Former Series Editors:
Gerhard Goos, Juris Hartmanis, and Jan van Leeuwen

More information about this series at http://www.springer.com/series/7409

Henrik Boström · Arno Knobbe
Carlos Soares · Panagiotis Papapetrou (Eds.)

Advances in Intelligent Data Analysis XV

15th International Symposium, IDA 2016
Stockholm, Sweden, October 13–15, 2016
Proceedings

Springer

Editors
Henrik Boström
Stockholm University
Stockholm
Sweden

Arno Knobbe
Leiden University
Leiden
The Netherlands

Carlos Soares
University of Porto
Porto
Portugal

Panagiotis Papapetrou
Stockholm University
Stockholm
Sweden

ISSN 0302-9743 ISSN 1611-3349 (electronic)
Lecture Notes in Computer Science
ISBN 978-3-319-46348-3 ISBN 978-3-319-46349-0 (eBook)
DOI 10.1007/978-3-319-46349-0

Library of Congress Control Number: 2016950907

LNCS Sublibrary: SL3 – Information Systems and Applications, incl. Internet/Web, and HCI

Printed on acid-free paper

This Springer imprint is published by Springer Nature
The registered company is Springer International Publishing AG
The registered company address is: Gewerbestrasse 11, 6330 Cham, Switzerland

Preface

We are proud to present the proceedings of the 15th International Symposium on Intelligent Data Analysis, which took place during October 13–15 in Stockholm, Sweden. The series started in 1995 and was held biennially until 2009. In 2010, the symposium re-focused to support papers that go beyond established technology and offer genuinely novel and game-changing ideas, while not always being as fully realized as papers submitted to other conferences.

IDA 2016 continued this approach and sought first-look papers that might elsewhere be considered preliminary, but contain potentially high-impact research. In addition, for the first time this year, IDA introduced an industrial challenge track. For the industrial challenge, researchers were invited to participate in a machine learning prediction challenge, where the task was to devise a prediction model for judging whether or not a vehicle faces imminent failure of a specific component, exploiting data collected from heavy Scania trucks in everyday usage.

The IDA symposium is open to all kinds of modelling and analysis methods, irrespective of discipline. It is an interdisciplinary meeting that seeks abstractions that cut across domains. IDA solicits papers on all aspects of intelligent data analysis, including papers on intelligent support for modelling and analyzing data from complex, dynamical systems.

Intelligent support for data analysis goes beyond the usual algorithmic offerings in the literature. Papers about established technology were only accepted if the technology was embedded in intelligent data analysis systems, or was applied in novel ways to analyzing and/or modelling complex systems. The conventional reviewing process, which favors incremental advances on established work, can discourage the kinds of papers that were selected for IDA 2016. The reviewing process addressed this issue explicitly: referees evaluated papers against the stated goals of the symposium, and any paper for which at least one program committee advisor wrote an informed, thoughtful, positive review was accepted, irrespective of other reviews. Indeed, this had a notable impact on what papers were included in the program.

We were pleased to see a very strong program. We received 75 submissions by 198 authors from 30 different countries, out of which 15 were accepted as regular papers, 12 as regular poster papers, and 4 as short papers (industrial challenge papers). All submissions were reviewed by three PC members and one PC advisor.

In addition, we were happy to accept two abstracts to the IDA horizon track:

- "Usable analytics at societal scale", by Daniel Gillblad
- "Cognitive Computing for the Automated Society", by Devdatt Dubhashi

We were honored to have the following distinguished invited speakers at IDA 2016:

- Samuel Kaski, Aalto University and University of Helsinki, Finland; on the topic "Bayesian Factorization of Multiple Data Sources"

- Sihem Amer Yahia, CNRS at LIG, Grenoble, France; on the topic "Worker-Centricity Could Be Today's Disruptive Innovation in Crowdsourcing"
- Foster Provost, New York University, USA; on the topic "The Predictive Power of Massive Data about Our Fine-Grained Behavior".

The conference was held at the Department of Computer and Systems Sciences of Stockholm University, Sweden.

We wish to express our gratitude to all authors of submitted papers for their intellectual contributions; to the program committee members and advisors and additional reviewers for their effort in reviewing, discussing, and commenting on the submitted papers, and to the members of the IDA steering committee for their ongoing guidance and support. We thank Isak Karlsson for running the conference website. Special thanks go to the industrial challenge chair, Tony Lindgren, for handling the submission and reviewing process of the industrial challenge papers. We gratefully acknowledge those who were involved in the local organization of the symposium: Lars Asker, Isak Karlsson, Jing Zhao, and Ram Gurung. We are grateful to our sponsors: Stockholm University, Scania AB, Vetenskapsrådet, Springer, The Artificial Intelligence Journal, and SERSC. We are especially indebted to KNIME, who funded the IDA Frontier Prize for the most visionary contribution presenting a novel and surprising approach to data analysis in the understanding of complex systems.

July 2016

Henrik Boström
Arno Knobbe
Carlos Soares
Panagiotis Papapetrou

Organization

General Chair

Panagiotis Papapetrou   Stockholm University, Sweden

Program Committee Chairs

Henrik Boström   Stockholm University, Sweden
Arno Knobbe   Leiden University, The Netherlands
Carlos Soares   University of Porto, Portugal

Local Chair

Lars Asker   Stockholm University, Sweden

Industrial Challenge Chair

Tony Lindgren   Stockholm University, Sweden

Sponsorship and Publicity Chairs

Lars Asker   Stockholm University, Sweden
Panagiotis Papapetrou   Stockholm University, Sweden

Advisory Chairs

Joost Kok   Leiden University, The Netherlands
Jaakko Hollmen   Aalto University, Finland
Matthijs van Leeuween   Leiden University, The Netherlands

Webmaster

Isak Karlsson   Stockholm University, Sweden

Program Committee Advisors

Stephen Swift   Brunel University, UK
Allan Tucker   Brunel University, UK
Michael Berthold   University of Konstanz, Germany
Matthijs van Leeuwen   Leiden University, The Netherlands
Elisa Fromont   University of Saint-Etienne, France

Arno Siebes	Universiteit Utrecht, The Netherlands
Hannu Toivonen	University of Helsinki, Finland
Nada Lavrac	Jozef Stefan Institute, Slovenia
Xiaohui Liu	Brunel University, UK
Elizabeth Bradley	University of Colorado, USA
Hendrik Blockeel	K.U. Leuven, Belgium
Frank Klawonn	Ostfalia University of Applied Sciences, Germany
Jaakko Hollmen	Aalto University School of Science, Finland
Tijl De Bie	Ghent University, Belgium

Program Committee

Wouter Duivesteijn	Ghent University, Belgium
Mykola Pechenizkiy	Eindhoven University of Technology, The Netherlands
Jefrey Lijffijt	Ghent University, Belgium
Lubos Popelinsky	Masaryk University, Czech Republic
Alexandra Poulovassilis	Birkbeck, University of London, UK
Saso Dzeroski	Jozef Stefan Institute, Slovenia
Christine Solnon	LIRIS CNRS UMR 5205/INSA Lyon, France
Nicos Pavlidis	Lancaster University, UK
Marc Plantevit	LIRIS - Université Claude Bernard Lyon 1, France
Maguelonne Teisseir	Cemagref - UMR Tetis, France
Albrecht Zimmermann	University of Normandy, France
George Magoulas	Birkbeck, University of London, UK
Ruggero G. Pensa	University of Turin, Italy
Andre Carvalho	USP, Brazil
Maarten Van Someren	University of Amsterdam, The Netherlands
Frank Takes	Leiden University, The Netherlands
Ricardo Cachucho	Leiden University, The Netherlands
Antonio Salmeron	University of Almeria, Spain
Wannes Meert	KU Leuven, Belgium
Joaquin Vanschoren	TU Eindhoven, The Netherlands
Indre Zliobaite	Aalto University, Finland
Martin Atzmueller	University of Kassel, Germany
Rudolf Kruse	University of Magdeburg, Germany
Paulo Cortez	University of Minho, Portugal
Brett Drury	LIAAD – INESC, Portugal
Jan N. van Rijn	Leiden University, The Netherlands
Vera Oliveira	University of Porto, Portugal
Fabrizio Angiulli	University of Calabria, Italy
Mohamed Nadif	University of Paris Descartes, France
Kaustubh Patil	MIT, USA
Nuno Escudeiro	ISEP - Instituto Superior de Engenharia do Porto, Portugal
Ana Aguiar	FEUP, Portugal
Roberta Siciliano	University of Naples Federico II, Italy

Kenny Gruchalla   NREL/CU-Boulder, USA
Miguel A. Prada   Universidad de Leon, Spain
Myra Spilliopoulou   Otto-von-Guericke-University of Magdeburg, Germany
Paula Brito   University of Porto, Portugal
Andreas Nuernberger   Otto-von-Guericke University of Magdeburg, Germany
Giovanni Montana   Imperial College, UK
Ricard Gavalda   Universitat Politècnica de Catalunya, Spain
Peter van der Putten   Leiden University and Pegasystems, The Netherlands
Loic Cerf   Universidade Federal de Minas Gerais, Brazil
Bernard De Baets   Ghent University, Belgium
Jose-Maria Pena   Universidad Politècnica de Madrid, Spain
Anton Dries   KU Leuven, Belgium
Johannes Furnkranz   TU Darmstadt, Germany
Alipio M. Jorge   University of Porto, Portugal
Antti Ukkonen   Finnish Institute of Occupational Health, Finland
Thibault Sellam   CWI, The Netherlands
Fabrizio Riguzzi   University of Ferrara, Italy
Gustavo Batista   University of Sao Paulo, Brazil
Ulf Brefeld   Leuphana University of Lüneburg, Germany
Niklas Lavesson   Blekinge Institute of Technology, Sweden
Jose A. Lozano   The University of the Basque Country, Spain
Jose Del Campo   Universidad de Málaga, Spain
Frank Hoppner   Ostfalia University of Applied Sciences, Germany
Ingrid Fischer   University of Konstanz, Germany
Ad Feelders   Universiteit Utrecht, The Netherlands
Wojtek Kowalczyk   Leiden University, The Netherlands
Bruno Cremilleux   Université de Caen, France
Maria Bielikova   Slovak University of Technology in Bratislava,
                         Slovakia
Javier Gonzalez   University of Sheffield, UK
Harm de Vries   Leiden University, The Netherlands
Irena Koprinska   The University of Sydney, Australia
Vitor Santos Costa   Universidade do Porto, Portugal
Jesse Read   Aalto University, Espoo, Finland
Cor Veenman   Netherlands Forensic Institute, The Netherlands
Peter Flach   University of Bristol, UK
Adolfo Martinez-Uso   Technical University of Valencia, Spain
Lawrence Hall   University of South Florida, USA
François Portet   University of Grenoble Alpes, France
Jose Balcazar   Universitat Politècnica de Catalunya, Spain
Tias Guns   KU Leuven, Belgium
Douglas Fisher   Vanderbilt University, USA
Norbert Jankowski   Nicolaus Copernicus University, Poland
Eirini Ntoutsi   Leibniz University Hanover, Germany

Sponsors and Supporters

- SCANIA
- The Swedish Research Council
- KNIME
- Springer
- The Artificial Intelligence Journal
- Stockholm University
- SERSC

Contents

DSCo-NG: A Practical Language Modeling Approach for Time
Series Classification 1
Daoyuan Li, Tegawendé F. Bissyandé, Jacques Klein,
and Yves Le Traon

Ranking Accuracy for Logistic-GEE Models 14
Nasser Davarzani, Ralf Peeters, Evgueni Smirnov, Joël Karel,
and Hans-Peter Brunner-La Rocca

The Morality Machine: Tracking Moral Values in Tweets 26
Livia Teernstra, Peter van der Putten, Liesbeth Noordegraaf-Eelens,
and Fons Verbeek

A Hybrid Approach for Probabilistic Relational Models Structure Learning ... 38
Mouna Ben Ishak, Philippe Leray, and Nahla Ben Amor

On the Impact of Data Set Size in Transfer Learning Using Deep
Neural Networks...................................... 50
Deepak Soekhoe, Peter van der Putten, and Aske Plaat

Obtaining Shape Descriptors from a Concave Hull-Based
Clustering Algorithm.................................... 61
Christian Braune, Marco Dankel, and Rudolf Kruse

Visual Perception of Discriminative Landmarks in Classified Time Series ... 73
Tobias Sobek and Frank Höppner

Spotting the Diffusion of New Psychoactive Substances over the Internet ... 86
Fabio Del Vigna, Marco Avvenuti, Clara Bacciu, Paolo Deluca,
Marinella Petrocchi, Andrea Marchetti, and Maurizio Tesconi

Feature Selection Issues in Long-Term Travel Time Prediction 98
Syed Murtaza Hassan, Luis Moreira-Matias, Jihed Khiari,
and Oded Cats

A Mean-Field Variational Bayesian Approach to Detecting Overlapping
Communities with Inner Roles Using Poisson Link Generation 110
Gianni Costa and Riccardo Ortale

Online Semi-supervised Learning for Multi-target Regression in Data
Streams Using AMRules 123
Ricardo Sousa and João Gama

A Toolkit for Analysis of Deep Learning Experiments............... 134
  Jim O'Donoghue and Mark Roantree

The Optimistic Method for Model Estimation..................... 146
  James Brofos, Rui Shu, and Frank Zhang

Does Feature Selection Improve Classification? A Large Scale
Experiment in OpenML...................................... 158
  Martijn J. Post, Peter van der Putten, and Jan N. van Rijn

Learning from the News: Predicting Entity Popularity on Twitter 171
  Pedro Saleiro and Carlos Soares

Multi-scale Kernel PCA and Its Application to Curvelet-Based Feature
Extraction for Mammographic Mass Characterization................. 183
  Sami Dhahbi, Walid Barhoumi, and Ezzeddine Zagrouba

Weakly-Supervised Symptom Recognition for Rare Diseases
in Biomedical Text...................................... 192
  Pierre Holat, Nadi Tomeh, Thierry Charnois, Delphine Battistelli,
  Marie-Christine Jaulent, and Jean-Philippe Métivier

Estimating Sequence Similarity from Read Sets for Clustering
Sequencing Data... 204
  Petr Ryšavý and Filip Železný

Widened Learning of Bayesian Network Classifiers.................. 215
  Oliver R. Sampson and Michael R. Berthold

Vote Buying Detection via Independent Component Analysis............ 226
  Antonio Neme and Omar Neme

Unsupervised Relation Extraction in Specialized Corpora
Using Sequence Mining.................................... 237
  Kata Gábor, Haïfa Zargayouna, Isabelle Tellier, Davide Buscaldi,
  and Thierry Charnois

A Framework for Interpolating Scattered Data Using Space-Filling Curves... 249
  David J. Weston

Privacy-Awareness of Distributed Data Clustering Algorithms Revisited 261
  Josenildo C. da Silva, Matthias Klusch, and Stefano Lodi

Bi-stochastic Matrix Approximation Framework for Data Co-clustering 273
  Lazhar Labiod and Mohamed Nadif

Sequential Cost-Sensitive Feature Acquisition....................... 284
  Gabriella Contardo, Ludovic Denoyer, and Thierry Artières

Explainable and Efficient Link Prediction in Real-World Network Data   295
  Jesper E. van Engelen, Hanjo D. Boekhout, and Frank W. Takes

DGRMiner: Anomaly Detection and Explanation in Dynamic Graphs   308
  Karel Vaculík and Luboš Popelínský

Similarity Based Hierarchical Clustering with an Application
to Text Collections   320
  Julien Ah-Pine and Xinyu Wang

Determining Data Relevance Using Semantic Types and Graphical
Interpretation Cues   332
  Eduardo Haruo Kamioka, André Freitas, Frederico Caroli,
  and Siegfried Handschuh

A First Step Toward Quantifying the Climate's Information Production
over the Last 68,000 Years.   343
  Joshua Garland, Tyler R. Jones, Elizabeth Bradley, Ryan G. James,
  and James W.C. White

HAUCA Curves for the Evaluation of Biomarker Pilot Studies with Small
Sample Sizes and Large Numbers of Features.   356
  Frank Klawonn, Junxi Wang, Ina Koch, Jörg Eberhard,
  and Mohamed Omar

Stability Evaluation of Event Detection Techniques for Twitter.   368
  Andreas Weiler, Joeran Beel, Bela Gipp, and Michael Grossniklaus

IDA 2016 Industrial Challenge: Using Machine Learning
for Predicting Failures   381
  Camila Ferreira Costa and Mario A. Nascimento

An Optimized k-NN Approach for Classification on Imbalanced Datasets
with Missing Data.   387
  Ezgi Can Ozan, Ekaterina Riabchenko, Serkan Kiranyaz,
  and Moncef Gabbouj

Combining Boosted Trees with Metafeature Engineering
for Predictive Maintenance.   393
  Vítor Cerqueira, Fábio Pinto, Claudio Sá, and Carlos Soares

Prediction of Failures in the Air Pressure System of Scania Trucks
Using a Random Forest and Feature Engineering   398
  Christopher Gondek, Daniel Hafner, and Oliver R. Sampson

Author Index   403

DSCo-NG: A Practical Language Modeling Approach for Time Series Classification

Daoyuan Li$^{(\boxtimes)}$, Tegawendé F. Bissyandé, Jacques Klein, and Yves Le Traon

Interdisciplinary Centre for Security, Reliability and Trust (SnT),
University of Luxembourg, Luxembourg, Luxembourg
{daoyuan.li,tegawende.bissyande,jacques.klein,yves.letraon}@uni.lu

Abstract. The abundance of time series data in various domains and their high dimensionality characteristic are challenging for harvesting useful information from them. To tackle storage and processing challenges, compression-based techniques have been proposed. Our previous work, Domain Series Corpus (DSCo), compresses time series into symbolic strings and takes advantage of language modeling techniques to extract from the training set knowledge about different classes. However, this approach was flawed in practice due to its excessive memory usage and the need for *a priori* knowledge about the dataset. In this paper we propose DSCo-NG, which reduces DSCo's complexity and offers an efficient (linear time complexity and low memory footprint), accurate (performance comparable to approaches working on uncompressed data) and generic (so that it can be applied to various domains) approach for time series classification. Our confidence is backed with extensive experimental evaluation against publicly accessible datasets, which also offers insights on when DSCo-NG can be a better choice than others.

1 Introduction

Time series data usually refer to temporally or spatially ordered data, which are abundant in numerous domains including health-care, finance, energy and industry applications. Besides their abundance, time series data are becoming increasingly challenging to efficiently store, process and mine useful information due to their high dimensionality characteristics. In order to tackle these challenges, researchers have proposed many approaches to model time series more efficiently. Compression-based techniques are especially promising and have been adopted in many recent studies, including dimensionality reduction [9,11,21] and numerosity reduction [24]. Symbolic Aggregate approXimation (SAX) [15] is an approach that is capable of both dimensionality and numerosity reduction. Among all time series data mining tasks, time series classification (TSC) has received great interests from researchers and practitioners thanks to its wide application scenarios including speech recognition, medical diagnosis, etc.

Our previous work, Domain Series Corpus (DSCo) [13] for TSC, takes advantage of SAX to compress real-valued time series data into text strings and builds per-class language models as a means of extracting representative patterns in the

© Springer International Publishing AG 2016
H. Boström et al. (Eds.): IDA 2016, LNCS 9897, pp. 1–13, 2016.
DOI: 10.1007/978-3-319-46349-0_1

training phase. To classify unlabeled samples, we compute the fitness of each symbolized sample against all per-class models by finding the best way to segment this sample and choose the class represented by the model with the best fitness score. We also prove that although DSCo works with approximated data, it can perform similarly to approaches that work with original uncompressed numeric data. One issue with DSCo, however, lies in its excessive memory usage when calculating the fitness score of one sample against language models, which makes it impractical for real-world applications.

In this paper, we set to improve DSCo's time and space complexity and propose a next generation of DSCo: DSCo-NG. We follow our initial intuition that time series data are similar to *sentences from different languages or dialects*, but apply a more efficient approach to find nuances of difference from these *languages*. Specifically, unlike in DSCo where we try to find the best way to recursively segment time series, DSCo-NG breaks time series into smaller segments of the same size, and this simplification of the classification process also leads to simplified language model inference in the training phase. Overall, the contributions of this paper are summarized as follows:

- We propose a new practical language modeling-based approach for time series classification, which has a linear time complexity and small memory footprint. Previously DSCo works optimally on a High Performance Computing (HPC) platform, e.g. ULHPC [20], while DSCo-NG can virtually run efficiently on any personal computers thanks to its low complexity.
- We have tested our approach extensively on an open archive which contains datasets from various domains, demonstrating by comparison with state-of-the-art approaches and first generation DSCo that DSCo-NG is both performant and efficient.
- We investigate the performance of DSCo-NG by scrutinizing the characteristics of datasets and provide insights in application scenarios when DSCo-NG could be a better choice than other approaches.
- We offer a new perspective for TSC: traditional TSC approaches compare instances against instances, which can be computationally inefficient when the training dataset is large, while our approach aggregates training sets into models and compares the fitness of instances to such models, making comparisons more efficient.

The remainder of this paper is organized as follows. Section 2 provides the necessary background information on time series classification as well as our first trial of using language modeling for TSC. Section 3 briefly surveys related research work to ours. Section 4 presents the details of our new improvements, while experiments and evaluation results are described in Sect. 5. Section 6 concludes the paper with directions for future work.

2 Background

In this section, we briefly introduce time series, TSC, SAX and DSCo. For a more detailed information on DSCo, the readers are encouraged to refer to [13].

Traditionally, time series data refers to temporally ordered data, e.g., data sequences that are related to time. However, data mining community [7] embraces a broader definition, relaxing the time aspect and incorporating any ordered sequences. For instance, images may also be transformed into time series representation [25]. In this paper, we define a time series $T = t_0, t_1, ..., t_{n-1}$, where t_i ($0 \leq i \leq n-1$) is a real-valued number and that T has a length of n, i.e., $|T| = n$.

TSC is a common category of tasks that involves learning from a training dataset and applying the learned knowledge to classify instances from a testing dataset, where instance classes or labels are often unknown or purposefully hidden. TSC tasks are commonly found in various application domains such as image and speech recognition, medical analysis, industrial automation, etc. Many techniques have been proposed for TSC, including k-Nearest Neighbors, shapelets [25], and bag-of-features [2]. In practice, the Nearest Neighbor (1NN) approach has been proven to work very well [1], especially when combined with a good time series distance metrics such as Dynamic Time Warping (DTW) [4] and Time Warp Edit Distance (TWED) [16,19].

In the literature of time series data mining, real-valued data are sometimes transformed into symbolic representations, so as to potentially benefit from the enormous wealth of data structures and algorithms made available by the text processing and bio-informatics communities. Besides, symbolic representation approaches make it easier to solve problems in a streamed manner [15]. Finally, many algorithms target discrete data represented by strings over floating point numbers. Symbolic Aggregate approXimation (SAX) [15] is one such technique that is popular among the community [14,17]. It can perform both dimensionality reduction and numeriosity reduction on time series and transforms real valued time series data into a string of alphabets.

Once time series data are compressed into strings, DSCo [13] builds per-class unigram and bigram language models which contains artificial words and phrases that have been extracted from the training dataset. Figure 1 illustrates how DSCo works. When classifying unlabeled instances, DSCo calculates the

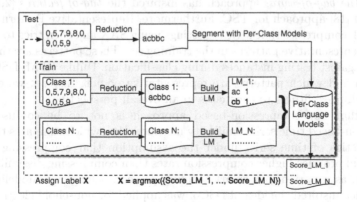

Fig. 1. Flow chart illustration of DSCo's training (in yellow) and testing process. (Color figure online)

fitness scores of one instance against all per-class language models, and the fittest model's class label will be assigned to this instance. DSCo builds on the simply intuition that time series signals are comparable to sentences and phrases from natural languages and dialects in the real-world: each dialect have their unique words and patterns, which is similar to distinguishable features in time series. In DSCo, we try to recursively segment an instance using a Viterbi algorithm until we find the best way to divide such instance with a given language model. Due to this intensive process, DSCo has an almost linear time complexity and space complexity of $O(m^2 n^2)$, where m is the number of instances in the training set and n is the length of time series.

3   Related Work

Due to TSC's wide application scenarios, there are a plethora of algorithms made available by the research community. An extensive review of time series mining has been done by Fu [7]. Here we only survey the works that are closely related to ours due to space limitation. Since DSCo-NG is a compression-based approach, we introduce related approaches that also takes advantage of time series compression techniques.

There are basically two methods for compressing time series, i.e., dimensionality reduction that works on the time axis and numerosity reduction that works on the value axis. Dimensionality reduction mechanisms include Piecewise Linear Representation (PLR) [8], Piecewise Aggregate Approximation (PAA) [9], and methods that keeps only perceptually important points (PIP) [6]. Our previous work [11] takes advantage of Discrete Wavelet Transform for dimensionality reduction. On the value axis, Xi et al. [24] have proposed using numerosity reduction to speed up TSC, and Lin et al. have proposed SAX [15], which converts real-valued data into a symbolic form. Note that it is possible to apply both dimensionality reduction and numerosity reduction using SAX.

Symbolic representation of time series has opened a new avenue for TSC since it makes it possible to borrow paradigms from the text mining community. For instance, the *bag-of-words* approach has inspired the *bag-of-features* [2,22] and SAX-VSM [18] approach for TSC. Furthermore, Representative Pattern Mining (RPM) [23] compresses time series to strings using SAX and then tries to identify the most representative patterns in the training set. These patterns are then used to match against testing instances during classification. Unlike RPM, DSCo does not try to find which patterns are representative or not. Instead, we evaluate testing instances' fitness to each class in an overall perspective.

Note that our compression-based approach is not to be confused with compression-based time series similarity measures [10], which compares the compression ratios of time series under the assumption that compressing similar series would produce higher compression rates than compressing dissimilar ones.

Finally, as a part of our smart building project, a language modeling approach, which inspired the idea of DSCo, was applied for household electric appliance profiling [12], where language models are used to classify and maintain profiles of different appliances.

4 Next Generation Domain Series Corpus for TSC

Since our approach is based on a simple intuition that if we abstract time series classes to *languages*, these *languages* will be descriptive so that it is able to differentiate instances or *sentences*. In practice, we firstly harvest descriptive language models of different patterns from a training corpus. Later in the classification phase, these language models are used to find out which instances are likely to be written in a corresponding language.

The main complexity of original DSCo lies in the classification process, where testing instances are recursively segmented in order to produce the best segmentation result using a language model, in DSCo-NG we try to break the testing instances into sub-sequences of the same length. Then we calculate the product of bigram probability of these sub-sequences. This scheme is inspired by the intuition that when using a sliding widow of size w to iterate over the training set, all possible unigrams and bigrams are already captured within the language model of a specific class. As a result, there is no need to use a sliding window of variable length during the classification process, thus reducing the classification complexity. To better illustrate how DSCo-NG works, we detail it in three steps in the subsections below.

4.1 Compressing Time Series into Texts

There are potentially many approaches that can compress time series data into texts. For instance, one may think of creating a mapping from range of values to alphabets. However, for the benefit of reusing existing mature techniques, we have leveraged SAX for this task. Recall that SAX is capable of both dimensionality and numerosity reduction, as long as the required length and cardinality parameters are specified. Previously with DSCo we arbitrarily reduced the dimensionality of all long (of size larger than 100) time series to 100, for the sake of computational efficiency. Here for DSCo-NG we do not conduct dimensionality reduction since DSCo-NG is efficient enough, allowing us to remove one parameter (or heuristics-based decision) from our processing pipeline. However, as our previous study [11] suggests, conducting dimensionality reduction can potentially increase overall classification accuracy.

4.2 Extracting Language Models

Once time series are compressed to texts, a language model can be extracted to summarize each time series class. Since the text representation does not have word boundaries, we need to create artificial words. To that end, we employ a sliding window mechanism that generates such words. In order to facilitate reader understanding, we reproduce the procedure from [13] in Algorithm 1. This algorithm collects all possible sub-strings of length w within a string, so that no descriptive segment is left uncaptured from the original time series. For example, we can break string `abcde` into the following 2-alphabet words: *ExtractWords(abcde, 2)* produces an output of $[ab, bc, cd, de]$.

Algorithm 1. Extract *words* from a string (S) using a sliding window (of length w).

1: **procedure** EXTRACTWORDS(S, w)
2:       *words* ← ∅
3:       **for** $i \leftarrow 0, GetLength(S) - w + 1$ **do**
4:           *word* ← $SubString(S, i, w)$ ▷ Sub-string of size w
5:           *words* ← *words* ∪ {*word*}
6:       **return** *words*

Next, we build ngram language models for each time series class in our training set, which is illustrated in Algorithm 2. Unlike DSCo that requires a minimum word length and a maximum word length to capture words, here we use a single length w. Note that the probability of ngrams are calculated independently, since different classes may have different number of training instances.

Algorithm 2. Build language models (LMs) from a list (SL) of ($string, label$) pairs.

1: **procedure** BUILDLM(SL, w)
2:       $LMs \leftarrow \varnothing$
3:       **for all** ($string, label$) ∈ SL **do**
4:           **if** $NGrams_{label} \notin LMs$ **then**
5:               $NGrams_{label} \leftarrow \varnothing$
6:           *words* ← $ExtractWords(string, w)$)
7:           **for all** $ngram$ ∈ $GetNGrams(words)$ **do**
8:               InsertOrIncreaseFreq($NGrams_{label}, ngram$)
9:           $LMs \leftarrow LMs \cup NGrams_{label}$
10:       ConvertFreqToProbability(LMs)
11:       **return** LMs

4.3   Classifying Unlabeled Instances

As mentioned earlier, classification in DSCo-NG is performed by checking which language model is the best fit for the tested sample. Specifically, we compare the sample's fitness scores to each model, which is calculated following the ngram statistical language model probability as shown in Eq. 1.

$$P(w_1, ..., w_m) = \prod_{i=1}^{m} P(w_i | w_1, ..., w_{i-1}) \approx \prod_{i=1}^{m} P(w_i | w_{i-(n-1)}, ..., w_{i-1}) \qquad (1)$$

In practice, bigrams ($n = 2$) and trigrams ($n = 3$) are most prominent [3]. We have opted for the bigram model due to its simplicity for both the language model extraction process and fitness score calculation, which is approximated as

$P(w_1, ..., w_m) \approx \prod_{i=1}^{m} P(w_i|w_{i-1})$. During the classification process, we need to break time series strings into words. Unlike original DSCo which breaks *sentences* into variable sized *words*, here we adopt the same sliding window size w as the uniform word length. As we shall show later, this simplified process yields similar classification accuracy but greatly reduces the complexity compared with DSCo.

4.4 Time and Space Complexity

During the preprocessing phase, SAX has a linear time and space complexity when transforming real-valued time series into text representation. When extracting language models in the training phase, each training sample is went through once and models are stored to external storage, resulting an $O(n)$ time and space complexity. Finally, the classification process go through testing samples constant times with language models loaded from external storage, yielding linear time complexity. Language models loaded to memory has a theoretic complexity of $O(\alpha^w)$ where α is the alphabet size used when using SAX to compress real-valued data, and w is the length of *artificial words*. In practice, language models seldom exceed a few megabytes, due to the fact that time series in a domain have a very limited number of *words*.

DSCo-NG's real advantage comes when the training set is large. Given a training set of m_1 time series of length n, when classifying a testing set of m_2 instances, traditional kNN approaches have to conduct $m_1 \times m_2$ pairwise comparisons. Even when using a linear similarity measure such as Euclidean distance, the overall time complexity goes up to $O(m_1 m_2 n)$. On the other hand, DSCo-NG would only have a computational complexity of $O(cm_2 n)$ where c is the number of classes and $c \ll m_1$, making DSCo-NG a magnitude faster than kNN. And this is indeed great improvement even compared with DSCo which has a time complexity of $O(cm_2 n w^2)$ and space complexity of $O(m_1^2 n^2)$.

5 Experimental Evaluation

In order to evaluate performance of our new approach, we have implemented DSCo-NG and tested it on an open dataset archive. To facilitate reproducibility, we have open sourced our implementation with full documentation and tutorials on GitHub[1]. We opt for testing with the UCR Time Series Classification Archive [5] for three reasons: (1) this archive has a large number of publicly accessible datasets; (2) these datasets are from a wide range of domains, from environmental monitoring to medical diagnosis; (3) it comes with precomputed classification accuracy rates for DTW-based 1NN, which is the most widely used similarity measure in the research community and has become the *de facto* state-of-the-art benchmark for TSC. In the experiments below we consider 39 datasets from the *Newly Added Datasets* sub-archive because of its uniform file format and structure.

---

[1] Repository is available at https://github.com/serval-snt-uni-lu/dsco.

5.1 Implementation and Setup

In theory, when calculating ngram probabilities, the larger n is, the more accurate these probabilities will be. However, in practice it is seldom the case, due to the lack of training data and the rise of complexity when n becomes larger. As a result, our implementation considers the bigram model with unigram fallback as a trade-off between efficiency and accuracy. Note that falling back to unigrams may not always work, when specific unigrams are missing from the training set. In this case it is necessary to employ a penalty mechanism to offset the influence of such unigrams. From our experience, these missing unigrams' probability could be set as a constant of low probability value, so that the missing probabilities do not overwhelm the existing ones and lead to inaccurate classification.

5.2 Parameter Optimization

Ideally, time series mining approaches should have as few parameters as possible, even parameter-free, so as to avoid presumption on data [10]. In reality it is extremely difficult to achieve. For instance, even the popular DTW distance requires a warping window size to be set in order to produce optimal results. In DSCo-NG, we essentially have two parameters: the cardinality of SAX alphabet when compressing real-valued data to text strings and the sliding window size or length of artificial words. Normally, approaches based on SAX have to specify both the cardinality and a PAA size to which time series are reduced. Since DSCo-NG does not necessarily need dimensionality reduction, we only need to fix for a suitable cardinality, i.e., a good alphabet size that keeps sufficient information during time series compression. To that end, we try to reduce time series using different cardinality values from 3 to 20, which is range supported by major SAX implementations. For length of artificial words, we also fix a range to 2 to 20 in order to avoid extremely long words, in order to limit the size of language models.

Figure 2 presents the classification accuracy from four datasets across different domains. As shown, although these four datasets have different characteristics in terms of training dataset size, time series length and number of classes, there is a clear trend when high classification accuracy is achieved. That is, generally good accuracy is achieved with small to medium SAX alphabet size and the alphabet size has more impact than the word length (imagine projecting the 3D plots to the 2D plane defined by the alphabet size and accuracy axis). This is extremely useful to narrow down the parameter space, even though in fact our parameter space is already small ($18 * 19 = 342$ combinations in total). Note that there are other methods available for finding the optimal parameters. For instance, in [23] the authors have adopted an algorithm named DIRECT. Thanks to the small parameter space and efficiency of DSCo-NG we employ a brute force approach for finding the best parameters for different datasets. Naturally, there is not a single parameter setting that guarantees good performance, since different datasets can be totally different in number of classes, size, time series lengths and variation amplitude. However, it is indeed possible to set the same parameters for datasets with similar characteristics.

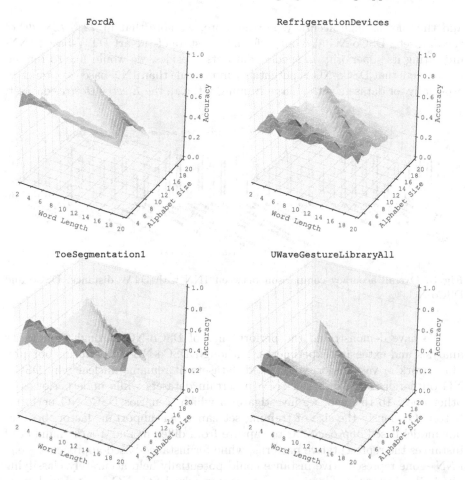

Fig. 2. 3D surface plots of classification accuracy with different parameters, darker blue indicates higher accuracy. (Color figure online)

5.3   Comparison of Classification Performance

Now that we have fixed the parameters for DSCo-NG, here we set to compare its performance with its predecessor DSCo and the state-of-the-art approach DTW-based 1NN classifier. As an intuitive and simple benchmark, we only consider the classification accuracy here because the accuracy results are available in the UCR archive and it is in general what the time series classification community compare with. Figure 3 presents the classification results. It clearly demonstrates that DSCo-NG outperforms its predecessor. In fact, in *90 % (35/39)* of the datasets, DSCo-NG is more or equally accurate compared with DSCo, indeed suggesting performance improvement in accuracy. This is probably due to the fact that DSCo tries to find the best way to segment time series; however, with insufficient training data this segmentation process will result in suboptimal segmentation and thus not as high accuracy. Furthermore, we note that in *72% (28/39)* of the datasets, DSCo-NG also outperforms the state-of-the-art DTW-based 1NN, indicating its superiority in specific datasets. Besides, we would like to remind the readers that DSCo-NG is potentially more scale than 1NN based approaches, especially for datasets with a large training set, e.g., the *ElectricDevices* dataset.

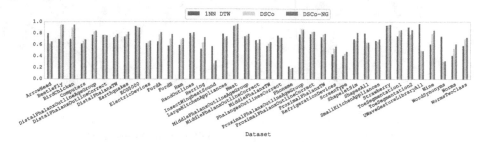

Fig. 3. Overall accuracy comparison between 1NN with DTW distance, DSCo and DSCo-NG.

We have demonstrated the performance of DSCo-NG through complexity analysis and extensive experiments. Although DSCo-NG outperforms our previous work in vast majority of tested datasets, it remains unclear why DSCo-NG outperforms DTW-based 1NN in certain datasets while underperforms in other ones. To this end, we investigate in which scenarios DSCo-NG performs better. Obviously the size of training set can be an important factor, because our model-based approach has to capture from different and a large number of instances the representative patterns, while for instance-based approaches – e.g. kNN – one representative instance could potentially help accurately classifying all similar instances. This is a major reason why DSCo-NG greatly underperforms 1NN for the *WordSynonyms* dataset, which has many (25) classes but very few (267) training instances. Besides, some classes in this dataset has as few as two instances, making the language model extraction highly inaccurate for DSCo-NG.

Besides training set size, in this study we found another important factor that lies in how small segments constitute a time series. Figure 4 shows why DSCo-NG does not perform well for *InsectWingbeatSound*: these two classes consist of similar segments installed in different positions of time series. Thus DSCo-NG will consider these segments as the same word unless we set an extremely long word length. Similarly, DSCo-NG underperforms for *UWaveGestureLibraryAll* because instances in this dataset are composed of three different segments.

Finally, we demonstrate with one example why DSCo-NG outperforms DTW-based 1NN. Consider the two classes from the *FordA* dataset as shown in Fig. 5. It is obvious that visually it is impossible for a human being to distinguish these two classes, because there are two many samples that are not properly aligned like in Fig. 4. As a result, for 1NN classifier, these samples could be distracting so

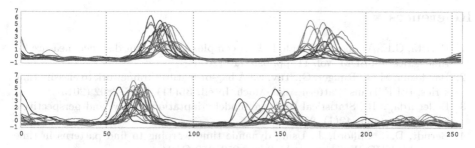

Fig. 4. All instances of two classes (1 and 5) from *InsectWingbeatSound*'s training set.

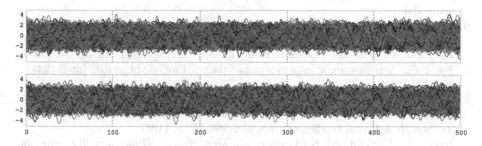

Fig. 5. All instances of two classes (−1 and 1) from *FordA*'s training set.

that it fails to find similar samples given a testing instance. However, DSCo-NG is able to aggregate samples within a class so that it finds the overall descriptive way to differentiate different classes.

6   Conclusions and Future Work

In this study we have improved our previous work DSCo and propose a new approach for TSC. Through complexity analysis and extensive experiments, we show that DSCo-NG is both efficient and performant when comparing with DSCo and the state-of-the-art DTW-based 1NN. Besides, DSCo-NG does not require datasets to be properly aligned, as a result it can save time and efforts preparing for time series data, and result in better classification accuracy with not properly aligned data. Finally, unlike DTW-based 1NN and similar approaches, DSCo-NG can work with data of variable length, which make it suitable for streaming applications.

Since DSCo-NG uses SAX to discretize real-valued time series to text representations, there can be other symbolization techniques to replace SAX and make DSCo-NG parameter-free. In the future, we plan to investigate such opportunities and study the impact of different symbolization techniques on the performance of DSCo-NG.

Acknowledgment. The authors would like to thank Paul Wurth S.A. and Luxembourg Ministry of Economy for sponsoring this research work.

References

1. Batista, G.E., Wang, X., Keogh, E.J.: A complexity-invariant distance measure for time series. In: SDM, vol. 11, pp. 699–710 (2011)
2. Baydogan, M.G., Runger, G., Tuv, E.: A bag-of-features framework to classify time series. IEEE Trans. Pattern Anal. Mach. Intell. **35**(11), 2796–2802 (2013)
3. Bellegarda, J.R.: Statistical language model adaptation: review and perspectives. Speech Commun. **42**(1), 93–108 (2004)
4. Berndt, D.J., Clifford, J.: Using dynamic time warping to find patterns in time series. In: KDD Workshop, vol. 10, pp. 359–370 (1994)
5. Chen, Y., Keogh, E., Hu, B., Begum, N., Bagnall, A., Mueen, A., Batista, G.: The UCR time series classification archive, July 2015. www.cs.ucr.edu/~eamonn/time-series_data/
6. Chung, F.L., Fu, T.C., Luk, R., Ng, V.: Flexible time series pattern matching based on perceptually important points. In: International Joint Conference on Artificial Intelligence Workshop on Learning from Temporal and Spatial Data, pp. 1–7 (2001)
7. Fu, T.C.: A review on time series data mining. Eng. Appl. Artif. Intell. **24**(1), 164–181 (2011)
8. Keogh, E.: Fast similarity search in the presence of longitudinal scaling in time series databases. In: Proceedings of the Ninth IEEE International Conference on Tools with Artificial Intelligence, pp. 578–584. IEEE (1997)
9. Keogh, E., Chakrabarti, K., Pazzani, M., Mehrotra, S.: Dimensionality reduction for fast similarity search in large time series databases. Knowl. Inf. Syst. **3**(3), 263–286 (2001)
10. Keogh, E., Lonardi, S., Ratanamahatana, C.A.: Towards parameter-free data mining. In: Proceedings of the Tenth ACM SIGKDD International Conference on Knowledge Discovery and Data Mining, pp. 206–215. ACM (2004)
11. Li, D., Bissyande, T.F., Klein, J., Le Traon, Y.: Time series classification with discrete wavelet transformed data: insights from an empirical study. In: The 28th International Conference on Software Engineering and Knowledge Engineering (2016)
12. Li, D., Bissyande, T.F., Kubler, S., Klein, J., Le Traon, Y.: Profiling household appliance electricity usage with n-gram language modeling. In: The 2016 IEEE International Conference on Industrial Technology, Taipei, pp. 604–609. IEEE (2016)
13. Li, D., Li, L., Bissyande, T.F., Klein, J., Le Traon, Y.: DSCo: a language modeling approach for time series classification. In: The 12th International Conference on Machine Learning and Data Mining, New York (2016)
14. Li, Y., Lin, J.: Approximate variable-length time series motif discovery using grammar inference. In: Proceedings of the Tenth International Workshop on Multimedia Data Mining, p. 10 (2010)
15. Lin, J., Keogh, E., Wei, L., Lonardi, S.: Experiencing SAX: a novel symbolic representation of time series. Data Min. Knowl. Disc. **15**(2), 107–144 (2007)
16. Marteau, P.F.: Time warp edit distance with stiffness adjustment for time series matching. IEEE Trans. Pattern Anal. Mach. Intell. **31**(2), 306–318 (2009)
17. Senin, P., et al.: GrammarViz 2.0: a tool for grammar-based pattern discovery in time series. In: Calders, T., Esposito, F., Hüllermeier, E., Meo, R. (eds.) ECML PKDD 2014. LNCS, vol. 8726, pp. 468–472. Springer, Heidelberg (2014). doi:10.1007/978-3-662-44845-8_37

18. Senin, P., Malinchik, S.: SAX-VSM: interpretable time series classification using SAX and vector space model. In: IEEE 13th International Conference on Data Mining, pp. 1175–1180. IEEE (2013)

19. Serrà, J., Arcos, J.L.: An empirical evaluation of similarity measures for time series classification. Knowl. Based Syst. **67**, 305–314 (2014)

20. Varrette, S., Bouvry, P., Cartiaux, H., Georgatos, F.: Management of an academic HPC cluster: the UL experience. In: Proceedings of the 2014 International Conference on High Performance Computing and Simulation (HPCS 2014), Bologna, Italy, pp. 959–967. IEEE, July 2014

21. Wang, Q., Megalooikonomou, V.: A dimensionality reduction technique for efficient time series similarity analysis. Inf. Syst. **33**(1), 115–132 (2008)

22. Wang, X., Mueen, A., Ding, H., Trajcevski, G., Scheuermann, P., Keogh, E.: Experimental comparison of representation methods and distance measures for time series data. Data Min. Knowl. Disc. **26**(2), 275–309 (2013)

23. Wang, X., Lin, J., Senin, P., Oates, T., Gandhi, S., Boedihardjo, A.P., Chen, C., Frankenstein, S.: RPM: representative pattern mining for efficient time series classification. In: Proceedings of the 19th International Conference on Extending Database Technology (2016)

24. Xi, X., Keogh, E., Shelton, C., Wei, L., Ratanamahatana, C.A.: Fast time series classification using numerosity reduction. In: Proceedings of the 23rd International Conference on Machine Learning, pp. 1033–1040. ACM (2006)

25. Ye, L., Keogh, E.: Time series shapelets: a new primitive for data mining. In: Proceedings of the 15th ACM SIGKDD International Conference on Knowledge Discovery and Data Mining, pp. 947–956. ACM (2009)

Ranking Accuracy for Logistic-GEE Models

Nasser Davarzani[1(✉)], Ralf Peeters[1], Evgueni Smirnov[1], Joël Karel[1],
and Hans-Peter Brunner-La Rocca[2]

[1] Department of Data Science and Knowledge Engineering, Maastricht University,
P.O.BOX 616, 6200 MD Maastricht, The Netherlands
{n.davarzani,ralf.peeters,smirnov,joel.karel}@maastrichtuniversity.nl
[2] Department of Cardiology, Maastricht University Medical Center,
Maastricht, The Netherlands
hp.brunnerlarocca@mumc.nl

Abstract. The logistic Generalized Estimating Equations (logistic-GEE) models have been extensively used for analyzing clustered binary data. However, assessing the goodness-of-fit and predictability of these models is problematic due to the fact that no likelihood is available and the observations can be correlated within a cluster. In this paper we propose a new measure for estimating the generalization performance of the logistic GEE models, namely ranking accuracy for models based on clustered data (RAMCD). We define RAMCD as the probability that a randomly selected positive observation is ranked higher than randomly selected negative observation *from another cluster*. We propose a computationally efficient algorithm for RAMCD. The algorithm can be applied for two cases: (1) when we estimate RAMCD as a goodness-of-fit criterion and (2) when we estimate RAMCD as a predictability criterion. This is experimentally shown on clustered data from a simulation study and a biomarkers' study.

Keywords: Clustered data · Generalized Estimating Equation · Goodness-of-fit · Predictability · Ranking accuracy

1 Introduction

Clustered data are common in biomedical, clinical, and social-science research [2,9,14]. They are defined as data with a clustered/grouped structure. A cluster (group) can consist of variable measurements of related subjects or repeated variable measurements for a single subject such that in either case the measurements may correlate.

To analyze clustered data, the correlation within clusters needs to be taken into account. To this end, Liang and Zeger [10] proposed an extension of the Generalized Linear Model (GLM) for clustered data with either dichotomous or continuous outcomes [16]. They introduced Generalized Estimating Equations (GEE) to estimate the parameters of the GLM model for dealing with correlated outcomes.

© Springer International Publishing AG 2016
H. Boström et al. (Eds.): IDA 2016, LNCS 9897, pp. 14–25, 2016.
DOI: 10.1007/978-3-319-46349-0_2

The GEE models are widely used for analysis of clustered data, particularly if outcomes are binary (see e.g., [8]). However, due to the fact that no likelihood is available and the residuals (observed outcome minus expected terms) are correlated within a cluster, there is no consensus how to evaluate the GEE models.

This paper addresses the problem of evaluating logistic GEE models. The problem has been considered by several authors (see e.g., [6,7,13]). As a result, several criteria and tests have been proposed for assessing the goodness-of-fit of logistic GEE models. However, most of them have their own shortcomings making impossible having a commonly accepted criterion or test. Below we briefly describe relevant work and then propose our solution.

Barnhart and Williamson [4] proposed a model-based and robust goodness-of-fit test for logistic-GEE models. The method is based on partitioning the space of covariates into distinct regions. The main disadvantage of this method is that applying this method might be problematic when many continuous covariates contribute to the model, or sample sizes are small.

Williamson et al. [15] proposed a Kappa-like classification statistic to assess the model fit of GEE models with categorical outcomes. The disadvantage of the statistic is that for two-class imbalanced data it usually tends to be close to zero (i.e., it states that the model is poorly fitted). Moreover, since no distribution of the statistic is given, interpretation of the statistic is not obvious.

One of the well-established goodness-of-fit statistics for GEE is an *quasilikelihood under the independence model information criterion* (QIC) [12] which is the extension of Akiake's information criterion (AIC) [3]. As a goodness-of-fit and model-selection criterion, the model with smaller QIC is preferred. Since QIC is a function of both quasilikelihood (that depends on the size of the working dataset) and the number of estimated parameters in the GEE model, it indicates the quality of a model relative to other models, fitted with the same data set. That is why it might have different ranges for different data sets. Therefore, QIC is not an applicable criterion for comparing the goodness-of-fit of GEE models for different data sets.

If we generalize the aforementioned goodness-of-fit test statistics and criteria for logistic GEE models, we can derive the following shortcomings: (a) difficulty of interpretation, (b) a relative range of the criterion values (i.e., the range depends on the number of subjects and number of covariates in the model), (c) restriction on the number and types of covariates in the model being evaluated, (d) bias in case of two-class imbalanced data, and (e) inapplicability to indicate the predictability of the model being evaluated.

To propose a criteria that does not suffer from problems (a)–(e), we observe that: (1) logistic GEE models are models trained on clustered data, and (2) logistic GEE models output probabilities of being positive for test observations. The latter implies that logistic GEE models can induce an ordering over those observations. Thus, logistic GEE models actually solve the bipartite ranking task for clustered data [1]. The task is as follows: given labeled clustered data, find an ordering on test observations so that positive observations are ranked higher than negative ones. The standard measure for the quality of that ordering is ranking accuracy. However, it is not applicable for the logistic GEE models, since it does not take into account the within-cluster correlation that might be present, and thus it is not valid.

In this paper we extend the concept of ranking accuracy for clustered data. We propose a new measure that we call *ranking accuracy for models based on clustered data* (RAMCD). It is defined as a probability that a randomly selected positive observation is ranked higher than randomly selected negative observation *from another cluster*. By the definition RAMCD employs the within-cluster correlation in the data used. It focuses on estimating the generalization performance of the logistic GEE models when ranking uncorrelated observations.

We show that RAMCD can be used as a goodness-of-fit criterion and a predictability criterion (i.e., it can be used for estimating the generalization performance of the logistic GEE models beyond training data). For the latter we propose a modification to standard k-fold cross validation method applicable for clustered data.

When comparing RAMCD with the presented standard goodness-of-fit test statistics and criteria for logistic GEE models we observe that RAMCD does not suffer from any of problems (a) to (e) (given above). The main reasons are that: (1) RAMCD is a probability that is easy to interpreted; (2) RAMCD does not impose any restriction on the models being evaluated; (3) RAMCD is not biased for binary imbalanced data (since it indicates class separation); and (4) RAMCD can be used as a goodness-of-fit criterion and a predictability criterion.

The rest of the paper is organized as follows. Section 2 briefly formalizes the bipartite ranking task for clustered data and logistic-GEE model. RAMCD is introduced in Sect. 3. Section 4 provides the experiments and Sect. 5 concludes the paper.

2   Bipartite Ranking Task and Logistic GEE Models

The bipartite ranking task assumes that we have n subjects. The i-th subject is represented by a cluster of m_i observations such that the tth observation, $t = 1, ..., m_i$ is given with p covariates $X_{it1}, ..., X_{itp}$ in $\mathbb{R}$ and a binary outcome variable Y_{it}. Hence, the i-th cluster is identified by X_i and Y_i, where $X_i = (X_{i1}, ..., X_{im_i})'$ in which $X_{it} = (X_{it1}, ..., X_{itp})$ is $1 \times p$ vector of covariates for observation t for subject i and $Y_i = (Y_{i1}, ..., Y_{im_i})'$ is $m_i \times 1$ vector of binary outcomes. For any $i \neq j$ we assume that the correlation between Y_i and Y_j equals 0.0 while the components of each Y_i may be correlated and the covariates may be either fixed or changing at every cluster level. Given n clusters X_i and Y_i for $i \in [1, n]$, the goal of bipartite ranking is to find a real-value ranking function that maps any observation X_{it} to real number. The ranking function can be used to induce ordering over the observations X_{it}.

The logistic-GEE model solves the bipartite ranking task, since it is essentially a ranking function for clustered data. It describes the relationship between the covariates and outcome variables with the following equation:

$$\log \left(\frac{\pi_{it}}{1 - \pi_{it}} \right) = \beta_0 + \beta X'_{it} \ , i = 1, ..., n, \ t = 1, ..., m_i, \tag{1}$$

where $\pi_{it} = \mathrm{E}(Y_{it} | X_{it})$, β_0 is the population averaged intercept term and $\beta = (\beta_1, ..., \beta_p)$ is the vector of population averaged (or marginal) coefficients.

The logistic-GEE model can be obtained by estimating the unknown regression coefficient vector $\gamma = (\beta_0, \beta)$. Estimating the coefficients can be done by solving the following generalized estimating equations [10]:

$$\sum_{i=1}^{n} \left(\frac{\partial \pi_i}{\partial \beta_h} \right)' V_i^{-1} (Y_i - \pi_i) = 0, \ h = 0, ..., p, \tag{2}$$

where, for $i = 1, ..., n$, $\pi_i = (\pi_{i1}, ..., \pi_{im_i})'$, $V_i = A_i^{1/2} R_i(\alpha) A_i^{1/2}$ is the working covariance matrix for Y_i, A_i, is a diagonal matrix $\mathrm{diag}[\pi_{i1}(1 - \pi_{i1}), ..., \pi_{im_i}(1 - \pi_{im_i})]$, α is an $m \times 1$ vector of unknown parameters, associated with the correlation between outcomes Y_{it} and Y_{is} of cluster i, $m = \max(m_1, ..., m_n)$, and $R_i(\alpha)$ is the working correlation matrix for Y_i.

We note that the working correlation matrix $R_i(\alpha)$, parameterized by α, might be defined in different ways depending on the nature of correlation between outcomes Y_{it} and Y_{is}. Zeger and Liang [16] proposed a method for estimating the parameter vectors α and γ in Eq. (3). The method operates by minimizing the weighted sum of squared residuals using IRLS, described in [11].

3 Ranking Accuracy for Models Based on Clustered Data

In this section we introduce the ranking accuracy for models based on clustered data (RAMCD). RAMCD is formally defined in Subsect. 3.1. The algorithm for computing RAMCD is provided in Subsect. 3.2 together with a complexity analysis. Subsect. 3.3 explains how the algorithm can be used for estimating RAMCD as a criterion of the model's goodness-of-fit and as a criterion for the model's predictability.

3.1 Definition

According Eq. (1) any logistic GEE model is essentially a scoring classifier. It outputs a score, a probability π_{it}, for any observation X_{it}. Given a test data of n number of clusters $\langle X_i, Y_i \rangle$, the probabilities π_{it} induce an ordering over the observations from the clusters. To judge the quality of the probabilities π_{it}, we judge the quality of the ordering, they induce, and compare that ordering with the binary outcome variables Y_{it}. The standard measure for such a comparison is ranking accuracy [1]. It is defined as a probability that a randomly selected positive observation is ranked higher than randomly selected negative observation. However, as it might be seen from the definition, the ranking accuracy does not take into account the within-cluster correlation that might be present and thus it is not valid for clustered data. This calls for a new special ranking accuracy applicable for models based on clustered data.

We introduce the ranking accuracy for models based on clustered data (RAMCD) by analogy. Consider a set of observations X_{it} where each cluster is present with exactly one observation. The number of such sets equals $\sum_{i=1}^{n} m_i$. The probabilities π_{it} induce an ordering for each of these sets. To compare these orderings with the binary outcome variables Y_{it} we introduce RAMCD. RAMCD is defined as a probability that a randomly selected positive observation is ranked higher than randomly selected negative observation *from another cluster*. By the definition RAMCD employs the within-cluster correlation in the data used and focuses on estimating the generalization performance of the logistic GEE models when ranking uncorrelated observations.

RAMCD is easy to interpret, since it is a probability (i.e., it ranges between 0 and 1). The value of 1.0 indicates that the orderings imposed correspond completely to the binary outcome variables Y_{it} in the clustered data, and the value of 0.0 shows that the orderings are reversed to that with value of 1.0. The value of 0.5 is the worst case. It indicates bad orderings that do not correspond at all to the outcome variables. However, we note that RAMCD of 0.5 does not always imply a random logistic GEE model (e.g., when the data is class-imbalanced).

Below we introduce the exact formula for RAMCD. We first introduce statistics imposed by the binary outcome variables Y_{it}. Following the RAMCD definition we determine for any positive observation X_{it} the number P_{it} of negative observations from other clusters:

$$P_{it} = \sum_{j=1, j \neq i}^{n} \sum_{t=1}^{m_j} I\{Y_{jt} = 0\} \tag{3}$$

where I is the indicator function. The number P_{it} can be interpreted as the number of pairs that consist of positive observation X_{it} and negative observation from any other cluster. It is the same for any positive observation in cluster i. This implies that the number P_i of pairs for all the positive observations in cluster i is equal to:

$$P_i = \sum_{t=1}^{m_i} P_{it} I\{Y_{it} = 1\} \tag{4}$$

and the total number P of pairs of observations over all the clusters imposed by the binary outcome variables Y_{it} is equal to:

$$P = \sum_{i=1}^{n} P_i \tag{5}$$

Once the statistics imposed by the binary outcome variables Y_{it} have been defined, we introduce statistics for comparing the orderings imposed by probabilities π_{it}. We assume that for any observation X_{it} we have a probability estimate

π_{it} provided by a logistic GEE model. We rank the observations $\boldsymbol{X}_{it}$ according to π_{it}. To judge whether a particular positive observation $\boldsymbol{X}_{it}$ from cluster i is ranked properly we compute the number CP_{it} of correct pairs produced by the ranking through combining with all negative observations $\boldsymbol{X}_{jt}$ from all other clusters j such that $j \neq i$. The number CP_{it} is given by:

$$CP_{it} = \sum_{j=1, j\neq i}^{n} \sum_{t_j=1}^{m_j} (I\{\pi_{it} > \pi_{jt_j}\} + \frac{1}{2}I\{\pi_{it} = \pi_{jt_j}\})I\{Y_{jt_j} = 0\} \qquad (6)$$

Number CP_{it} does not stay the same for each positive observation in cluster i. Hence, the number CP_i of all correct pairs produced by combining all the positive observations $\boldsymbol{X}_{it}$ from cluster i with all the negative observations $\boldsymbol{X}_{jt}$ over all the clusters j given that $j \neq i$ is equal to:

$$CP_i = \sum_{t=1}^{m_i} CP_{it}I\{Y_{it} = 1\} \qquad (7)$$

and the number CP of all the correct pairs produced by the ranking is:

$$CP = \sum_{i=1}^{n} CP_i \qquad (8)$$

Thus, formally our RAMCD with respect to the ranking produced is defined equal to:

$$RAMCD = \frac{CP}{P} \qquad (9)$$

3.2   Algorithm

Below in Fig. 1 we provide an algorithm for RAMCD. Given data with n number of clusters $\langle \boldsymbol{X}_i, \boldsymbol{Y}_i \rangle$, and a vector $\boldsymbol{\pi}_i$ of observation probabilities π_{it} for each cluster $\langle \boldsymbol{X}_i, \boldsymbol{Y}_i \rangle$, the algorithm computes RAMCD induced by the observation probabilities π_{it} w.r.t. outcome variable Y_{it}. The main steps are as follows. First, the algorithm computes the statistics imposed by the binary outcome variables Y_{it}: it computes number P_i for each cluster i (see formula (4)) and total number P (see formula (5)). Then, the algorithm computes statistics necessary for comparing the orderings imposed by probabilities π_{it}. For that purpose the observations $\boldsymbol{X}_{it}$ over all the clusters are sorted according to π_{it} in decreasing order of magnitude into list L_π. The algorithm scans the sorted list L_π to compute numbers CP_{it}, CP_i, and CP (initially set equal to 0). For list scanning it keeps a counter C_i for all the clusters $i \in [1, n]$ that represents the number of all correct pairs that start with a positive observation from cluster i and end with a negative observation from another cluster given that both observations have not been visited in list L_π. Therefore, C_i is initialized equal to $\frac{P_i}{m_i}$ which is the number of pairs derived by combining a positive observation from cluster i with all possible negative observations from other clusters.

Algorithm *RAMCD*
Input: n number of clusters $\langle X_i, Y_i \rangle$,
    Vector π_i of observation probabilities π_{it} for each cluster $\langle X_i, Y_i \rangle$.
Output:
    RAMCD induced by the observation probabilities π_{it} with respect to outcome variable Y_{it}.
    for $i := 1$ to n **do**
        Compute P_i using formula (4);
    Compute P using formula (5);
    Sort all the observations X_{it} according to π_{it} in decreasing order of magnitude into list L_π.
    for $i := 1$ to n **do**
        $CP_i = 0$;
        $C_i = \frac{P_i}{m_i}$;
    $CP = 0$;
    for each observation X_{it} in L_π **do**
        if $Y_{it} = 0$ **then**
            for $j = 1$ to n **do**
                if $j \neq i$ **then**
                    $C_j = C_j - 1$;
        else
            $CP_{it} = C_i$;
            $CP_i = CP_i + CP_{it}$;
    for $i := 1$ to n **do**
        $CP = CP + CP_i$;
    return $\frac{CP}{P}$.

Fig. 1. Algorithm for computing ranking accuracy for models based on clustered data.

After the initialization the algorithm sequentially visits the observations X_{it} in the sorted list L_π. For each observation X_{it} the actions taken depends on the output variable Y_{it}. If the observation is negative ($Y_{it} = 0$), then the algorithm decrements the counter C_j for each cluster j different from cluster i. This is to indicate that all the positive observations X_{jt} with probability π_{jt} that is lower than π_{it} cannot form a correct pair with observation X_{it} according to the ordering imposed on L_π. If the observation is positive ($Y_{it} = 1$), then the algorithm assigns the counter value C_i to the number CP_{it} and then this number is added to number CP_i according to formula (7). Once all the numbers CP_i have been computed, the algorithm computes number CP (see formula (8)) and then outputs the RAMCD (see formula (9)).

The algorithm for RAMCD is computationally efficient. Its space complexity is $O(nm)$, where n is the number of clusters and m is the size of the clusters. This complexity is due to the sorted list L_π that has to be explicitly maintained by the algorithm. The time complexity is $O(nm log_2(nm))$ and it coincides with the time complexity of the sorting algorithm[1]. We note that the time complexity of scanning list L_π is linear in the size of the list (nm) and that is why it does not influence the asymptotic time complexity.

---

[1] We assume the usage of efficient sorting algorithms like merge sort.

3.3   Goodness of Fit and Predictability

The ranking accuracy for models based on clustered data (RAMCD) can be used as a criterion of model's goodness-of-fit and as a criterion for model's predictability. If a logistic GEE model has been trained and tested on the same data, then RAMCD is a goodness-of-fit criterion. In this case RAMCD estimates how the logistic GEE model fits the data when only uncorrelated observations are taken into account.

If a logistic GEE model has been tested using k-fold cross validation on the data, then RAMCD is a predictability criterion. However, the randomization part of the cross validation has to be controlled such that observations within any cluster are being selected for only one folder. In this way we do not introduce additional bias when computing probabilities π_i due to the within-cluster correlation. This guarantees that the algorithm estimates RAMCD that indicates the predictability of the GEE model beyond training data when only uncorrelated observations are taken into account.

4   Experiments

In this section we present the experiments with RAMCD and QIC on simulated data and biomarker data. The experiments are employed to compare these two criteria.

4.1   Experiments with Simulated Data

This subsection presents two experiments with RAMCD of logistic GEE models on a simulated data. The simulated data is described by 30 time-dependent covariates $(X_1, X_2, \ldots, X_{30})$. It contains 500 clusters with maximum sizes of $m = 10$ and autoregressive working correlation structure of order 1 with correlation of 0.25.

The first experiment is in the context of the goodness-of-fit test. We compare RAMCD and QIC in a function of the GEE model complexity. For that purpose we add the covariates X_1 to X_{30} one by one into the GEE model and each time plot the RAMCD and QIC in Fig. 2. The Figure shows that the RAMCD and QIC follow similar trends in a function of the model complexity. There exists however some fluctuations of RAMCD when it is close to 0.5. In these cases GEE models are under-fitted and exhibit random performance which is not captured by QIC.

The second experiment is in the context of model selection: we employ RAMCD for forward feature selection when it is used as a goodness-of-fit criterion and when it is used as a predictability criterion. In the first case RAMCD is estimated on the simulated data and it is denoted as RAMCD. In the second case RAMCD is estimated using one-cluster-out cross validation on the simulated data and it is denoted as RAMCD-CV. In both cases we compare the results of the model selection with those obtained by QIC.

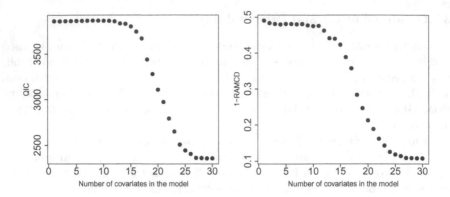

Fig. 2. QIC and RAMCD as functions of GEE model complexity.

The process of forward feature selection is sequential; i.e., the covariates are added one by one. It is guided by a hill-climbing search which for RAMCD (QIC) adds that covariate that maximizes (minimizes) the RAMCD (QIC) of the resulted GEE model. The process stops when further improvement is not possible.

Table 1. Forward feature selection for logistic-GEE model using RAMCD, RAMCD-CV, and QIC. Each box represents the selected covariate and the value of selection criterion (RAMCD, RAMCD-CV, or QIC). The bold variables are those that are not selected.

Step	RAMCD	RAMCD-CV	QIC	Step	RAMCD	RAMCD-CV	QIC
1	X_{18} (0.590242)	X_{18} (0.581824)	X_{18} (3791.772)	16	X_{25} (0.877261)	X_{25} (0.873539)	X_{25} (2464.293)
2	X_{17} (0.621768)	X_{17} (0.614062)	X_{17} (3732.281)	17	X_{27} (0.881865)	X_{27} (0.878116)	X_{27} (2425.637)
3	X_{16} (0.649819)	X_{16} (0.643470)	X_{16} (3658.207)	18	X_{26} (0.886264)	X_{26} (0.882337)	X_{26} (2387.744)
4	X_{19} (0.672773)	X_{19} (0.666522)	X_{19} (3587.769)	19	X_9 (0.889523)	X_9 (0.885488)	X_9 (2360.825)
5	X_{15} (0.696743)	X_{15} (0.690881)	X_{15} (3502.061)	20	X_{29} (0.890200)	X_{29} (0.885943)	X_{29} (2357.297)
6	X_{20} (0.718667)	X_{20} (0.713321)	X_{20} (3422.789)	21	X_2 (0.890596)	X_6 (0.886166)	X_6 (2355.108)
7	X_{13} (0.743073)	X_{13} (0.737916)	X_{13} (3325.778)	22	X_6 (0.890973)	X_{28} (0.886391)	X_{28} (2353.878)
8	X_{14} (0.759801)	X_{14} (0.754877)	X_{14} (3241.496)	23	X_{28} (0.891342)	X_2 (0.886536)	X_2 (2352.896)
9	X_{22} (0.778013)	X_{22} (0.773413)	X_{22} (3141.077)	24	X_5 (0.891657)	X_5 (0.886586)	X_5 (2352.300)
10	X_{21} (0.798329)	X_{21} (0.794059)	X_{21} (3028.665)	25	X_{30} (0.891823)	$\mathbf{X_4}$ (0.886572)	$\mathbf{X_7}$ (2352.430)
11	X_{12} (0.818305)	X_{12} (0.814269)	X_{12} (2905.000)	26	X_4 (0.891994)	$\mathbf{X_{30}}$ (0.886520)	$\mathbf{X_4}$ (2352.941)
12	X_{23} (0.834424)	X_{23} (0.830658)	X_{23} (2801.273)	27	X_7 (0.892142)	$\mathbf{X_7}$ (0.886462)	$\mathbf{X_{30}}$ (2353.607)
13	X_{24} (0.848698)	X_{24} (0.844904)	X_{24} (2698.514)	28	X_1 (0.892164)	$\mathbf{X_1}$ (0.886271)	$\mathbf{X_8}$ (2355.205)
14	X_{11} (0.861617)	X_{11} (0.858049)	X_{11} (2600.898)	29	X_8 (0.892181)	$\mathbf{X_8}$ (0.886086)	$\mathbf{X_1}$ (2356.971)
15	X_{10} (0.870470)	X_{10} (0.866909)	X_{10} (2527.606)	30	$\mathbf{X_3}$ (0.892134)	$\mathbf{X_3}$ (0.885836)	$\mathbf{X_3}$ (2359.079)

The results of model selection for RAMCD, RAMCD-CV, and QIC are provided in Table 1. The Table shows that RAMCD-CV and QIC are rather consistent: they lead to the same ordered set of covariates on the simulated data when the process of feature selection stops. This means that RAMCD-CV and

QIC result in the same GEE model. However, if we continue to add covariates after the stopping condition, the RAMCD-CV and QIC become less consistent. As expected, the values of RAMCDs are higher than those of RAMCD-CVs at each step which results in a bigger set of selected covariates. In this context we note that RAMCD is less consistent with RAMCD-CV and QIC than those two measures together.

4.2 Experiments with Biomarkers' Data

This subsection presents a model-selection (biomarker selection) process guided by RAMCD-CV on the data from the TIME-CHF study [5]. The TIME-CHF study (The Intensified versus standard Medical therapy in Elderly patients with Congestive Heart Failure) includes 499 patients aged 60 years or older, with left ventricular ejection fraction (LVEF) $< 45\%$ and NYHA II or more, from 15 centers in Switzerland and Germany. Patients were followed for 6 pre-specified visits after baseline, 1st, 3rd, 6th, 12th and 18th month. Six biomark-ers, PREA (prealbumin), SST2 (soluble ST2), IL6 (Interleukin-6), hsCRP (high sensitivity C-reactive protein), GDF15 (growth differentiation factor 15), SFLT (soluble fms-like tyrosine kinase-1,) and BPsyst (Systolic blood pressure) and LVEF were measured at every visit and dosages of a heart failure (HF) drug Loop (Loop diuretics per se) were available on a daily basis. Patients were fol-lowed up for 19 months and the outcome variable for i^{th} patient at month t, Y_{it}, $i = 1, ..., 499$, $t = 1, ..., 19$, takes the value of one if the patient experienced HF hospitalization or death at the given month, otherwise zero. In this setup, more weight is given to the outcome death (weight 2 for death and 1 for the other observations).

The medication covariate Loop is down-sampled to monthly values by tak-ing the average drug dosage during the previous month. Since the biomarkers, BPsyst and LVEF have been recorded just in six visits; obviously for these six measurements, the covariates gets the exact value, and between these six vis-its we used last observation carried forward method (LOCF) and put the value of the covariates of the previous visit. There exist eight fixed covariates that measured only at the baseline; Age, Gender (1 = male, 0 = female), Coronary artery disease (CAD), Kidney-disease, Diabetes, Anemia, Charlsonscore (Charl-son comorbidity score) and Rales, where CAD, Kidney-disease and Diabetes are binary variable that indicates whether the patients are suffering from these dis-eases or not (1 = yes, 0 = no) and Rales (1 = abnormal lung sounds, 0 = normal lung sounds).

The goal of the study is to select the best subset of covariates (biomark-ers) to explain the variation of the probability of HF hospitalization and death. To this end, we apply a forward feature selection process using the proposed RAMCD-CV and QIC as model-selection criteria to find the best GEE model. Table 2 shows the selected covariates for the GEE model at every step of for-ward selection process based on RAMCD-CV and QIC. Both criteria lead to the same selected subset of covariates (GDF15, SST2, CAD, Loop, hsCRP,

Table 2. Selected covariates at each step of forward selection method using RAMCD-CV and QIC as model selection criteria.

RAMCD-CV-Covariates	RAMCD-CV	p-values	QIC-Covariates	QIC	p-values
Intercept	0.5	0.000000	Intercept	1917.96	0.000000
GDF15	0.738816	0.000002	GDF15	1753.53	0.000002
SST2	0.767409	0.000001	SST2	1691.76	0.000001
CAD	0.775987	0.003211	Loop	1660.43	0.000060
Loop	0.784891	0.000060	CAD	1649.28	0.003211
hsCRP	0.790796	0.031393	BPsyst	1645.50	0.018600
Age	0.794360	0.020222	Age	1639.86	0.020222
BPsyst	0.795732	0.018600	hsCRP	1635.74	0.031393
Rales	0.796831	0.073626	Rales	1635.28	0.073626

Age, BPsyst and Rales), however, the selected subsets were obtained in different orders for each criterion. The estimated coefficients and corresponding p-values of selected covariates, when using RAMCD-CV as a model-selection, are presented in Table 2.

5   Conclusion

In this paper we proposed RAMCD as a new measure for estimating the generalization performance of logistic GEE models. RAMCD was defined as a probability that a randomly selected positive observation is ranked higher than randomly selected negative observation *from another cluster*. We showed that RAMCD focuses on estimating the generalization performance of the logistic GEE models when ranking uncorrelated observations. We proposed a computationally efficient algorithm for RAMCD and showed that it can be applied for two cases: (1) when we estimate RAMCD as a goodness-of-fit criterion and (2) when we estimate RAMCD as a predictability criterion. The algorithm was experimentally tested on clustered data from a simulation study and a biomarkers' study. The experiments showed that RAMCD is consistent with the QIC criterion.

We compared RAMCD with the standard goodness-of-fit test statistics and criteria for logistic GEE models: we observed that RAMCD does not suffer from any of their problems. The main reasons are that: (1) RAMCD is a probability that is easy to interpreted; (2) RAMCD does not impose any restriction on the models being evaluated; (3) RAMCD is not biased for binary imbalanced data (since it indicates class separation); and (4) RAMCD can be used as a goodness-of-fit criterion and a predictability criterion.

Finally, we note although RAMCD has been initially designed for the logistic GEE models, it is applicable to any model for bipartite ranking based on clustered data. This is due to the fact that RAMCD employs model's probabilities and data labels; i.e., it does not use any internal information from the model being tested. Thus, we conclude that RAMCD is a general measure for models for bipartite ranking based on clustered data.

References

1. Agarwal, S., Graepel, T., Herbrich, R., Har-Peled, S., Roth, D.: Generalization bounds for the area under the ROC curve. J. Mach. Learn. Res. **6**, 393–425 (2005)
2. Ahsan, H., Chen, Y., Parvez, F., Zablotska, L., Argos, M., Hussain, I., Momotaj, H., Levy, D., Cheng, Z., Slavkovich, V., Van Geen, A.: Arsenic exposure from drinking water and risk of premalignant skin lesions in Bangladesh: baseline results from the health effects of arsenic longitudinal study. Am. J. Epidemiol. **163**(12), 1138–1148 (2006)
3. Akaike, H.: A new look at the statistical model identification. IEEE Trans. Autom. Control **19**(6), 716–723 (1974)
4. Barnhart, H.X., Williamson, J.M.: Goodness-of-fit tests for GEE modeling with binary responses. Biometrics **54**(2), 720–729 (1998)
5. Brunner–La Rocca, H.P., Buser, P.T., Schindler, R., Bernheim, A., Rickenbacher, P., Pfisterer, M., TIME-CHF-Investigators: Management of elderly patients with congestive heart failuredesign of the Trial of Intensified versus standard Medical therapy in Elderly patients with Congestive Heart Failure (TIME-CHF). Am. Heart J. **151**(5), 949–955 (2006)
6. Evans, S.R., Hosmer Jr., D.W.: Goodness of fit tests for logistic GEE models: simulation results. Commun. Stat. Simul. Comput. **33**(1), 247–258 (2004)
7. Evans, S., Li, L.: A comparison of goodness of fit tests for the logistic GEE model. Stat. Med. **24**(8), 1245–1261 (2005)
8. Hanley, J.A., Negassa, A., Forrester, J.E.: Statistical analysis of correlated data using generalized estimating equations: an orientation. Biometrics **157**(4), 364–375 (2003)
9. Lafata, J.E., Pladevall, M., Divine, G., Ayoub, M., Philbin, E.F.: Are there race/ethnicity differences in outpatient congestive heart failure management, hospital use, and mortality among an insured population? Med. Care **42**(7), 680–689 (2004)
10. Liang, K.Y., Zeger, S.L.: Longitudinal data analysis using generalized linear models. Biometrika **73**(1), 13–22 (1986)
11. McCullagh, P.: Quasi-likelihood functions. Ann. Stat. **11**(1), 59–67 (1983)
12. Pan, W.: Akaike's information criterion in generalized estimating equations. Biometrics **57**(1), 120–125 (2001)
13. Pulkstenis, E., Robinson, T.J.: Two goodness-of-fit tests for logistic regression models with continuous covariates. Stat. Med. **21**(1), 79–93 (2002)
14. Titler, M.G., Jensen, G.A., Dochterman, J.M., Xie, X.J., Kanak, M., Reed, D., Shever, L.L.: Cost of hospital care for older adults with heart failure: medical, pharmaceutical, and nursing costs. Health Serv. Res. **43**(2), 635–655 (2008)
15. Williamson, J.M., Lin, H.M., Barnhart, H.X.: A classification statistic for GEE categorical response models. J. Data Sci. **1**, 149–165 (2003)
16. Zeger, S.L., Liang, K.Y.: Longitudinal data analysis for discrete and continuous outcomes. Biometrics **42**(1), 121–130 (1986)

The Morality Machine: Tracking Moral Values in Tweets

Livia Teernstra[1,2](✉), Peter van der Putten[1], Liesbeth Noordegraaf-Eelens[2], and Fons Verbeek[1]

[1] Media Technology, Leiden University, Leiden, The Netherlands
{p.w.h.van.der.putten,f.j.verbeek}@liacs.leidenuniv.nl
[2] Faculty of Social Science, Erasmus University Rotterdam, Rotterdam, The Netherlands
livia@helloliefje.com, eelens@euc.eur.nl

Abstract. This paper introduces The Morality Machine, a system that tracks ethical sentiment in Twitter discussions. Empirical approaches to ethics are rare, and to our knowledge this system is the first to take a machine learning approach. It is based on Moral Foundations Theory, a framework of moral values that are assumed to be universal. Carefully handcrafted keyword dictionaries for Moral Foundations Theory exist, but experiments demonstrate that models that do not leverage these have similar or superior performance, thus proving the value of a more pure machine learning approach.

Keywords: Text classification · Moral values · Social technologies

1 Introduction

There has been growing interest in social sciences research to leverage intelligent data analysis to automatically gather and analyze large amounts of data. A potentially interesting yet relatively unexplored area is ethics, which so far has been approached more theoretically rather than empirically, especially with machine learning methods. The instantaneous and opinionated nature of Internet media such as Twitter provides an immediate outlet for emotions, opinions, information and interactions, loaded with moral perspectives [14]. Accordingly, Twitter is a promising data source for interdisciplinary research on ethics. However, most social science research examines the diffusion of information rather than the content [1,7,17]. Even when content is analysed, this has mostly been focused on commercial or political motivations [2,18]. Likewise within intelligent data analysis, social media monitoring is a popular topic, but it is typically limited to sentiment or opinion mining for business applications, and is lacking theoretical social science foundations. Hence, there is room for an approach that combines morality research with social network content analysis.

The main purpose of this study is to provide an overview of The Morality Machine, a proof of concept system that detects and monitors moral sentiment

H. Boström et al. (Eds.): IDA 2016, LNCS 9897, pp. 26–37, 2016.
DOI: 10.1007/978-3-319-46349-0_3

in Twitter communications, using a text classification approach. It is based on an ethical framework from social psychology called Moral Foundations Theory (MFT), which assumes universal moral foundations exist that can be used to categorize and study ethical problems and discourse [8,11].

As an example debate, this study will explore public opinion on austerity measures in the Eurozone, and specifically the discussion of the Greek exit of the Euro (the 'Grexit'). Austerity is a good topic to explore because it is often discussed in the context of moral hazard [10]. Some state that by bailing out Southern European nations that have shown lack of fiscal discipline, it is encouraging such behaviour rather than criticising it, and that the irresponsible behaviour of these governments is the root cause of the European financial crisis [4]. Conversely, others point out that richer countries have been main beneficiaries of economical support to poorer countries in the past, and that all EU countries have a duty to look after each other and protect the integrity of the EU. Consequently, the Grexit discussion is framed in a moral light, where 'good' and 'bad' nations and policies are distinguished, and there is no shortage of opinions. This austerity dispute will be used to contextualise the methodology since it has the potential to engage all moral foundations.

Related work that classify text into moral foundations typically use dictionary based techniques, meaning that large word lists grounded in psychological theory must first be created and validated manually, as opposed to being discovered automatically by machine learning [2,21,22]. When using these dictionaries, frequencies of morally related words generate moral loadings for texts [5,6,18]. However, the relative importance of these frequencies for detecting certain moral foundations are not derived from evidence. Thus, machine learning algorithms are useful since they can automatically determine lexical indicators for each foundation, without the need to create a dictionary beforehand. Additionally, lexical indicators for each moral foundation can be gleaned from the algorithm, which can be used for further research into moral expressions.

To our knowledge, this is the first study that aims to detect and monitor Moral Foundations using a machine learning approach. It is also the first study to examine moral expressions of the public regarding the Grexit. It uses generally accepted machine learning techniques to explore moral expressions in a natural real world setting, through the use of the Twitter platform. Specifically, this study will determine if supervised machine learning models are able to classify Tweets into moral foundations at an acceptable accuracy, potentially without relying on handcrafted dictionaries.

The remainder of this paper is structured as follows; Sect. 2 provides background on Moral Foundation Theory and text classification. Our methodology is outlined in Sect. 3 and experiments and results are described in Sect. 4. The paper ends with a discussion (Sect. 5) and conclusion (Sect. 6).

2 Related Work

Early ideas in moral reasoning originate in Greek Philosophy. Contemporary moral research asserts that the backbone of our moral decisions lie in a combination of biological and environmental factors [18]. These inert, deep-seated motivations can serve different social functions. This section provides an overview of Moral Foundations Theory (MFT), a framework of assumed universal moral values, and applies it to the case study of the Grexit. It also examines previous research on content analysis using MFT.

2.1 Moral Foundations Theory

The assertion behind MFT is that intrinsic, cognitive responses in individuals can be used to explain the variation in human moral reasoning across cultures [12]. Hence, the theory posits that there is an innate and universal morality which transcends cultural boundaries. This universal morality can be categorised into different foundations, which can be thought of as 'moral building blocks'. Each foundation is fostered within cultures, which serves the purpose of constructing narratives, virtues and institutions. The fostering of foundations differs between groups, where some may emphasize one foundation over another [8]. The foundations can be held simultaneously by individuals and societies, and may conflict with one another.

In the context of this research, six foundations will be used to classify Twitter data. Although there are normally five foundations which form the basis of MFT, a sixth (Liberty - Oppression) has previously been included in the model for other politically driven studies, so we included it [8]. The foundations are briefly described in Table 1, along with example Tweets. The moral foundation which drives opinions can stem from society at large, smaller communities, or individual moral preferences. As such, this study asserts no preference for a specific moral standpoint, as its main focus is learning to classify Grexit Tweets into moral foundations, as a case study for empirical ethics. Also by definition a framework is framed by an underlying theory, which should not be seen as objective or value free. MFT provides a useful framework to distinguish ethical statements, but we do not want to imply it is the only valid one. See for example [20] for a critical review.

2.2 Moral Foundations Text Analysis

In social sciences, dictionary based approaches are predominantly used for text classification. A Moral Foundations Dictionary (MFD) is also available. This dictionary gives linguistic indications for the five basic moral foundations (hence 'liberty' is excluded). The MFD was created for use with the Linguistic Inquiry Word Count (LIWC) program [9]. LIWC is one of the most widely used social science tools for text analysis and is also commonly used for Tweet classification [6,22]. Yet, there is no current research which uses the MFD with LIWC to detect moral foundations in Tweets.

Instead, textual analytics using the MFD with LIWC has been applied in analysis of long texts such as news articles and web blogs, where rhetorical moral assessments were assigned to each text [6,18]. The analysed texts are authored by opinion leaders, such as news media or bloggers, rather than the general public.

Table 1. Descriptions of moral foundations

Foundation	Description	Tweets
Care, Harm	The desire to cherish protect others, identification of a victim and sympathy with him	European control of the IMF is helping Greece
		Greece runs out of funding options despite Euro zone reprieve
Fairness, Cheating	The notions of justice and rights, applied to shared rules in a community. Relates to reciprocal altruism	Greece forced to sell assets and cut spending to pay back debts to EU
		It's easy for the Dutch to go hard on 'Greece'
Loyalty, Betrayal	Relating to 'in-groups'; friends, family, community, as well as showing virtues of patriotism	If I had to choose between #Greece and #Germany, I know which way I'd go...
		Greece may stay in the Eurozone for the time being there are no guarantees it can become a responsible member
Authority, Subversion	Submission to and respect for legitimate authority and traditions	Greece says Euro zone approves reform plan
		German elites are willing to let the Euro crash to guarantee their own political survival
Sanctity, Degradation	Stems from feelings of disgust and contamination. Relating to the virtue that 'the body is a temple', and should not be defiled	There really is no space inside the Euro for a radical left government
		The four-month extension on the Greek debt lowers the risk of Greece leaving the Euro zone
Liberty, Oppression	The resentment of tyranny and desire for autonomy. This is often in tension with the foundation of 'Authority'	Greece needs a path out of the Euro
		Greece really might leave the Euro

For example, research on the Ground Zero Mosque showed that blog authors showed more lexical similarity among virtuous terms for the foundations care, fairness and authority [6]. One can then gather that expression of the other foundations may be constructed differently amongst cultural groups. Due to the differences in textual expressions of moral opinion, dictionary based approaches can be problematic when drawing conclusions about moral reasoning. And given that the dictionary is hand built rather than learned, it is very dependent on it being correct and complete. All in all, the use of MFT and MFD in text analysis is in its infancy, and there is notable room for improvement.

3   Experiments and Results

In this section we provide an overview of the experiments and results, and then review each step of the process and the accompanying results in more detail.

3.1   Overall Procedure

Tweets were collected, and for a random sample frequent keywords were generated as well as bigrams, and bigrams and Tweets were labelled. This gave us 3 data sets: just raw data, raw data with bigrams and raw data with the moral foundations dictionary (MFD). We created two variants of each, one with and one without stop words removed. Skipping stop word removal worked best, so on this data we then carried out learning curve experiments to assess the impact of training set size. For the best performing variants we ran an additional five fold cross validation test. The best model was then deployed to the full data set minus the labeled Tweets to illustrate how the model can be used to track moral sentiment on new data.

3.2 Data Collection

In order to gather initial public reactions to Eurozone meetings, English language Tweets with keywords 'Euro' and 'Greece' were collected from three specific times in 2015, using a custom built streaming Twitter data collector. The search term 'Grexit' was omitted, as it is more prominent in the financial sector, so it excludes Tweets from those who are not familiar with the term. Moreover, 'grexit' tends to carry a certain connotation, focusing only on Greece leaving the Eurozone, rather than economic issues as a whole. The exact dates, number of Tweets and events are outlined in Table 2. Each week of data collection yielded between 4000 and 7000 Tweets, resulting in a total of 18,986 Tweets. The duplicate entries were then removed (including re-Tweets), leaving only unique Tweets (N = 8,292). Note all our coding was done with in Python including the Python Natural Language ToolKit [3].

Table 2. Data collection time periods

Data set	Date Range	N	Event
1	24/02/2015 to 03/03/2015	7,037	Eurozone Finance ministers agreed to extend the Greek bailout for another 4 months
2	28/04/2015 to 04/05/2015	4,856	Eurozone Finance ministers meet to discuss reform packages from Athens
3	11/05/2015 to 23/05/2015	7,066	Athens announces repayments to International Monetary Fund to avoid default

3.3 Data Preparation

Tweets in our system are primarily represented as distributions across sets of keywords (bag of words) and these distributions are then fed into the classifiers. A baseline set of keywords to be used is the Moral Foundations Dictionary (MFD). In our machine learning approach we can already improve over basic MFD label counting because the relationship between MFD keywords and moral foundations classes is learned. In addition we generate keyword sets from the data.

First, data was changed to lower case and hexadecimal codes for emojis were removed, leaving plain text for coding and analysis. Also URLs were replaced with the code 'URL' in order to determine the frequency of link sharing, rather than the most popularly shared links. Next, generated frequency counts for the 100 most common words and 100 most common bi-grams (pairs of consecutive words) were produced for efficiency. Optionally, once the relevance of the data was confirmed by Ethics scholars, a list of common stop words was applied. Stop words contain the most common words in a language and corpus. Removal of these words often yields more accurate predictions in linguistic processing and classification [19]. The most common words were examined without removing stop words, then the most frequently Tweeted words in the data set were added to a standard stop word list, including 'URL', 'greece' and 'euro'. We kept the raw version of the data and keyword sets as well.

The next step was to manually label a random selection of 2000 Tweets with the correct moral foundation. The codes were initially based on the MFD, where related words and synonyms were used to guide classification. Beginning with a dictionary-based approach was useful in order to obtain a more tangible picture of lexical indicators for each of the foundations. However, since the MFD didn't include liberty, a list of synonyms for this foundation was created. Then, detailed descriptions of each of the foundations were used to better understand the nuances in each foundation, as outlined in the work of Graham et al. [8] and Haidt [11]. So the combination of specific, related words as well as detailed descriptions of the foundations were used to code the Tweets.

Manual labeling of Tweets is a challenging task, given the inherent ambiguity of some Tweets, the short length of Tweets, the potential of multiple moral foundations being covered in Tweets and use of writing styles such as irony, sarcasm, satire or mere trolling. We choose not to filter out hard to label Tweets as this could bias the sample, nor did we want to include an 'unknown' category, as it would limit the usefulness of the model for monitoring. We considered approaching it as a multi-label problem, however in our view there were far more cases where the Tweet was simply hard to label due to ambiguity than that there was sufficient evidence to conclude that multiple foundations were being addressed, also given the short lengths of Tweets in contrast to the longer texts (blog posts, articles) studied in related work. For similar reasons we discounted an approach where we would have scored the Tweets on the various dimensions to a particular degree. So we kept it simple by manually labeling each Tweet with a single label. These other approaches are indeed interesting areas for future research, but we decided to generate baseline results first.

The most frequent class occurred in 21 % of cases, thus a majority vote baseline model has an accuracy of 21 %. Two coders also labeled a set of bigrams (N = 112) to determine the degree which coders could agree on moral classes. The coders agreed on 66 % of the classifications. It is acknowledged that coding bigrams more difficult than coding Tweets, yet it gives an indication of intercoder agreement in classifying moral foundations with little contextual information. Therefore, any accuracy higher than 21 % is an improvement of the classifier over selection of the most frequently occurring class, and any accuracy around 66 % would show that the classifier is matching human classification of bigrams.

4   Modeling and Evaluation

Previous research using the MFD was conducted on long texts, examining moral loadings and linguistic relations between these texts [6,18]. Since Tweets are short, single-label output (one classification per Tweet) was chosen over multiple labels. We used Multinomial Naive Bayes (NB) and Maximum Entropy (ME) as classification algorithms [15]. The most relevant key difference for this study relates to the independence of features, where NB assumes conditional independence and ME can exploit contextual information (such relationships between words) for classification. Despite the fact that the independence assumption is typically violated, Naive Bayes has shown in general to be a robust classification method, especially for noisy, high variance problems [16].

The data was split into a training (N = 1,300) and a test set (N = 700). Classifiers were built on the raw data (no stop word removal) and the clean data (stop word removal). NB showed higher overall accuracy (raw = 65 %, clean = 64 %) than ME (raw = 57 %, clean = 55 %). Removing stop words did not seem to increase classifier accuracy for either algorithm. To study the impact of training set size, we trained classifiers on raw data training sets of increasing size, with increments of 100 Tweets, up until a maximum of 1300, whilst keeping the test set constant. We also varied the feature set between the raw features, raw features with the MFD and raw features with the bigrams. The results in Fig. 1 show that NB performance is not significantly improved by adding the dictionary, and performance drops if bigrams are added. Detailed results for ME are omitted for brevity, but ME performs best with the addition of bi-grams, achieving 57 % accuracy, and for training set sizes of 300 instances or more, NB outperforms ME. Under almost all conditions, the NB classifier outperformed ME, shown in Fig. 2 (best feature set for each). Over time, the learning curves of both classifiers flattens. It is therefore expected that additional training data will not improve classifier accuracy.

These results were confirmed by a 5-fold cross-validation comparison, where the mean accuracy for NB was 64.7 % (SD = 0.03, p = .000) compared with the ME mean accuracy of 54.2 % (SD = 0.02, p = .000). The difference in classifier accuracy is significant (T = 13.9, p = .000). Overall, the NB classifier is 10 % points more accurate than ME in classifying Tweets into moral foundations.

Confusion matrices, precision, recall and F measures are provided in Tables 3 and 4 for this model (ME results are omitted for brevity). These tables refer to a subsample of Tweets used for training and testing the model. The most frequently correctly classified foundation was care (N = 108), followed by authority (N = 87). Liberty was the least often correctly classified foundation (N = 38). Despite care being most frequently classified correctly, the precision, recall and

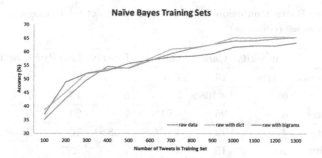

Fig. 1. NB classifier test set accuracy for different training set sizes

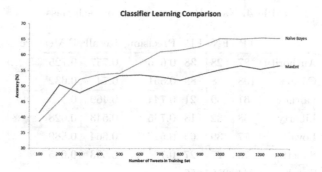

Fig. 2. NB and ME classifiers test set accuracy for different training set sizes

F-measures in Table 4 show otherwise. Taking the relative accuracy into account, authority was the most accurately classified ($F = 0.73$), followed by sanctity ($F = 0.66$) and care ($F = 0.63$). Fairness was the least accurate ($F = 0.58$). Therefore, this model overall works best in identifying Tweets stemming from the foundation of authority.

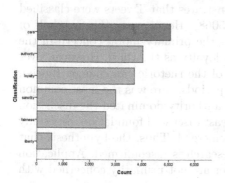

Fig. 3. Classification of all Tweets

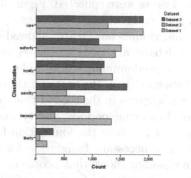

Fig. 4. Classification of Tweets per time period

Table 3. Naive Bayes Confusion Matrix, comparing actual frequencies (rows) and predicted frequencies (columns)

	Authority	Care	Fairness	Liberty	Loyalty	Sanctity
Authority	<87>	13	2	.	9	4
Care	6	<108>	7	5	13	7
Fairness	8	25	<61>	7	23	6
Liberty	4	12	4	<38>	10	2
Loyalty	10	15	5	1	<77>	8
Sanctity	10	23	3	.	14	<73>

Table 4. Naive Bayes accuracy for each class

	TP	FN	FP	Precision	Recall	F-Measure
Authority	87	28	38	0.696	0.757	0.725
Care	108	38	88	0.551	0.74	0.632
Fairness	61	69	21	0.744	0.469	0.575
Liberty	38	32	13	0.745	0.543	0.628
Loyalty	77	39	69	0.527	0.664	0.589
Sanctity	73	27	27	0.73	0.593	0.655
Total	444	256	256			

4.1 Deployment

The most accurate algorithm, with the least training time required (NB, raw data, no MFD) was trained with all labeled data ($N = 2000$). Following learning, the model was used to classify the remaining Tweets ($N = 16,986$). Deployment of the model enabled analysis of changes in moral concerns following key meetings regarding the Greek exit of the Eurozone. There were 3 different time frames where Tweets were collected. Figure 3 demonstrates that Tweets were classified most frequently in the care category ($N = 5068$). Hence, over the first half of 2015, individuals on Twitter showed care as the primary moral concern in the Grexit debate, authority as the second, and loyalty as the third. However, over time, the predominant moral underpinning of the rhetoric can change. Indeed, Fig. 4 shows that in the first and third time periods, care was the most common concern, whereas in the second time period, authority dominated the discussion overall. In all time periods, liberty was the least discussed foundation, especially in data set 2, where the foundation barely emerged. Thus, the hypothesis that liberty is a necessary foundation for this research is disconfirmed. Application of the classifier shows that people on Twitter are not primarily concerned with liberty or oppression of any party in this debate.

The running means of the Tweets made through the data sets shown in Figs. 5, 6 and 7. These means show the discourse over the number of Tweets,

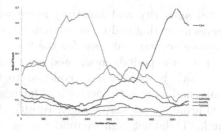

Fig. 5. Running Mean for data set 1   **Fig. 6.** Running Mean for data set 2

disregarding the time and day they were sent. This compensates for different time zones and allows time for news to disseminate. In Fig. 5, care has two dominant peaks, despite initial discussion referring to authority. Towards the end of the week, loyalty and fairness was behind the discussion. Figure 6 shows that in the first Tweets of data set 2, authority is a key concern, but is replaced with care in the later Tweets. In the final data set (see Fig. 7) there are multiple points of interest. The first peak shows that authority drove the early Tweets, followed by loyalty. At the end of the discussion, care became the dominant foundation. Liberty was not a relative point of concern in any data set.

Fig. 7. Running Mean for data set 3

One key finding is that the data shows that public discussion is not in line with analysts moral view of the situation, as shown especially in Figs. 6 and 7. Economic analysts tend to approach Grexit discussion from angles of fairness and loyalty, such as the potential loyalty of Greece shifting beyond the European Union if they were to leave [13]. The public discussion frequently centers on the foundation of care, referring to helping Greece with extensions or bailouts. Loyalty was indeed more present than fairness, but clearly care and authority were salient moral foundations especially in data sets 2 and 3.

5 Discussion

The results show that the NB classifier is a good starting point for attributing moral foundations to Tweets. The three most accurately classified foundations (care, authority and sanctity) agree with previous research [6]. The learning curves show that coding more than 2,000 Tweets for training a classifier will not improve accuracy, at least for the feature sets and ground truth used.

For time periods monitored care is the primary moral concern of the public, which is somewhat in contrast to the dominant economic views that are concerned more with loyalty and fairness.

Perhaps most surprisingly, results also showed that addition of words from the MFD did not improve model accuracy. Therefore, the usefulness of the MFD in a frequency based classification approach is called into question. If using this dictionary is desired in future research, improvements to the dictionary should be made by including words identified as the most informative features following training the NB algorithm. However, the efforts in improvement of the MFD may only have marginal implications for model accuracy. It may be prudent to discontinue the MFD, since these dictionaries are costly to build and maintain, and a pure machine learning approach has similar accuracy and uses less assumptions.

6   Conclusion

This study presents several experiments to determine if machine learning methods can be used to accurately detect moral foundations in Tweets regarding the Grexit. A Naive Bayes (NB) model trained on raw data was 10 % points more accurate than the a Maximum Entropy (ME) model, with best results achieved on raw data without bigram or Moral Foundations Dictionary (MFD) attributes. Specifically, the fact that the NB model doesn't require the handcrafted MFD is an interesting result.

At this point, it is difficult to compare with other moral foundation classification research, as thus far none have used a machine learning approach. However, the accuracy of the NB model is comparable to the agreement of moral classification between humans for bigrams (64.7 % compared with 66 %, respectively). Moreover, the model is roughly 3 times more accurate than the ZeroR measure of 21.4 %. Hence, using a NB classifier is a good starting point for categorization of Tweets into their dominant moral foundations.

References

1. Adamic, L.A., Glance, N.: The political blogosphere and the 2004 US election: divided they blog. In: Proceedings of the 3rd International Workshop on Link Discovery, pp. 36–43. ACM (2005)
2. Agarwal, A., Xie, B., Vovsha, I., Rambow, O., Passonneau, R.: Sentiment analysis of Twitter data. In: Proceedings of the Workshop on Languages in Social Media, pp. 30–38. Association for Computational Linguistics (2011)
3. Bird, S., Klein, E., Loper, E.: Natural Language Processing with Python. O'Reilly Media, Sebastopol (2009)
4. Bond, J.: It aint over till the fat lady sings. Sens-Public (2012). http://www. sens-public.org/article979.html

5. Clifford, S., Jerit, J.: How words do the work of politics: moral foundations theory and the debate over stem cell research. J. Politics **75**(03), 659–671 (2013)
6. Dehghani, M., Sagae, K., Sachdeva, S., Gratch, J.: Linguistic analysis of the debate over the construction of the Ground Zero Mosque. J. Inform. Technol. Politics **11**, 1–14 (2014)
7. Freelon, D.: On the interpretation of digital trace data in communication and social computing research. J. Broadcast. Electron. Media **58**(1), 59–75 (2014)
8. Graham, J., Haidt, J., Koleva, S., Motyl, M., Iyer, R., Wojcik, S.P., Ditto, P.H.: Moral foundations theory: the pragmatic validity of moral pluralism. Adv. Exp. Soc. Psychol. **47**, 55–130 (2013)
9. Graham, J., Haidt, J., Nosek, B.A.: Liberals and conservatives rely on different sets of moral foundations. J. Pers. Soc. Psychol. **96**(5), 1029 (2009)
10. Grauwe, P.: The eurozone as a morality play. Intereconomics Rev. Eur. Econ. Policy **46**(5), 230–231 (2011)
11. Haidt, J.: The righteous mind: why good people are divided by politics and religion. Vintage, New York (2012)
12. Haidt, J., Joseph, C.: Intuitive ethics: how innately prepared intuitions generate culturally variable virtues. Daedalus **133**(4), 55–66 (2004)
13. Lazarou, A.: Greece: The many faces of Yanis Varoufakis. Green Left Weekly (104) (2015)
14. Lazer, D., Pentland, A.S., Adamic, L., Aral, S., Barabasi, A.L., Brewer, D., Christakis, N., Contractor, N., Fowler, J., Gutmann, M., et al.: Life in the network: the coming age of computational social science. Science **323**(5915), 721 (2009). (New York, NY)
15. Manning, C.D., Schütze, H.: Foundations of Statistical Natural Language Processing. MIT Press, Cambridge (1999)
16. van der Putten, P., van Someren, M.: A bias-variance analysis of a real world learning problem: the CoIL Challenge 2000. Mach. Learn. **57**(1), 177–195 (2004)
17. Ratkiewicz, J., Conover, M., Meiss, M., Gonçalves, B., Patil, S., Flammini, A., Menczer, F.: Truthy: mapping the spread of astroturf in microblog streams. In: Proceedings of the 20th International Conference Companion on World Wide Web, pp. 249–252. ACM (2011)
18. Sagi, E., Dehghani, M.: Measuring moral rhetoric in text. Soc. Sci. Comput. Rev. **32**(2), 132–144 (2014)
19. Saif, H., Fernández, M., Alani, H.: Automatic stopword generation using contextual semantics for sentiment analysis of Twitter. In: CEUR Workshop Proceedings, vol. 1272 (2014)
20. Suhler, C.L., Churchland, P.: Can innate, modular foundations explain morality? Challenges for Haidt's moral foundations theory. J. Cogn. Neurosci. **9**, 2103–2116 (2011)
21. Taboada, M., Brooke, J., Tofiloski, M., Voll, K., Stede, M.: Lexicon-based methods for sentiment analysis. Comput. Linguist. **37**(2), 267–307 (2011)
22. Tumasjan, A., Sprenger, T.O., Sandner, P.G., Welpe, I.M.: Predicting elections with Twitter: what 140 characters reveal about political sentiment. ICWSM **10**, 178–185 (2010)

A Hybrid Approach for Probabilistic Relational Models Structure Learning

Mouna Ben Ishak[1](✉), Philippe Leray[2], and Nahla Ben Amor[1]

[1] LARODEC Laboratory, ISG, Université de Tunis, Tunis, Tunisia
mouna.benishak@gmail.com
[2] DUKe Research Group, LINA Laboratory UMR 6241,
University of Nantes, Nantes, France

Abstract. Probabilistic relational models (PRMs) extend Bayesian networks (BNs) to a relational data mining context. Just like BNs, the structure and parameters of a PRM must be either set by an expert or learned from data. Learning the structure remains the most complicated issue as it is a NP-hard problem. Existing approaches for PRM structure learning are inspired from classical methods of learning the BN structure. Extensions for the constraint-based and score-based methods have been proposed. However, hybrid methods are not yet adapted to relational domains, although some of them show better experimental performance, in the classical context, than constraint-based and score-based methods, such as the Max-Min Hill Climbing ($MMHC$) algorithm. In this paper, we present an adaptation of this latter to relational domains and we made an empirical evaluation of our algorithm. We provide an experimental study where we compare our new approach to the state-of-the art relational structure learning algorithms.

Keywords: Probabilistic relational model · Relational structure learning · Relational Max-Min Hill Climbing

1 Introduction

Statistical relational learning (SRL) has emerged as a field of machine learning that enables effective and robust reasoning about relational data structures [8]. Probabilistic relational models (PRMs) [10,17] are an extension of Bayesian networks (BNs) [16] which allow to work with relational database representation rather than propositional data representation. PRMs are interested in manipulating structured representation of the data, involving objects described by attributes and participating in relationships, actions, and events. The probability model specification concerns classes of objects rather than simple attributes.

In order to be used, PRMs have to be constructed either by an expert or using learning algorithms. PRM learning implies finding a graphical structure as well as a set of conditional probability distributions that fit the best way to the relational training data. PRM structure learning remains the most challenging

© Springer International Publishing AG 2016
H. Boström et al. (Eds.): IDA 2016, LNCS 9897, pp. 38–49, 2016.
DOI: 10.1007/978-3-319-46349-0_4

issue, as it is considered as a NP-Hard problem [7]. Only few works have been proposed to learn PRMs [6] or almost similar models [11,13,14] from relational data. Proposed algorithms are inspired from standard BNs learning approaches. Those latter are divided into three families, namely, constraint-based, score-based and hybrid approaches [5]. PRM structure learning approaches are adaptations of either constraint-based or score-based approaches. However, it has been shown that, for BNs, some hybrid approaches provide better experimental results than constraint-based and score-based methods [18]. In this paper we present a new hybrid algorithm to learn the structure of a PRM from a complete relational dataset. Our proposal is an adaptation of the Max-Min Hill Climbing ($MMHC$) algorithm [18]. We call it Relational Max-Min Hill Climbing algorithm, $RMMHC$ for short. Also, we provide an experimental study where we compare the $RMMHC$ algorithm to state-of-the-art methods. The remainder of this paper is as follows: Sect. 2 presents useful background and discusses related work. Section 3 details the $RMMHC$ algorithm. Section 4 provides the empirical study. Finally, Sect. 5 concludes and outlines some perspectives.

2 Background

We start by providing a brief recall on PRMs and presenting methods to learn BNs and PRMs structure from data.

2.1 Probabilistic Relational Models

A PRM is defined through two components: a graphical one, a dependency structure defined over the attributes of a relational structure (i.e., an entity-relationship model or a relational schema) containing classes ans class attributes, and a numerical component that quantifies probabilistic dependencies between variables of the relational structure.

Relational model. A relational structure consists of a set of classes $\mathcal{X} \equiv \mathcal{E} \cup \mathcal{R}$, where $\mathcal{E}$ is a set of entity classes and $\mathcal{R}$ is a set of relationship classes. Each $R \in \mathcal{R}$ links a set of entity classes $R(E_i \ldots E_j)$. Each $X \in \mathcal{E} \cup \mathcal{R}$ has a set of attributes denoted by $\mathcal{A}(X)$. Every attribute takes on a range of values $\mathcal{V}(X.A)$.

A relational skeleton σ is a partial specification of an instance of a relational structure. It specifies the set of class objects that exist in a domain and the relations that hold between them.

Example 1. An example of a relational structure is depicted in Fig. 1(a), with three classes $\mathcal{X} = \{Movie, Vote, User\}$. $\mathcal{E} = \{Movie, User\}$. $\mathcal{R} = \{Vote\}$ The entity class $User$ has three attributes $\mathcal{A}(User) = \{Gender, Age, Occupation\}$. The linked entities of the relationship $Vote$ are $Movie$ and $User$ (Dotted links).

Figure 1(b) shows an example of a relational skeleton for the relational schema of Fig. 1(a). It consists of three $User$ objects and five $Movie$ objects. User $user1$ has voted for two movies $M = movie1$ and $movie2$.

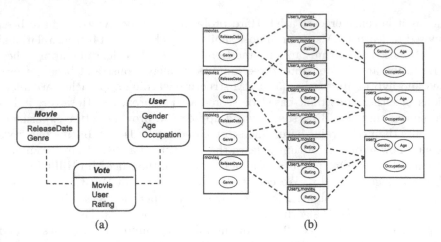

Fig. 1. Example of a relational structure (a) and a relational skeleton (b) for the movie domain inspired from the MovieLens dataset http://grouplens.org/datasets/movielens/

Probabilistic model. A PRM $\mathcal{M} = (\mathcal{S}, \Theta)$ brings together the strengths of probabilistic graphical models and the relational representation of data. A dependency structure $\mathcal{S}$ is constructed by adding probabilistic dependencies between class attributes, $\forall X.A \in \mathcal{A}(X)$, there is a set of parents $Pa(X.A) = \{U_1, \ldots, U_l\}$. The numerical component is composed of the conditional probability distributions (CPD) of the attributes in the context of their parents in the dependency structure $P(X.A|Pa(X.A))$. Probabilistic dependencies may be intra or inter classes, this depends on the path that connects the child to its parent. Several paths may be found depending on the way how the relational structure has been traversed. Friedman et al. [6] specify the path between the parent and child variables using a slot chain. Heckerman et al. [9] refer to as constraint and Maier et al. [13] call it relational path.

Moreover, depending on the cardinality (i.e., the number of items an entity can participate in a relationship), it is possible for an attribute object to have multiple parents objects (i.e., a *Many* cardinality). This number of parents is finite but not known in advance and it varies from one object to another. Whereas, there is only one CPD shared among all objects of a given parent attribute $X.A$. To address this issue, the notion of aggregation has been adopted from database theory: *An aggregate γ takes a multiset of values of some ground type, and returns a summary of it. γ can be the MAX, MIN, MODE, etc.*

Each parent U_i has then the form $X.B$ if it is a simple attribute in the same class. $X.K.B$ or $\gamma(X.K.B)$ otherwise, where K is a path and γ is an aggregation function. Aggregators are needed if the path contains at least a *Many* cardinality.

Example 2. Figure 2(a) shows a PRM for the relational structure of Fig. 1(a). *User.Occupation* has two parents from the same class *User*. *Vote.Rating* has two parents: *Vote.User.Gender* from the *User* class and *Vote.Movie.Genre* from

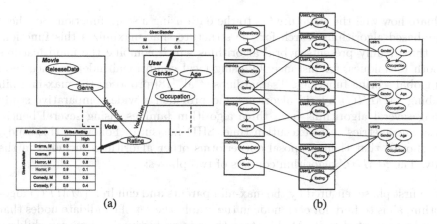

Fig. 2. Example of a probabilistic relational model (a) and a ground graph (b) for the movie domain of Fig. 1

the *Movie* class. *Vote.Movie.genre* → *Vote.rating* is an example of a probabilistic dependency derived from a path of length one where *Vote.Movie.genre* is the parent and *Vote.rating* is the child as shown by Fig. 2(a). Also, varying the path length may give rise to other dependencies. For instance, using a path of length three, we can have a probabilistic dependency from $\gamma(Vote.User.User^{-1}.Movie.genre)$ to *Vote.rating*. In this case, *Vote.rating* depends probabilistically on an aggregate value of all the genres of movies rated by a particular user. $User.User^{-1}.Movie$ is the path and $User^{-1}$ specifies the path part which involves the use aggregators.

Given a PRM $\mathcal{M}$ and a relational skeleton σ, we can construct a ground Bayesian network (GBN) by applying the probabilistic dependencies specified in $\mathcal{M}$ to the object attributes of σ. The CPD for each $x.A$ is inherited from the CPD $P(X.A|Pa(X.A))$ defined in the PRM. An example of the graphical structure of a GBN is shown by Fig. 2(b).

2.2 From BN to PRM Structure Learning

A wealth of literature has been produced that seeks to understand and provide methods for BN structure learning from data [5]. Some of the proposed approaches have been extended to learn from relational domains. In this section we start by a brief survey on BN structure learning approaches, then we present existing approaches for PRM structure learning.

BN structure learning is known as an NP-Hard problem [3]. BN structure learning methods are divided into three main families. The first family tackles this issue as a constraint satisfaction problem. Constraint-based algorithms look for independencies (dependencies) in the data, using statistical tests then, try to find the most suitable graphical structure with this information. The second family treats structure learning as an optimization problem. They evaluate how well the structure fits to the data using a score function. So, these Score-based algorithms search for the structure that maximizes this function. The third family presents hybrid algorithms which combine the main features of both techniques, for instance, by using local conditional independence tests and global scoring functions. Tsamardinos et al. [18] proposed the max-min hill climbing ($MMHC$) hybrid algorithm and provided a wide comparative study among several algorithms from three algorithm families, using several benchmarks and metrics (e.g., execution time, SHD measure). Following this study, they showed that their proposal outperforms other algorithms included in the study. The $MMHC$ algorithm consists of two phases:

- The first phase, ensured by the max-min parents and children ($MMPC$) algorithm, aims to find, for each node in the graph, the set of candidate nodes that can be connected to it. At this stage there is no distinction between children and parents nodes and links orientation is not of interest. $MMPC$ discovers the set of candidate parents and children (CPC) for a target variable T. It consists of a raw neighborhood identification step ensured by the $\overline{MMPC}$ algorithm and an additional symmetrical correction step, where $MMPC$ removes from each set $CPC(T)$ each node X for which $T \notin CPC(X)$. $\overline{MMPC}$ consists of a forward phase where for each variable T of the graph, a set of variables are added to $CPC(T)$, and a backward phase whose role is to remove false dependencies detected in the forward phase. Dependency is measured using an association measurement function such as mutual information or χ^2.
- The second phase allows the construction of the graph $\mathcal{G}$ using the greedy search heuristic constrained to the set of candidate parents and children of each node resulting from the first phase.

PRM structure learning aims at finding the dependency structure $\mathcal{S}$ for a given relational structure and a relational observational dataset that instantiates this structure. As we have seen in Sect. 2.1, paths may be arbitrary large and give rise to complicated models. So that a user specified value, a maximum path length (K_{max}), is required to limit the length of possible paths that one can cross in the model. Only few works have been proposed to learn PRM structure from relational data [6,11,13,14]. These latter are inspired from classical methods for BN structure learning.

Friedman et al. [6] proposed the Relational Greedy Hill-Climbing Search (RGS) algorithm. For each path length $k \in \{0, K_{max}\}$ RGS defines a hypothesis space of potential PRM structures (i.e., neighbors) it is willing to consider, using the add_edge, $delete\_edge$ and $reverse\_edge$ operators. Then, it computes the score of each neighbor, and keeps the graph that has the best score, until it reaches a structure that has the highest score in the list of neighbors. As score function, they used a relational extension of the Bayesian Dirichlet (BD) score [4]. In this process, the neighborhood search space could be super-exponential.

Maier et al. proposed two constraint-based approaches. The first is a relational extension of the PC algorithm to learn PRM structure from relational data [14]. Yet, unlike the PC algorithm which is sound and complete the RPC

algorithm did not satisfy these criteria. The second approach comes to refine
the RPC algorithm [13]. They proposed the relational causal discovery (RCD)
algorithm and proved that this approach is sound and complete for causally sufficient relational data. The RCD algorithm performs on two phases. In the first
phase, given a maximum path length, RCD starts by providing the set of all
potential dependencies. Then continues by removing conditional independences
found using conditional independence tests. Because of asymmetry caused by
the use of aggregate functions, RCD verifies whether a statistical association is
detected between two variables in both directions and it leaves the dependency
if a statistical association exists in at least one direction, but omits this information about orientation. In the second phase, RCD determines the orientation
of the dependencies discovered previously. Orientation rules are similar to those
used by the PC algorithm. In [11] the authors proposed a refined version of the
RCD algorithm in term of time complexity and space.

Algorithm 1 $\overline{RMMPC}$

Require: $schema$: A relational model, $\mathcal{D}$: A database instance
  $Current\_Path\_length$: A path length, T: A target attribute
Ensure: CPC: The set of parents and children of T, $CPC(T) = CPC(T)^{sym} \cup$
  $CPC(T)^{asym}$
  1: $Pot_{list} = Generate\_potential\_list(T, Current\_Path\_length)$
     % Phase I: Forward
  2: **repeat**
  3:   $\langle F, assocF \rangle = MaxMinHeuristic(T, CPC(T), Pot_{list})$
  4:   **if** $assocF \neq 0$ **then**
  5:     **if** $Current\_path\_length = 0$ OR $does\_Not\_Contains\_Many\_Relationship(F)$
         then
  6:       $CPC(T)^{sym} = CPC(T)^{sym} \cup F$
  7:     **else**
  8:       $CPC(T)^{asym} = CPC(T)^{asym} \cup F$
  9:     **end if**
 10:     $CPC(T) = CPC(T)^{sym} \cup CPC(T)^{asym}$
 11:     $Pot_{list} = Pot_{list} \backslash F$
 12:   **end if**
 13: **until** CPC has not changed or $assocF = 0$ or $Pot_{list} = \emptyset$
     % Phase II: Backward
 14: **for all** $A \in CPC(T)$ **do**
 15:   **if** $\exists S \subseteq CPC(T), s.t. Ind(A; T|S)$ **then**
 16:     $CPC(T) = CPC(T) \backslash \{A\}$
 17:   **end if**
 18: **end for**

Hybrid approaches combine both techniques and some algorithms, such as the
$MMHC$, experimentally outperforms the classical approaches. Yet, no hybrid
algorithm has been proposed for PRMs. In the next section, we will provide a
new hybrid approach to learn PRM structure from relational data. Our proposal
is a relational extension of the $MMHC$ algorithm detailed at Sect. 2.2, that we
refer to as relational max min hill climbing ($RMMHC$).

3 $RMMHC$: The Relational Max Min Hill Climbing Algorithm

$RMMHC$ preserves the same phases as the $MMHC$ algorithm (cf. Sect. 2.2). The neighborhood identification phase, ensured by the $RMMPC$ algorithm, handles asymmetry caused by the use of aggregators and leads to a partially oriented neighborhood (cf. Sect. 3.1). This latter is then used to simplify the global structure identification phase (cf. Sect. 3.2).

3.1 Relational Max Min Parents and Children: $RMMPC$

Neighborhood identification: $\overline{RMMPC}$. The $\overline{RMMPC}$ algorithm aims to find the list of neighbors of a target attribute T, that consists of either children or parents of T, from a set of potential variables. For BNs, $\overline{MMPC}$ does not make a difference between a node in the graph structure and a variable, and the potential set of parents and children of a node T is $\mathcal{V}\backslash T$, where $\mathcal{V}$ is the set of BN nodes. While, a PRM is a meta-model used to describe the overall behavior of a system in a relational domain. For a PRM, and due to the horizon of crossed paths, the number of potential variables is not fixed. Thus, we have to make the difference between an attribute and a variable:

Algorithm 2 $RMMPC$

Require: *schema*: A relational model, $\mathcal{D}$: A database instance, $Current\_Path\_length$: A path length, T: A target attribute
Ensure: CPC: The set of parents and children of T, $CPC(T) = CPC(T)^{sym} \cup CPC(T)^{asym}$
1: **if** $Current\_Path\_length = 0$ **then**
2:     $CPC(T)^{sym} = \emptyset, CPC(T)^{asym} = \emptyset, CPC(T) = \emptyset$
3: **end if**
4: $CPC(T) = \overline{RMMPC}(schema, \mathcal{D}, T, Current\_Path\_length)$
5: **for all** $A \in CPC(T)$ **do**
6:     **if** $Current\_Path\_length = 0$ **then**
7:         $CPC(A)^{sym} = \emptyset, CPC(A)^{asym} = \emptyset, CPC(A) = \emptyset$
8:     **end if**
9:     $CPC(A) = \overline{RMMPC}(schema, \mathcal{D}, A, Current\_Path\_length)$
10:     **if** $A \in CPC(T)^{sym}$ AND $T \notin CPC(A)^{sym}$ **then**
11:         $CPC(T) = CPC(T)\backslash\{A\}$
12:     **end if**
13: **end for**

- An attribute is characterized by its name, domain, a set of possible aggregators and the class that it belongs to. A child is an attribute.
- A variable is characterized by its name, domain, the class that it belongs to, a specific aggregator type and the path that it is derived from. A parent is a variable and its path starts from the class to which the child belongs. This notion is defined in [13] as a canonical dependency.

Consequently, each parent is a variable, while each target is an attribute. When searching the $CPC(T)$, T is a target attribute. $CPC(T)$ consists of the candidate parents and children of T, and $|CPC(T)|$ depends on the length of the traversed path $k \in \{0 \ldots K_{max}\}$. For each value of k, a subset of potential parents and children can be generated. As the final generated $CPC(T)$ list may be very large, we adopt the same strategy as [6] and we proceed by phases. That is, suppose that we want to provide the list of children and parents of each attribute T given a maximum path length k_{max}, the neighborhood identification will be done on $k_{max} + 1$ phases. At phase 0, we will search for the set of parents and children of attribute T from the same class as T, at phase 1, we will search for the set of parents and children of attribute T in classes related to T class using paths of length one. At phase 2, we will go through further classes and search for the set of parents and children of attribute T in classes related to T class by traversing paths of length 2 and so on. The neighborhood identification, for one specified value of path length, is described by Algorithm 1. The *Generate_potential_list* method aims to identify the list of potential parents and children of a target attribute T given a path length k. Its result is a set of potential variables of the form $X_T.A$ for intra-class dependencies and $X_T.K.Y.A$ or $\gamma(X_T.K.Y.A)$ for inter-class dependencies, where K is a path of length k. More details about this method can be found in [1].

On the other hand, as some dependencies may require aggregators, there is an inherent asymmetry and this list of candidate dependencies is closely related to the path composition. So that, we propose to divide the neighborhood list, CPC, into two sub-lists. Formally, $CPC(T) = CPC(T)^{sym} \cup CPC(T)^{asym}$, where:

- $CPC(T)^{sym}$: The set of potential children and parents of target attribute T coming either from the same class as T, with path length equal to 0 or from paths that do not contain any *Many* relationship.
- $CPC(T)^{asym}$: The set of potential variables coming from the other paths. In this case, A could only be a potential parent of T [7].

As for the standard case [18], $MaxMinHeuristic$ selects the variables that maximize the $MinAssoc$ with target attribute T conditioned to the subset of the currently estimated $CPC(T) = CPC(T)^{sym} \cup CPC(T)^{asym}$.

Symmetrical correction: $RMMPC$. The $RMMPC$ algorithm (Algorithm 2) comes to refine the result of Algorithm 1 by applying a symmetrical correction to the $CPC(T)$ provided by $\overline{RMMPC}$. As $CPC(T)$ consists of two subsets, the symmetrical correction depends on the concerned subset.

- For each $A \in CPC(T)^{sym}$, we must verify that $T \in CPC(A)^{sym}$, otherwise, A has to be removed from $CPC(T)^{sym}$. This symmetrical correction is equivalent to the symmetrical correction of standard $MMPC$.
- For each $A \in CPC(T)^{asym}$, we cannot apply the symmetrical correction since the SQL queries involved in such a case are not equivalent and the resulting datasets on which we will apply statistical tests are not the same. However, $\forall A \in CPC(T)^{asym}$, A can only be a parent of T. By this way, we can deduce the dependency direction, directly from the first phase of $RMMHC$.

A detailed toy example on the various steps of this phase can be found in [1].

3.2   Global Structure Identification

The global structure identification is performed using a score-based algorithm only on the set of variables derived from the first local search phase. We choose to work with the RGS procedure, using the relational Bayesian score. In this case, $Pot_K(X.T)$ consists of the CPC list of attribute $X.T$ found on the local search step. As this set contains two subsets, the choice of the operator to be performed during the neighbors generation process depends on the concerned subset:

Algorithm 3 $RMMHC$

Require: *schema:*   A   relational   model, $\mathcal{D}$:   A   database   instance, k_{max}: $Maximun\_Path\_Length$
Ensure: The local optimal dependency graph $\mathcal{S}$
   % *Local search*
1: **for** $Current\_Path\_length = 0$ to k_{max} **do**
2:    **for all** T **do**
3:       $CPC(T) = RMMPC(schema, \mathcal{D}, T, Current\_Path\_length)$
4:    **end for**
5: **end for**
   % *Global search*
6: $\mathcal{S} = RGS(schema, \mathcal{D}, CPC)$

- For $CPC(T)^{sym}$: each $A \in CPC(T)^{sym}$ can be either a child or a parent of $X.T$ so all the operators, namely, add_edge, $delete\_edge$ and $reverse\_edge$ can be tested.
- For $CPC(T)^{asym}$: each $A \in CPC(T)^{asym}$ is a potential parent of $X.T$ so only the add_edge and $delete\_edge$ operators can be tested.

The global search step is expensive in term of complexity, since the size of the generated neighborhood may increase rapidly. $RMMHC$ performs the local search procedure in phases until reaching the K_{max} value. The result of this search procedure will be the CPC list of all variables for all path lengths. This partially directed result allows to further reduce the size of the search space during greedy search. It is used as input to the global search procedure that will be run only one time. The overall process is as presented by Algorithm 3.

3.3   Time Complexity of the Algorithms

The $MMPC$ algorithm consists of the $\overline{MMPC}$ algorithm of complexity $\mathcal{O}(|Pot_{list}|.2^{|CPC|})$ and an additional symmetrical correction. Thus, its overall complexity is $\mathcal{O}(|Pot_{list}|^2 .2^{|CPC|})$. At each iteration of the classical greedy search algorithm, the number of possible local changes is bounded by $\mathcal{O}(\mathcal{V}^2)$, where $\mathcal{V}$ is the number of nodes in the graph [18].

Our $\overline{RMMPC}$ algorithm presents the same steps as for the standard case, augmented with the $Generate\_potential\_list$ procedure which is of complexity $\mathcal{O}(N^k)$, where N is the number of classes and k is the current path length. Thus, its time complexity, at each k value, $k \in \{0 \ldots K_{max}\}$ remains equal to

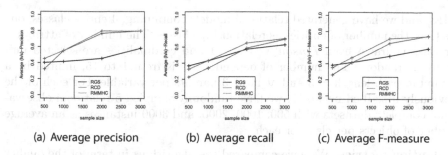

(a) Average precision   (b) Average recall   (c) Average F-measure

Fig. 3. The average values of Precision, Recall and F-Measure with respect to the sample size

$\mathcal{O}(|Pot_{list}|.2^{|CPC|})$. Thus, augmented with the symmetrical correction, the time complexity of the $RMMPC$ algorithm is $\mathcal{O}(|Pot_{list}|^2.2^{|CPC|})$. For RGS, we have to iterate on attributes and for each attribute, we have to iterate on the list of all its potential parents. Let us consider β the number of potential parents that could be reached, then the number of possible local changes is bounded by $\mathcal{O}(\beta.\mathcal{V})$. Note that $\beta = |CPC|$ when the RGS is called after a local search step performed using $RMMPC$ algorithm. In $RMMHC$ algorithm, the local search step has been augmented with an outer loop presenting the current path length to consider at each iteration. Thus the final complexity of the local search is $\mathcal{O}(K_{max}.|Pot_{list}|^2.2^{|CPC|})$.

4 Experiments

We will compare the $RMMHC$ algorithm to the state-of-the-art approaches, namely, the RGS and RCD algorithms (cf. Sect. 2.2). The RCD is supposed to correct the theoretical problems of RPC and an experimental study on these two approaches can be found in [13]. Thus the RPC algorithm is excluded from the comparative study. In term of specific implementations, we have re-implemented the RGS algorithm and used our version in the experimental study. We have used the source code of the RCD algorithm available in[1]. As both RCD and $RMMHC$ use statistical independence tests, we have implemented the linear regression test to fit the RCD implementation and we have used it to perform statistical tests during the local search phase of $RMMHC$. To judge conditional independence, we have run both RCD and $RMMHC$ using a threshold $\alpha = 0.05$.

Networks and Datasets. Unlike standard Bayesian networks, where a set of ground truth models (i.e., benchmarks) is available to perform experimentations, there is no such models defined in the context of PRMs. Consequently, we have used our generating process, already described in [2] to generate gold models and relational database instances. We have followed the same experimental protocol

---

[1] https://kdl.cs.umass.edu/display/public/Relational+Causal+Discovery.

as [13] and we have generated relational models containing: 4 entity classes, one less than the number of entities as relationship classes. The number of attributes per class is drawn from $Poison(\lambda = 1) + 1$ and cardinalities are selected uniformly at random. The number of dependencies is from 1 to 15, limited by a maximum path length $= 3$ and at most 3 parents per variable. For each of the previously described networks, we have randomly sampled 5 relational observational complete datasets with 500, 1000, 2000, and 3000 instances as an average number of objects per class for each.

Evaluation metrics. We have compared the algorithms in term of the quality of reconstruction. Using the Precision, Recall and F-score measurement defined in [13].

Experimental results. Figure 3 presents the experimental results in term of Precision, Recall and F-score. RGS presents the worst result for all sample sizes ≥ 1000. $RMMHC$ outperforms RGS and RCD in term of Precision for all sample sizes and it presents the best Recall and F-score values for sample sizes ≥ 1000. For small sample size ($=500$), $RMMHC$ and RGS have similar results, followed by the RCD algorithm. Figure 3(a) shows that for sample size ≥ 1000, beyond 50 % of the dependencies retrieved by $RMMHC$ are relevant. Figure 3(b) shows that for sample size ≥ 1000, $RMMHC$ was able to find beyond 40 % of the relevant dependencies. Both values are increased by raising the sample size.

5   Conclusion

We proposed a first hybrid approach to learn PRMs structure from relational observational data. Our $RMMHC$ algorithm is based on a local search phase that allows to handle asymmetry and leads to a partially oriented neighborhood. This latter is used as input to simplify the global structure identification phase, optimize the search space and consequently enhance the scalability. We have also presented a first comparative study of state-of-the-art relational structure learning approaches and experiments showed that our approach presents good results in term of quality of reconstruction. However, this work is just the beginning for several challenging research tasks.

$RMMHC$ can be improved to deal with more complex structural uncertainty [7], or it can be adapted to learn PRM extensions [15]. Another avenues for future research is combining other theories to learn the model structure [12]. Also, one interesting perspective consists on the use of some prior knowledge, derived from knowledge representation frameworks such as ontologies, as input to the learning process.

References

1. Ben Ishak, M.: Probabilistic relational models: learning and evaluation. Ph.D. dissertation, Université de Nantes, Ecole Polytechnique; Université de Tunis, Institut Supérieur de Gestion de Tunis (2015)
2. Ben Ishak, M., Leray, P., Ben Amor, N.: Probabilistic relational model benchmark generation: principle and application. Intell. Data Anal. Int. J. **20**, 615–635 (2016)
3. Chickering, D.M., Geiger, D., Heckerman, D.: Learning Bayesian networks is NP-hard. Technical report, MSR-TR-94-17, Microsoft Research (1994)
4. Cooper, G.F., Herskovits, E.: A Bayesian method for the induction of probabilistic networks from data. Mach. Learn. **9**, 309–347 (1992)
5. Daly, R., Shen, Q., Aitken, S.: Learning Bayesian networks: approaches and issues. Knowl. Eng. Rev. **26**, 99–157 (2011)
6. Friedman, N., Getoor, L., Koller, D., Pfeffer, A.: Learning probabilistic relational models. In: Proceedings of the International Joint Conference on Artificial Intelligence, pp. 1300–1309 (1999)
7. Getoor, L., Koller, D., Friedman, N., Pfeffer, A., Taskar, B.: Probabilistic relational models. In: Getoor, L., Taskar, B. (eds.) Introduction to Statistical Relational Learning. MIT Press, Cambridge (2007)
8. Getoor, L., Taskar, B.: Introduction to Statistical Relational Learning. MIT Press, Cambridge (2007)
9. Heckerman, D., Meek, C., Koller, D.: Probabilistic models for relational data. Technical report, Microsoft Research, Redmond, WA (2004)
10. Koller, D., Pfeffer, A.: Probabilistic frame-based systems. In: Proceedings of AAAI, pp. 580–587. AAAI Press (1998)
11. Lee, S., Honavar, V.: On learning causal models from relational data. In: Proceedings of the 30th AAAI Conference on Artificial Intelligence (AAAI 2016), pp. 3263–3270 (2016)
12. Li, X.-L., He, X.-D.: A hybrid particle swarm optimization method for structure learning of probabilistic relational models. Inf. Sci. **283**, 258–266 (2014)
13. Maier, M., Marazopoulou, K., Arbour, D., Jensen, D.: A sound and complete algorithm for learning causal models from relational data. In: Proceedings of the Twenty-Ninth Conference on Uncertainty in Artificial Intelligence, pp. 371–380 (2013)
14. Maier, M., Taylor, B., Oktay, H., Jensen, D.: Learning causal models of relational domains. In: Proceedings of the Twenty-Fourth AAAI Conference on Artificial Intelligence, pp. 531–538 (2010)
15. Marazopoulou, K., Maier, M., Jensen, D.: Learning the structure of causal models with relational and temporal dependence. In: Proceedings of the Thirty-First Conference on Uncertainty in Artificial Intelligence, pp. 572–581 (2015)
16. Pearl, J.: Probabilistic Reasoning in Intelligent Systems. Morgan Kaufmann, San Francisco (1988)
17. Pfeffer, A.J.: Probabilistic reasoning for complex systems. Ph.D. dissertation, Stanford University (2000)
18. Tsamardinos, I., Brown, L.E., Aliferis, C.F.: The max-min hill-climbing Bayesian network structure learning algorithm. Mach. Learn. **65**, 31–78 (2006)

On the Impact of Data Set Size in Transfer Learning Using Deep Neural Networks

Deepak Soekhoe[✉], Peter van der Putten, and Aske Plaat

LIACS, Leiden University, P.O.Box 9512, 2300 RA Leiden, The Netherlands
d.p.soekhoe@umail.leidenuniv.nl,
{p.w.h.van.der.putten,a.plaat}@liacs.leidenuniv.nl

Abstract. In this paper we study the effect of target set size on transfer learning in deep learning convolutional neural networks. This is an important problem as labelling is a costly task, or for new or specific classes the number of labelled instances available may simply be too small. We present results for a series of experiments where we either train on a target of classes from scratch, retrain all layers, or subsequently lock more layers in the network, for the Tiny-ImageNet and MiniPlaces2 data sets. Our findings indicate that for smaller target data sets freezing the weights for the initial layers of the network gives better results on the target set classes. We present a simple and easy to implement training heuristic based on these findings.

Keywords: Deep learning · Convolutional neural networks · Transfer learning · Learning curves · AlexNet

1 Introduction

Current deep learning research achieves state-of-the-art performance in image classification tasks [6,13,15,18]. Modern models make use of deep convolutional neural networks (CNN) such as AlexNet [8]. However, training these models on large data sets such as ImageNet [1] can take up a significant amount of time, and the number of labelled examples per class available may be limited, so learning from scratch has its downsides. One approach to overcome this problem is to use transfer learning. The objective of transfer learning is to use knowledge of a source task and *transfer* that to a new target task [10]. It provides considerable benefits over learning from scratch (i.e. from a random initialisation of the weights). One obvious advantage is that a model can learn more efficiently since it starts with a pre-initialised weight matrix.

In their study, Yosinski *et al.* [17] trained AlexNet on the ImageNet data set and found that the first three layers in a CNN contain generic and reusable features. Beyond the third layer, the features gradually become more specific with respect to the source data set. However, the authors did not take into account the size of the target data set, on which the model with the transferred features will be trained.

© Springer International Publishing AG 2016
H. Boström et al. (Eds.): IDA 2016, LNCS 9897, pp. 50–60, 2016.
DOI: 10.1007/978-3-319-46349-0_5

The size of the target data set plays an important role, since it affects how much impact transfer learning will have on the performance. Thus, it is logical to ask how well extracted features generalise to smaller data sets. It would be helpful to know at what data set size transfer learning would be still beneficial. More specifically, at what layer is the model still able to generalize to a small data set size? Therefore, it is of both academic and practical interest to investigate at what *target data set* size transfer learning can still provide any additional value. Furthermore, Yosinski *et al.* only used the ImageNet data set [17]. It would be interesting to find out whether the transfer learning properties are different when using a data set from a different domain.

In this work we will expand the study by [17], and measure the effect of target data set size on the transferability of parameters in convolutional neural networks. Our main contribution is to quantify the extent to which features are able to generalise to the target data set when we systematically reduce its size. We will investigate this for each individual layer by evaluating the accuracy as a function of the data set size. We will have three variants of this. First, we will obtain a base score, without applying any form of transfer learning. In the second condition we will completely fine-tune all the layers of the network. In the third one, we will freeze the transferred features per individual layer. We will investigate this for different sizes of the target set. Moreover, we will test this on two different subsets of data sets, each with a different domain, ImageNet and Places2.

The remainder of this paper is organised as follows. Section 2 provides an overview of related work in deep learning and transfer learning. Section 3 describes our experimental setup and result, which addresses the data pre-processing steps we took, details about our feature transfer process and information regarding the training of the networks. In Sect. 4 we elaborate on our results and report our main findings, and conclude the paper in Sect. 5.

2   Related Work

Several studies have investigated the generalizability of features and have proven the success of transfer learning [4,9]. A popular strategy for transfer learning is fine-tuning, by training a linear classifier on top of the final layer of a CNN. Zeiler *et al.* [18] examined this by pre-training a CNN on ImageNet, and then training a linear classifier on three target data sets, PASCAL VOC 2012, Caltech-101 [3] and Caltech-256 [5]. They varied the target data set size, as well as the layer from which the classifier is trained on. They found that the model generalizes extremely well to Caltech-101 and Caltech-256, however less so to PASCAL. Nonetheless, the study proved the benefits of applying transfer learning. Similarly, good results were yielded in [11] using this approach of transfer learning. The authors pre-trained on ImageNet in combination with a SVM classifier, and use Pascal VOC and MIT-67 Indoor Scenes as target tasks.

In [2] the researchers investigated how well features transfer to different domain target problems, and they investigated at what layer in the network this is most optimal. They first trained AlexNet on the ImageNet data set, and tested these features on a basic object recognition task using the Caltech-101 [3] data set. Second, they tested the network on domain adaptation, where there is a small amount of data is available, using the Office database [12]. Thirdly, they tested how well their model performs on a more fine grained data set, using the Caltech-UCSD birds data set [16]. Since the images in this data set are very similar to each other, this is a rather difficult image classification task. Finally, the authors tested their model on the SUN-397 Large-Scale Scene Recognition database. This task is quite different from the source task, where the task was to classify objects. The objective of the SUN-397 data set is to classify scenic categories. In every experiment the authors improved the benchmark scores, indicating that the features learned from ImageNet provide substantial generalisable properties.

Our research is a direct extension of the work by Yosinski *et al.* [17]. They investigated how transferable features are between layers in the AlexNet architecture. To this end they trained two networks, N_1 and N_2, each on a random split of the ImageNet data set containing half of the data, split A and split B. After both networks were trained on their respective splits, the features of the first layer from network N_1, the base, were *transferred* to the first layer of network N_2, the target. The remaining layers in network N_2 were randomly initialised. Finally, network N_2 gets trained on the B partition of the ImageNet data set. Thus, what happens is that network N_2 does not train from scratch, but rather, it uses the pre-initialised features from network N_1. The researchers do this for layer one to seven in the network, transferring both from A to B as well as from B to A. They found that the features in the first three layers are fairly general and could be transferred and boost performance. However, features in deeper layers of the network are more specific to the source task and therefore, transferring them worsens the performance.

We hypothesize that a transfer learning approach by fixing the first layers is more valuable if the target set is smaller, and that for larger data sets updating all layers will give better results, and validate this on data sets from two different image classification domains.

3   Experiments and Results

In this section we present an overview of our experimental set up and results. We will start with a general overview of the approach, and then provide further detail in further sub sections.

3.1   Overall Approach

We will transfer features from a CNN trained on a source task, to a target task, i.e. data sets with disjunct outcome classes. We will consider the scenario where the target data set is the same size, as well as smaller in size as the source data set. The latter condition is the conventional setting in transfer learning [10].

Hypothetically the value of transfer learning should increase with smaller transfer data sets. Moreover, for each scenario we will investigate the first case where we will fine-tune all the layers with the transferred features. In the second case we will transfer the features but freeze the network weights in the first layers.

The CNN architecture we will use is AlexNet, developed by Krizhevsky *et al.* [8] (see Fig. 1), which was the winning model in the ImageNet Large Scale Visual Recognition Challenge 2012. The model consists of five convolutional layers and three fully connected layers. The first two convolutional layers are followed by a max pooling layer and a normalization layer respectively. The fifth convolutional layer is followed only by a max pooling layer. The first two fully connected layers contain 4,096 neurons. The final fully connected layer contains 1000 neurons for the target class scores. It is interesting to note that the authors used Rectified Linear Units (ReLUs) as activation functions instead of the regular sigmoid. Moreover, they applied a regularization technique called dropout to reduce overfitting [14].

3.2 Data Pre-processing

In our experiments, we use a subset of the ImageNet data set [1], **Tiny-ImageNet**. This data set contains 100,000 images with 200 classes, where each class contains 500 images, each of size 64×64 pixels. The data set contains images of a wide range of objects such as cats, parking meters, cliffs and rugby balls. The validation set contains 10,000 separate images.

Moreover, we extend the work by Yosinski *et al.* [17] by also repeating the experiments on a second data set, **MiniPlaces2**. This is a scaled down version of the larger MIT Places database [19]. The data set is made up of images with settings such as a food court, golf course, an office, and ice skating rink. It contains 100,000 images with 100 classes. Each class consists of 1000 pictures of size 128×128 pixels, however we resize them to 64×64 pixels to keep the image size consistent with Tiny-ImageNet. Again, the validation set contains 10,000 images. In Fig. 2 we present several image classes of both data sets to underline the difference between the two domains.

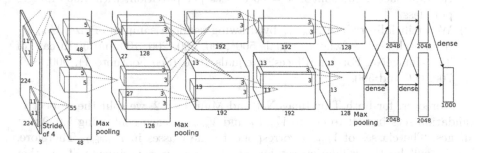

Fig. 1. AlexNet architecture (illustration taken from [8]).

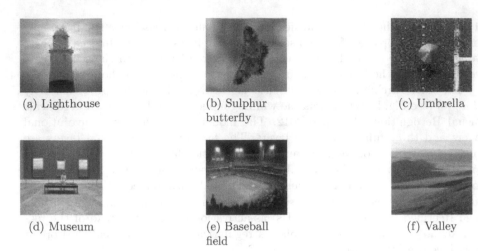

(a) Lighthouse

(b) Sulphur butterfly

(c) Umbrella

(d) Museum

(e) Baseball field

(f) Valley

Fig. 2. *Top:* a sample of training images from the Tiny-ImageNet data set. *Bottom:* a sample of training images from the MiniPlaces2 data set.

To measure the effect of data set size on the generalizability of features, we transfer the features from a source task to a target task, where the latter has a variable size. We will test this on a subset of the ImageNet and Places2 data set. We use a subset of the data sets rather than training on the full data sets of ImageNet and Places2 (respectively containing 1.2 million and 8.1 million images for training) due to computational limitations. We denote our target data set as N_{target}. Moreover, we define the data set splits with a variable size as M_{target_i} where $M_{target_i} \subseteq N_{target}$. To obtain M_{target_i} from N_{target} we execute the following procedure:

(1) We randomly split the entire data set into a source and a target partition, N_{source} and N_{target} respectively, where each partition contains 50,000 images. In both the source and target partition the images are equally distributed over $k = 100$ classes with 500 images per class for Tiny-ImageNet. In MiniPlaces2 the split is $k = 50$ classes per partition, with 1,000 images per class.
(2) We artificially reduce N_{target} by drawing random samples of size M_{target_i} from each class k, where i equals 500[1],400, 300, 200, 100 and 50 in case of Tiny-ImageNet.For MiniPlaces2 i equals 1000 (See Footnote 1), 900, 800, 700, 600 and 500.

Moreover, for both Tiny-ImageNet and MiniPlaces2, we split the respective validation sets in half to create V_{target} and V_{source}, each containing 5,000 test images. The classes of V_{target} correspond to the classes in N_{target}. Therefore, V_{target} will be the validation set for M_{target_i} in our experiments. The other

---

[1] Note that in the case where $i = 500$ and $i = 1,000$ we do not reduce N_{target} for Tiny-ImageNet and MiniPlaces2 respectively.

validation half, V_{source}, contains classes corresponding to N_{source}. In sum, we train our model on M_{target_i}, and evaluate it on a separate validation set V_{target}, to obtain our accuracy a.

3.3 Transferring Features

To create a model from which we can transfer the features, we first train our network on N_{source}. The parameters of the source model are stored in a Caffemodel object (see Sect. 3.4), which we use to transfer the parameters from the source model to the target model.

To obtain our baseline score we do not apply any transfer learning at all, and let the model train on the given training set. In our first experiment we fine-tune the network by transferring all the features from the source task to the model, and continue with backpropagation on the new task.

However, since we are also interested in at what layer l of the network features are able to generalize, we transfer the features from the source to the target task, one layer at a time. AlexNet has eight layers in total. Therefore, we transfer from layer $l = 1$, up until layer $l = 7$. When we transfer the parameters to the target model, we keep them fixed. That is to say, we do not update the parameters by gradient descent. The remaining $8 - l$ layers of the network we randomly initialize and let the errors backpropagate through the layers.

Finally, to get a mean accuracy score, we run the experiments again by following the same procedure, but now use N_{source} as N_{target} and vice versa.

3.4 Training

To conduct our experiments, we use the Caffe deep learning framework developed at UC Berkeley [7]. We make use of a single Nvidida GTX Titan X graphics card to enable Caffe in GPU mode, to speed up our training time. We use the AlexNet reference model which is included in Caffe. Detailed information about the model architecture can be found in [8]. Moreover, we follow the same training regime as specified by the reference model.

Furthermore, in terms of data augmentation we take a random crop in the training phase and use random mirroring as specified by Caffe. In the test phase we take a center crop of the images. Since our input images are 64×64, we change the crop size to 57, rather than upscaling the images to 256×256 and applying the default crop size of 227. Thus, we stay consistent with the ratio used in the AlexNet reference model. Moreover, we subtract the image mean from each image.

Finally, to determine for how many iterations we should train the models, we trained on N_{source} of both data sets and validated on the respective V_{source}, without applying any form of transfer learning.

We found that the model began to overfit on the training data around 10,000 iterations (see Figs. 3 and 4). Therefore, we found it reasonable for subsequent experiments to let each model run for 10 K iterations in order to measure the positive effect of transfer learning. Moreover, the more we reduce N_{target}, the faster the model will reach the point of overfitting, which is evidenced by the decreasing accuracy of the *base* training conditions across our experiments.

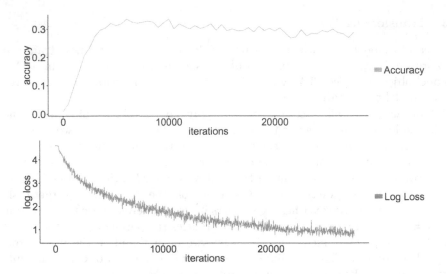

Fig. 3. *Top:* the accuracy on V_{source} after training on N_{source} of the Tiny-ImageNet data set after 25 K iterations. This split contains 100 classes, with 500 images per class. *Bottom:* the log loss over the training set with the identical split.

3.5   Results Tiny-ImageNet

In Fig. 5 we see the results of transfer learning on different data set sizes. The plot shows the accuracy on the validation set after 10 K iterations of training. The first two conditions are the base case and fine-tune all. The condition *base* indicates we did not apply transfer learning. Condition *FTall* means we fine-tuned through all the layers, and the notation *SnT* denotes up until which layers we freeze the transferred features from the source in the target model. For instance, *S3T* implies we transferred the first three feature layers from the model trained on N_{source} to the model trained on M_{target_i}. The final seven scores are the accuracies where we transfer the parameters per layer from the source, and freeze that particular layer. We notice an effect of data set size on the accuracy of the baseline score. As we decrease the data set size, we find that the accuracy decreases as well. In Fig. 5 we observe that the accuracy worsens as we keep more layers fixed when transferring parameters from the source task.

3.6   Results MiniPlaces2

As can be seen from Fig. 6, even though this is a task from a different domain, the results follow a pattern very similar to Tiny-ImageNet. With smaller target

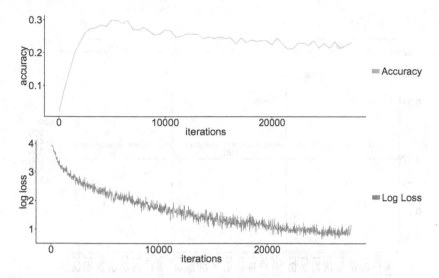

Fig. 4. *Top:* the accuracy on V_{source} after training on N_{source} of the MiniPlaces2 data set after 25 K iterations. This split contains 50 classes, with 1000 images per class. *Bottom:* The log loss over the training set with the identical split.

set sizes the benefits of locking the first few layers increases. Only for $M_{target_{1000}}$ the graphs seem to indicate that training from scratch is better, but this is truly just a baseline. In a real deployment one would probably expect that the source classes also still need to be recognized, and performance of tuning all layers is still lower then locking some of the initial layers.

4   Discussion

Our results reveal that data set size affects the accuracy in transfer learning with deep convolutional neural networks. The first effect we notice is on the baseline case (to repeat, just training the network with randomly initialized weights). We can see that the model starts to overfit on the training data when we artificially reduce the data set size, which leads to a steady decline in accuracy on both Tiny-ImageNet as well as MiniPlaces2. This can be explained by a sub-optimal parameter configuration as a result of overfitting on a small data set size.

Furthermore, fine-tuning all the layers only appears to have a positive effect with smaller data sets for Tiny-ImageNet where i in M_{target_i} ranges from 400 until 50, and MiniPlaces2 where M_{target_i} equals 500. This is an interesting result, as a network for which all layers can be adapted still benefits from potentially valuable initialization of the weights. We speculate that the source features are important for the target data set splits as well. Thus, the effect of initializing the model with parameters obtained from a model trained on a larger data set clearly shows its advantage. Moreover, we notice a visible spike in accuracy in all our graphs, when we transfer parameters from the first two layers. Likewise, there

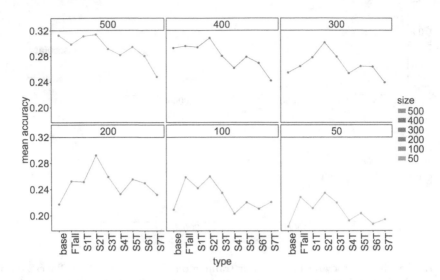

Fig. 5. Mean accuracy obtained after training on the target splits of Tiny-ImageNet where i in M_{target_i} equals 500, 400, 300, 200, 100 and 50 and validating on V_{target}. Note that we ran the same experiments again, but used N_{source} as N_{target} and vice versa. Thus, we obtained our mean accuracies by averaging the scores.

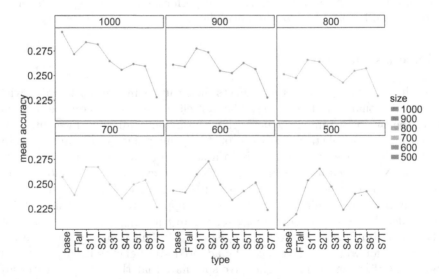

Fig. 6. Mean accuracy after training on the target splits of MiniPlaces2, where i in M_{target_i} equals 1000, 900, 800, 700, 600 and 500 and validating on V_{target}.

is a considerable decline in accuracy when transferring four layers, compared to transferring the first three layers.

The results in Fig. 5 generally follow the findings of the study by Yosinski *et al.* [17]. As we transfer more and more features (layers) from the source task, the accuracy initially goes up but then decreases. This can be attributed to feature specificity with regards to the source task. However, we observe a second positive spike in the accuracy at layer $l = 5$ in nearly all of our experiments. This result is quite surprising since the features have become substantially specific to the source, and yet generalize well to the new task. Evidently, the transferred features from the source task in this layer hold the same, or even superior, representational power compared to the features solely learned from a target data set.

All these results can be summarized into a fairly straightforward heuristic. For the first n instances of a new class, freeze the first l layers of the network. Once you have obtained more than n instances for new class, training can simply affect all layers. Obviously the values for n and l depend on the data and task at hand, in our experiments freezing the first 3 layers until 300 (Tiny-ImageNet) and respectively 900 (MiniPlaces2) instances per class gave the best results.

Our study could have benefited from having more samples per data point, by running repeated experiments. Since the initialization of the parameters happens at random, the parameters might converge at different local minima each time the model is run. This could effect the accuracy score in the test phase. Our results still indicate that transferring features from a larger source data set to a smaller target data set adds value by reducing the risk of overfitting, and improves performance.

5 Conclusion

In this paper we investigated the effect of data set size on the generalizability of features in deep convolutional neural networks. To this end, we transferred features from a pre-trained network to a new network. We systematically reduced the size of the target training set and trained our new network on these splits with the pre-initialized features. In support for a general rule of thumb heuristic, we found that freezing the first two to three layers of features results in a significant performance boost over the baseline score, especially for smaller target set sizes under a thousand instances per class.

References

1. Deng, J., Dong, W., Socher, R., Li, L.J., Li, K., Fei-Fei, L.: Imagenet: a large-scale hierarchical image database. In: IEEE Conference on Computer Vision and Pattern Recognition, CVpPR 2009, pp. 248–255. IEEE (2009)
2. Donahue, J., Jia, Y., Vinyals, O., Hoffman, J., Zhang, N., Tzeng, E., Darrell, T.: Decaf: a deep convolutional activation feature for generic visual recognition. arXiv preprint arXiv:1310.1531 (2013)

3. Fei-Fei, L., Fergus, R., Perona, P.: Learning generative visual models from few training examples: an incremental Bayesian approach tested on 101 object categories. Comput. Vis. Image Underst. **106**(1), 59–70 (2007)
4. Girshick, R., Donahue, J., Darrell, T., Malik, J.: Rich feature hierarchies for accurate object detection and semantic segmentation. In: Proceedings of the IEEE conference on Computer Vision and Pattern Recognition, pp. 580–587 (2014)
5. Griffin, G., Holub, A., Perona, P.: Caltech-256 object category dataset (2007)
6. He, K., Zhang, X., Ren, S., Sun, J.: Deep residual learning for image recognition. arXiv preprint arXiv:1512.03385 (2015)
7. Jia, Y., Shelhamer, E., Donahue, J., Karayev, S., Long, J., Girshick, R., Guadarrama, S., Darrell, T.: Caffe: convolutional architecture for fast feature embedding. In: Proceedings of the ACM International Conference on Multimedia, pp. 675–678. ACM (2014)
8. Krizhevsky, A., Sutskever, I., Hinton, G.E.: Imagenet classification with deep convolutional neural networks. In: Advances in Neural Information Processing Systems, pp. 1097–1105 (2012)
9. Oquab, M., Bottou, L., Laptev, I., Sivic, J.: Learning and transferring mid-level image representations using convolutional neural networks. In: Proceedings of the IEEE Conference on Computer Vision and Pattern Recognition, pp. 1717–1724 (2014)
10. Pan, S.J., Yang, Q.: A survey on transfer learning. IEEE Trans. Knowl. Data Eng. **22**(10), 1345–1359 (2010)
11. Razavian, A., Azizpour, H., Sullivan, J., Carlsson, S.: CNN features off-the-shelf: an astounding baseline for recognition. In: Proceedings of the IEEE Conference on Computer Vision and Pattern Recognition Workshops, pp. 806–813 (2014)
12. Saenko, K., Kulis, B., Fritz, M., Darrell, T.: Adapting visual category models to new domains. In: Daniilidis, K., Maragos, P., Paragios, N. (eds.) ECCV 2010, Part IV. LNCS, vol. 6314, pp. 213–226. Springer, Heidelberg (2010)
13. Simonyan, K., Zisserman, A.: Very deep convolutional networks for large-scale image recognition. arXiv preprint arXiv:1409.1556 (2014)
14. Srivastava, N., Hinton, G., Krizhevsky, A., Sutskever, I., Salakhutdinov, R.: Dropout: a simple way to prevent neural networks from overfitting. J. Mach. Learn. Res. **15**(1), 1929–1958 (2014)
15. Szegedy, C., Liu, W., Jia, Y., Sermanet, P., Reed, S., Anguelov, D., Erhan, D., Vanhoucke, V., Rabinovich, A.: Going deeper with convolutions. In: Proceedings of the IEEE Conference on Computer Vision and Pattern Recognition, pp. 1–9 (2015)
16. Welinder, P., Branson, S., Mita, T., Wah, C., Schroff, F., Belongie, S., Perona, P.: Caltech-ucsd birds 200 (2010)
17. Yosinski, J., Clune, J., Bengio, Y., Lipson, H.: How transferable are features in deep neural networks? In: Advances in Neural Information Processing Systems, pp. 3320–3328 (2014)
18. Zeiler, M.D., Fergus, R.: Visualizing and understanding convolutional networks. In: Fleet, D., Pajdla, T., Schiele, B., Tuytelaars, T. (eds.) ECCV 2014, Part I. LNCS, vol. 8689, pp. 818–833. Springer, Heidelberg (2014)
19. Zhou, B., Lapedriza, A., Xiao, J., Torralba, A., Oliva, A.: Learning deep features for scene recognition using places database. In: Advances in Neural Information Processing Systems, pp. 487–495 (2014)

Obtaining Shape Descriptors from a Concave Hull-Based Clustering Algorithm

Christian Braune[✉], Marco Dankel, and Rudolf Kruse

Otto-von-Guericke-University of Magdeburg,
Universitätsplatz 2, 39106 Magdeburg, Germany
{christian.braune,marco.dankel,rudolf.kruse}@ovgu.de

Abstract. In data analysis clustering is one of the core processes to find groups in otherwise unstructured data. Determining the number of clusters or finding clusters of arbitrary shape whose convex hulls overlap is in general a hard problem. In this paper we present a method for clustering data points by iteratively shrinking the convex hull of the data set. Subdividing the created hulls leads to shape descriptors of the individual clusters. We tested our algorithm on several data sets and achieved high degrees of accuracy. The cluster definition employed uses a notion of spatial separation. We also compare our algorithm against a similar algorithm that automatically detects the boundaries and the number of clusters. The experiments show that our algorithm yields the better results.

Keywords: Density based clustering · Convex hulls · Concave hulls · Noise removal · Automatic detection of cluster number

1 Introduction

Clustering is a fundamental problem in data analysis [14]. Its key function is to sort data points into an (unknown) amount of groups. Groups are usually disjunct, but this is not always necessary [2]. If points may partially belong to more than one cluster, we speak of fuzzy or possibilistic clustering [10,13]. The distinction between the two is mainly whether the membership degrees sum up to one or not. Hierarchical (agglomorative) clustering on the other hand starts with singleton clusters which are subsequently joined and merged to form a hierarchy of clusters from which the final clustering result may be extracted.

In this paper we will present an algorithm that *learns* the shape of the clusters in a data set by refining and splitting the hull of the data set. Let $\mathcal{X} \subset \mathbb{R}^d$ be a d-dimensional data set with data points $p_i \in \mathbb{R}^d$ in general position, i.e. no $d + 1$ points are co-hyperplanar. In the two-dimensional case this simply means that no three points lie on the same line. Should the data points not be in general position, we can still induce such a situation by adding a minuscule amount of gaussian noise to any of the degenerating points. This property is mainly necessary in our paper to ensure the runtime complexity of the convex hull algorithms and the uniqueness of the hull itself.

© Springer International Publishing AG 2016
H. Boström et al. (Eds.): IDA 2016, LNCS 9897, pp. 61–72, 2016.
DOI: 10.1007/978-3-319-46349-0_6

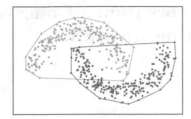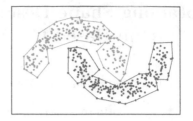

Fig. 1. Left we see the convex hulls of two non-convex clusters. Both hulls cover a lot more space than necessary, while the polygons on the right side appear to be a much better cluster shape descriptor.

The convex hull $\mathcal{H}$ of a set $\mathcal{X}$ of points is the set of all convex linear combinations of two points in the data set. On the one hand it is the smallest *convex set*, that contains $\mathcal{X}$. On the other hand it is the maximal set of points that can be constructed from the data points by convex linear combinations of points. Clusters as subsets of the data set may be described by their respective convex hulls which – in general – is not the best descriptor (see Fig. 1). The problem here is that clusters may actually not fill the space described by the convex hull and adding concavities to the hull may lead to a better description of the cluster shape. A *concave hull* is much less well defined since technically any set $\mathcal{H}^* \subseteq \mathcal{H}$ could qualify as concave hull as long as all p_i are still contained in $\mathcal{H}^*$. If in the following we speak of the convex hull, we explicitly mean the polygon that describes the border of the set $\mathcal{H}$ or $\mathcal{H}^*$.

In this paper we present a way of constructing a concave hull for each cluster contained in the data set by iteratively refining the convex hull of the complete data set. The algorithm is capable of finding the number of clusters on its own. The next section contains an overview about other similar approaches to constructing concave hulls. We also review some clustering algorithms that construct convex hulls to cluster data. In Sect. 3 we describe our algorithm and how we developed it while Scct. 4 contains the evaluation of the algorithm and an interpretation of the results. The paper concludes with some critical remarks on the algorithms performance and some future work which we plan to implement later.

2   Related Work

Calculating the convex hull has been studied extensively over the past decades (e.g. [1, 3, 9, 12]). There are already optimal algorithms for the construction of the convex hull considering different criteria for optimality. Concave hulls are also often called *characteristic shapes* in literature [6]. Many algorithms rely on the computation of a Delaunay tesselation [4] of the data set. All edges that would belong to the convex hull of the data set also belong to the Delaunay tesselation. Removing unwanted edges (e.g. edges that are longer than the longest edge of the minimum spanning tree), can lead to a polygon, that resembles the shape of the data well [6]. The caveat here is that usually only a single polygon is used to describe the data. This is not helpful if we want to find clusters. In [7] the concept of α-shapes was developed. Starting from the Delaunay tesselation edges are removed if there exists any circle (in the two-dimensional case) that contains the two points forming the edge and at least a third point. If no such circle can be found, the edge is kept and becomes part of the alpha shape.

The algorithm presented in [18] uses a similar approach to our method as it starts from the convex hull and iteratively replaces edges that are too long by more favorable candidates. The authors use the angle between the new edges and the old one as their criterion of choice. For this every point that is currently not and has never been on the hull polygon becomes a candidate. This increases the necessary search space. At the end a single polygon is found (c.f. Fig. 6.12 in [18]). Though this is certainly one possible way of representing the data set, notably in the depicted case several individual polygons would be more suitable than the narrow bridges that are currently connecting the individual clusters – especially since they sometimes leave rather large gaps.

In [17] the notion of concave hulls is used to describe clusters. However, it is not used to find clusters but to describe them, after a clustering algorithm has found dense, connected regions in the data space. This would actually be the opposite way of our approach, since we attend to find the clusters by finding their concave hulls. If our method was to be applied in this way, the major difference would be how new edges are chosen for the hull: the authors of [17] chose a modified gift wrapping approach (c.f. [12]) to choose an edge. However, we use a best-candidate approach to find a point that will become part of the new hull.

Similarly [15] has the goal to find a better representation of the clusters than a mere prototype to better compute the membership degrees of a fuzzy partitioning. They do not use all data points that belong to a cluster but rather only a subset that adequately describes the cluster's shape.

3   CLASH: Clustering Along Split Hulls

Similar to other concave hull methods shown in Sect. 2, the algorithm proposed by us will start off with the convex hull of the data set initially. This hull will be refined to adapt to the data set and, by doing this, the data set will be split naturally into different subsets. More accurately, our algorithm is composed of two major principles:

1. **Hull Refinement**
   i.e. iteratively replacing the currently longest edge of the hull with a *more suitable* edge path to describe the data, and
2. **Recursively Splitting the Data Set**
   into different clusters if necessary, i.e. splitting the current hull path into two separate, closed paths as soon as it converges into one point multiple times.

Algorithm 1. Refining an edge

Require: *hull* given as list of edges,
  $e_l = (\boldsymbol{p}_1, \boldsymbol{p}_2)$ is the longest edge currently in *hull*

1: **function** REFINE($\mathcal{P}$, *hull*, e_l)
2: $\boldsymbol{p}_m \leftarrow$ middle point on e_l
3: $\boldsymbol{n} \leftarrow$ normal vector on e_l pointing into the hull
4: $\boldsymbol{r} \leftarrow$ direction vector along e_l
5: $\mathcal{P}_{rel} \leftarrow$ all points $\boldsymbol{p} \in \mathcal{P}$ ▷ All relevant points above e_l
6:     with $(\boldsymbol{p} - \boldsymbol{p}_m) \cdot \boldsymbol{n} > 0$
7: $\mathcal{P}_{split} \leftarrow$ all points $\boldsymbol{p} \in \mathcal{P}_{rel}$ ▷ All possible split points
8:     with $\left| \frac{1}{|\boldsymbol{n}|} (\boldsymbol{p} - \boldsymbol{p}_m) \cdot \boldsymbol{r} \right| \leq \frac{1}{2} |e_l|$
9: $\boldsymbol{p}_{split} \leftarrow$ The point in $\mathcal{P}_{split}$ closest to p_m ▷ The split point

10: $\mathcal{P}_{left} \leftarrow$ all points $\boldsymbol{p} \in \mathcal{P}$ ▷ all points left of the split point
11:     with $(\boldsymbol{p} - \boldsymbol{p}_m) \cdot \boldsymbol{r} \leq 0$
12: $\mathcal{P}_{right} \leftarrow$ all points $\boldsymbol{p} \in \mathcal{P}$ ▷ all points right of the split point
13:     with $(\boldsymbol{p} - \boldsymbol{p}_m) \cdot \boldsymbol{r} > 0$

14: $hull_{new} \leftarrow hull[\boldsymbol{p}_2 \ldots \boldsymbol{p}_1]$ ▷ Current hull without e_l
15: **for** *edge* in convex_hull($\mathcal{P}_{left}$)$[\boldsymbol{p}_1 \ldots \boldsymbol{p}_{split}]$ **do**
16:     append edge to $hull_{new}$
17: **for** *edge* in convex_hull($\mathcal{P}_{right}$)$[\boldsymbol{p}_{split} \ldots \boldsymbol{p}_2]$ **do**
18:     append edge to $hull_{new}$
19: **return** $hull_{new}$

These steps will be explained in detail below. As an optional third step, in order to avoid overfitting of the hull to the data set, the resulting concave hulls may be simplified subsequently. This may be done, e.g., by deleting unnecessary edges.

3.1 Hull Refinement

Given an existing hull of the data (Fig. 2 ff. on the next page), our algorithm will iteratively select the longest edge $e_l = (\boldsymbol{u}, \boldsymbol{v})$ and replace it with another edge or a series of other edges that fit the data more accurately.

To do this, we look for an inner split point, which will become part of the new hull (Fig. 3). The data set is then split into two sub-sets: the data points that lie above the longest edge and left of the split point, and those points above the longest edge and right of the split point (Fig. 4). Since the convex hull usually is given in counter- clockwise order, we know that all points lie left of the hull. But as soon as the hull is not convex anymore, points may also lie on the right side (below or behind the longest edge, if we look into the direction of the normal vector of our current edge).

To find a good split point, our algorithm will first calculate the middle point $\boldsymbol{p}_m$ along e_l. From there we determine the set of all data points on the inner side of the longest edge, within the rectangular, inward-bound stripe upon e_l.

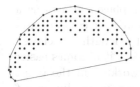

Fig. 2. Initial point set with convex hull.

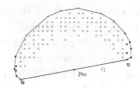

Fig. 3. Points on stripe above longest edge e_l.

Fig. 4. Left and right point set $\mathcal{X}_1$ and $\mathcal{X}_2$.

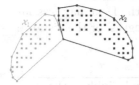

Fig. 5. Convex hulls of $\mathcal{X}_1$ and $\mathcal{X}_2$.

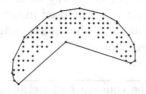

Fig. 6. Merged hull after only one iteration.

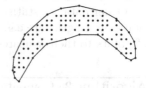

Fig. 7. Final hull.

This set is guaranteed to be non-empty since if there were no points in the stripe we would be either pointing into the wrong direction or there would be an edge that starts and ends outside of the stripe, thus making it a longer edge. This is in opposition to the assumption that e_l is the longest edge. From the points in $\mathcal{P}_{rel}$ we choose the point p_{split}, that lies closest to p_m.

For the left and right subsets according to this split point, a new partial convex hull is computed which then replaces e_l (Fig. 5). Thus, the new concave hull of the data set will consist of the former hull of the data set without the edge e_l, the series of edges along the convex hull of the left subset – starting from the left end vertex u of e_l up to the split point p_{split} – and the series of edges along the convex hull of the right subset – starting from p_{split} up to the right end vertex v of e_l (Fig. 6). Figure 7 shows the resulting hull after the first iteration.

This process is repeated until a termination criterion is met. For this, we define a minimum length θ and replace any edge shorter than this threshold. To find an appropriate minimum we make use of different data set statistics that characterize the overall density of the data set.

For this paper, the edge lengths of the Delaunay tessellation serve as such statistic: We assume that the more longer edges there are in this triangulation, the sparser is the data set around these edges, and vice versa. Thus, hull edges that would be considered an outlier within the Delaunay tessellation edge lengths are supposedly too long to appropriately describe the data set – and thus they should be subject to refinement.

In our algorithm edges which are $q = 3$ times longer than the p-th percentile of the edge lengths of the Delaunay tessellation will be replaced. q acts as a hyperparameter here helping to adjust the value of p. Of course we could always set $q = 1$ and just choose an appropriate p that represents the length just chosen by this combination. Our experiments however indicated that it becomes easier adjusting the value of p in the range $[0.5, 0.7]$ than e.g. working in the range $[0.95, 0.99]$. By adjusting the percentile parameter p, we automatically adjust our definition of long edges in relation to the data set density. The proper choice of p however is a less concern than one might think at first glance. The smaller the chosen value the more edges will be removed and the more ragged the resulting hull will be. In Sect. 3.3 we will present a way to smooth such hulls.

Other data set statistics that could be used instead of the Delaunay tessellation edge lengths are e.g. Minimum Spanning Tree edge lengths, or the distance of points to their k-nearest neighbors.

Algorithm 2. Computing the concave hull iteratively and dividing the data recursively

---

Require: $\mathcal{X}$ being a set of points,
           θ being the minimum edge length threshold

 1: **function** CONCAVEHULLS($\mathcal{X}$)
 2:     $hull \leftarrow$ convex hull of $\mathcal{X}$
 3:     **while** there is an edge longer than θ, and $|hull| \geq 3$ **do**
 4:         $e_l \leftarrow$ longest edge in $hull$
 5:         $hull \leftarrow$ REFINE($\mathcal{X}, hull, e_l$)
 6:         **if** a point p occurs twice in $hull$ **then**
 7:             split $hull$ into $hull_1$ and $hull_2$
 8:             $\mathcal{X}_1 \leftarrow$ points within $hull_1$,
 9:             $\mathcal{X}_2 \leftarrow$ points within $hull_2$
10:             $hulls_1 \leftarrow$ CONCAVEHULLS($\mathcal{X}_1, hull_1$)
11:             $hulls_2 \leftarrow$ CONCAVEHULLS($\mathcal{X}_2, hull_2$)
12:             **return** $hulls_1, hulls_2$
13:     **if** $|hull| \geq 3$ **then**
14:         **return** $hull$
15:     **else**
16:         **return** Nothing

3.2   Splitting into Multiple Smaller Hulls

When performing the steps above multiple times without meeting the termination criterion, we may encounter situations where the hull becomes so thin that it passes through one point multiple times. This would yield non-favorable results. Thus, once a point occurs twice on the hull, we split the hull into two closed hulls using this point. Along with these two hulls, the data set will be divided into two sub-sets, too.

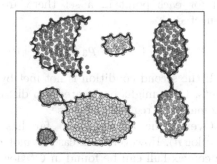

Fig. 8. Result for the `spirals` (left) and the `aggregation` (right) data set.

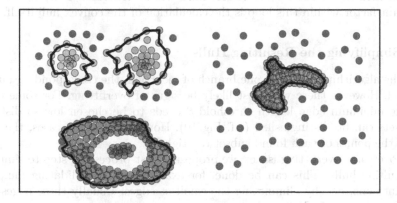

Fig. 9. Result for the `compound` data set.

The algorithm will then process both sub-sets separately and refine both hulls independently from each other. If, after some steps, another point occurs twice on one of those hulls, the corresponding sub-set will be split again. This is done for each subset recursively until either the termination criterion of a minimum edge length is met, or the sub-set consists of less than three points.

If there are less than three points in a data set, it is omitted since it is too small to calculate a hull that is not degenerated.

This way, the algorithm is automatically able to separate the data set into subsets while simultaneously computing their hulls. This will work for all non-overlapping clusters of points whose distance is greater than the minimum edge length threshold θ. More accurately, two point sets $\mathcal{X}_1$ and $\mathcal{X}_2$ will be separated properly by our algorithm, if

$$\forall\,(\boldsymbol{p}_1, \boldsymbol{p}_2) \in \mathcal{X}_1 \times \mathcal{X}_2: \quad \|\boldsymbol{p}_1 - \boldsymbol{p}_2\|_2 > \theta$$

and for each point in a set there are (at least) two points to which it is connected:

$$\forall\ \boldsymbol{p}_1 \in \mathcal{X}_i : \ \exists\ \boldsymbol{p}_2,\ \boldsymbol{p}_3 \in \mathcal{X}_i : \quad \|\boldsymbol{p}_1 - \boldsymbol{p}_2\|_2 \ \leq\ \theta\ \wedge\ \|\boldsymbol{p}_1 - \boldsymbol{p}_3\|_2 \ \leq\ \theta.$$

If the second condition is not met by some individual points within a set of points, these might be located too distant from the rest of the set and will be automatically regarded as noise.

Overall the algorithm so far has a worst-case runtime complexity of $\mathcal{O}(n^2 \log n)$. However, this case cannot occur in any practical setting. Usually the convex hull can be found in $\mathcal{O}(n \log k)$, where k is the number of points on the convex hull and n is the total number of points. If the parameters are chosen poorly and the concave hull found by our algorithm completely vanishes (c.f. Fig. 13), then the while loop in Algorithm 2 will be traversed $\mathcal{O}(n)$ times. The dominant factor within this loop is the calculation of the convex hull itself.

3.3   Simplifying the Resulting Hulls

Once the algorithm completes one branch of recursion, the corresponding hull is finalized. However there will most likely be signs of overfitting. For some data sets the minimum edge length threshold θ needs to be chosen low so distinct point sets can be distinguished (c.f. Fig. 10). However, in these cases, the hull will fit the points of each found subset too tightly.

To overcome overfitting issues, we propose a post-processing step to simplify the resulting hulls. This can be done, for example, by consolidating multiple edges into one, e.g. by eliminating the smallest edges. Usually these edges are those that distort the hull.

4   Evaluation

To evaluate our algorithm we generate several instances of different data sets with similar parameters. In the first, simple cases we generated blob-shaped clusters within a fixed bounding box and constant standard deviation. For each trial we generated a fixed amount of clusters, ranging from 3 to 6. For every such data set we used the same parameters for our algorithm (1000 points per data set, no hull simplification enabled, points only belong into a cluster, if they are contained in the polygon defined by the concave hull). For these data sets the ground truth is known from the data generation process and we can test the

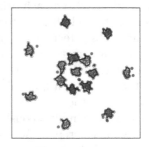

Fig. 10. Result for the R15 data set.

algorithm's results against this ground truth by using the adjusted rand index (ARI, [11]), and the V-measure [19]. The results for using $p = 0.95$ and the

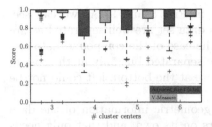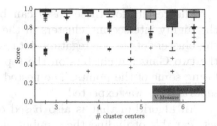

Fig. 11. Results for 1000 data sets with different numbers of clusters and different detection parameters (left: 0.95, right: 0.5, multiplier: 3)

$p = 0.5$ percentiles can be found in Fig. 11. These data set are usually clustered very well by any clustering algorithm and we use this test only to show, that our algorithm is not worse than others.

However, in the case of three clusters we can see some especially bad results with ARI scores between 0.4 and 0.6. If we take a closer look at the data set that caused these scores we see that these data sets contain clusters, that overlap to a great extent (see Fig. 12) – a situation that is hardly (if at all) solvable by any clustering algorithm.

Fig. 12. Two examplary data sets, where the standard setting used in the experiments fails. In the left data set, clusters are located too close to each other to be properly separated. In the right data set two clusters almost have the same center and the third cluster is located similarly close.

We also ran our algorithm on some well-known example data sets to show that it is capable of coping with several kinds of obstacles. In the spirals dataset algorithms have to cope with clusters whose convex hulls are highly intersecting. Any clustering algorithm based upon centroids will usually fail here. The aggregation data set contains several structures that do not show proper spatial separation of the clusters (lower left and right side of Fig. 8). The compound data set (Fig. 9) has been chosen since it contains one structure surrounded by noise with uniform density, two badly separated *gaussian* clusters and a nested structure.

The results for these cases can be seen in Figs. 8 and 9. As we can see, the highly intersecting clusters in the spirals data set are well separated as are the clusters of the aggregation data set. In the case of the compound data set, the two Gaussian clusters on the top left are separated but only at the cost of losing some of the points. The nested structure at the bottom left could not be separated as was expected.

An algorithm that is also based on the geometric information of the data set, capable of finding the number of clusters on its own, and that only needs few, self-tuning parameters list TRICLUST [16]. TRICLUST is largely based upon algorithms such as ASCDT [5], NSCABDT [20] or AUTOCLUST [8] with the distinction that it also generates border points for the clusters found and is thus better comparable to our method. For the experiments we use the parametrization proposed in [16]. The results of this algorithm compared to ours can be seen in Table 1 (V-measure only) In all cases our algorithm performs better – in most cases significantly better. The problem with TRICLUST here is, that it tends to find too few clusters if the data set is structured in a way that the Delaunay tesselation does not contain any edges with length that can be considered outliers. In the R15 data set (Fig. 10) all points are placed into a single cluster, while in the spirals data set, only two clusters are found – one containing less than 20 points (out of 3432). Our algorithm can find the correct number of clusters for the R15 data set if p (or θ) is chosen appropriately low. Otherwise the edges along the central clusters will not be removed and only eight clusters will be found.

Table 1. Comparison of V-Measure scores for TRICLUST and our algorithm on some examplary data sets.

Data set	TRICLUST	CLASH
aggregation	0.9422	0.9788
compound	0.2121	0.8603
R15	0.0000	0.9593
spirals	0.0034	0.9926

If we consider the results on data sets that do not contain any structure at all, we can see that the results differ slightly. In the case of high density noise with a uniform density distribution (blue noise), the algorithm finds one single structure. In the same setting but with lower density, clusters are found. The only remaining hulls are artifacts that have not been properly eliminated by the algorithm. When drawing points from a uniform distribution, some points are usually not that well separated and some artificial structures can be found. This is reflected by the results of our algorithm, that finds several small clusters which could be eliminated by further reducing the edge length threshold. All of these results can be seen in Fig. 13.

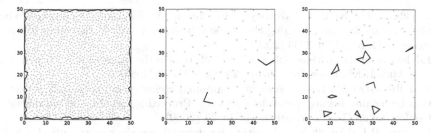

Fig. 13. Result for **high density** and **low density** blue noise (left and middle) and **uniform** noise (right).

5 Conclusion and Future Work

In this paper we presented an algorithm that is capable of clustering a diverse range of data sets with different properties according to the spatial separation of the cluster structures. By shrinking the convex hull to adapt to the shape of the individual clusters these are detected. The results (both visual and numerical) show, that our algorithm detects clusters with high accuracy. Only in the case where clusters are not well separated these are merged. Albeit slower than some other algorithms, our method not only labels points and finds the number of clusters on its own. It also generates a natural cluster shape description which can be used in various ways. E.g. in the future we plan to use the hull description found by our algorithm as a natural border for the clusters from which a fuzzy partitioning can be calculated.

The algorithm has been tested on two-dimensional data sets only. The extension of the clustering principal to more dimensions is straightforward. Instead of the longest edge we would have to look for the convex hull's facet with the largest hypervolume. A $(d-1)$-dimensional subspace perpendicular to this facet would sub-divide the data set and facets are replaced by the facets of the subsets convex hull to connect with the split point.

References

1. Barber, C.B., Dobkin, D.P., Huhdanpaa, H.: The quickhull algorithm for convex hulls. ACM Trans. Math. Softw. (TOMS) **22**(4), 469–483 (1996)
2. Braune, C., Besecke, S., Kruse, R.: Density based clustering: alternatives to DBSCAN. In: Celebi, E.B. (ed.) Partitional Clustering Algorithms, pp. 193–213. Springer, Heidelberg (2015)
3. Chan, T.M.: Optimal output-sensitive convex hull algorithms in two and three dimensions. Discrete Comput. Geom. **16**(4), 361–368 (1996)
4. Delaunay, B.: Sur la sphere vide. Izv. Akad. Nauk SSSR, Otdelenie Matematicheskii i Estestvennyka Nauk, **7**(793–800), 1–2 (1934)
5. Deng, M., Liu, Q., Cheng, T., Shi, Y.: An adaptive spatial clustering algorithm based on Delaunay triangulation. Comput. Environ. Urban Syst. **35**(4), 320–332 (2011)

6. Duckham, M., Kulik, L., Worboys, M., Galton, A.: Efficient generation of simple polygons for characterizing the shape of a set of points in the plane. Pattern Recogn. **41**(10), 3224–3236 (2008)
7. Edelsbrunner, H., Kirkpatrick, D.G., Seidel, R.: On the shape of a set of points in the plane. IEEE Trans. Inf. Theory **29**(4), 551–559 (1983)
8. Estivill-Castro, V., Lee, I.: Autoclust: automatic clustering via boundary extraction for mining massive point-data sets. In: Proceedings of the 5th International Conference on Geocomputation, pp. 23–25 (2000)
9. Graham, R.L.: An efficient algorithm for determining the convex hull of a finite planar set. Inf. Process. Lett. **1**(4), 132–133 (1972)
10. Höppner, F., Klawonn, F., Kruse, R., Runkler, T.: Fuzzy Cluster Analysis: Methods for Classification, Data Analysis and Image Processing. Wiley, London (1999)
11. Hubert, L., Arabie, P.: Comparing partitions. J. Classif. **2**(1), 193–218 (1985)
12. Jarvis, R.A.: On the identification of the convex hull of a finite set of points in the plane. Inf. Process. Lett. **2**(1), 18–21 (1973)
13. Klawonn, F., Kruse, R., Winkler, R.: Fuzzy clustering: more than just fuzzification. Fuzzy Sets Syst. **281**, 272–279 (2015)
14. Kruse, R., Borgelt, C., Klawonn, F., Moewes, C., Steinbrecher, M., Held, P.: Computational Intelligence: A Methodological Introduction. Springer Science & Business Media, London (2013)
15. Liparulo, L., Proietti, A., Panella, M.: Fuzzy clustering using the convex hull as geometrical model. Adv. Fuzzy Syst. **2015** (2015). Article No. 6
16. Liu, D., Nosovskiy, G.V., Sourina, O.: Effective clustering and boundary detection algorithm based on Delaunay triangulation. Pattern Recogn. Lett. **29**(9), 1261–1273 (2008)
17. Moreira, A., Santos, M.Y.: Concave hull: a k-nearest neighbours approach for the computation of the region occupied by a set of points. In: Proceedings of the International Conference on Computer Graphics Theory and Applications, GRAPP 2007 (2007)
18. Rosén, E., Jansson, E., Brundin, M.: Implementation of a fast and efficient concave hull algorithm. Technical report, University of Uppsala, Sweden (2014)
19. Rosenberg, A., Hirschberg, J.: V-measure: a conditional entropy-based external cluster evaluation measure. In: EMNLP-CoNLL, vol. 7, pp. 410–420 (2007)
20. Yang, X., Cui, W.: A novel spatial clustering algorithm based on Delaunay triangulation. In: International Conference on Earth Observation Data Processing and Analysis, p. 728530. International Society for Optics and Photonics (2008)

Visual Perception of Discriminative Landmarks in Classified Time Series

Tobias Sobek and Frank Höppner[✉]

Department of Computer Science, Ostfalia University of Applied Sciences,
38302 Wolfenbüttel, Germany
f.hoeppner@ostfalia.de

Abstract. Distance measures play a central role for time series data. Such measures condense two complex structures into a convenient, single number – at the cost of loosing many details. This might become a problem when the series are in general quite similar to each other and series from different classes differ only in details. This work aims at supporting an analyst in the explorative data understanding phase, where she wants to get an impression of how time series from different classes compare. Based on the interval tree of scales, we develop a visualisation that draws the attention of the analyst immediately to those details of a time series that are representative or discriminative for the class. The visualisation adopts to the human perception of a time series by adressing the persistence and distinctiveness of landmarks in the series.

1 Motivation

One can think of many different properties of time series that may or may not contribute to the similarity of two series. The literature offers a broad variety of similarity measures to address different properties, which were extensively tested in a classification setting (e.g. 1-nearest neighbour classifier) [6]. This is helpful if some black-box decision has to be made, but to gain insights into the similarity of time series, a distance measure or 1NN-classifier is almost pointless, as it provides no summary or model, let alone a visual representation of what makes a series more likely to belong to class A than B.

With this work we aim at a visual tool to highlight features in (labelled) time series that discriminate series from different classes, thereby supporting an analyst in understanding and interpreting the series. To achieve this we rely on landmarks in the series, such as minima and maxima, as these properties are also well perceived when visually inspecting series, but are dealt with counter-intuitively by prominent measures such as dynamic time warping (DTW) as we will see below. The identification of discriminative properties offers the opportunity to identify and appreciate even small features, which may not influence a distance measure to a significant extent. The difficulty is, however, to identify and match such features, as they may be affected by effects such as dilation, translation, scaling, noise, etc.

© Springer International Publishing AG 2016
H. Boström et al. (Eds.): IDA 2016, LNCS 9897, pp. 73–85, 2016.
DOI: 10.1007/978-3-319-46349-0_7

In the next section we briefly review DTW, because it is the most prominent elastic distance measure for time series; we discuss in particular why distance measures such as DTW may miss important structural properties (landmarks). In Sect. 3 we present an approach to identify, match and compare features of time series at multiple scales, which leads us directly to a visualisation of discriminative features. The method is evaluated experimentally in Sect. 4. Conclusions will be given in Sect. 5.

2   Related Work

By $\mathbf{x} = (x_1, \ldots, x_m) \in \mathbb{R}^m$ we denote a time series that consists of m values indexed from 1 to m. For the sake of simplicity, we refer to the indices i as the points in time when x_i was measured. While there is considerable work about the visualisation of temporal data [1], there is only little about the visualization of a set of labelled series. Several authors, e.g. [2], use multidimensional scaling or projection techniques to visualize a set of series in a scatter plot (one dot representing one series) by means of some distance measures (such as DTW). These approaches do not aim at showing individual or distinctive properties of the series (in contrast to this work), but to give a general overview.

2.1   Similarity Measures vs. Landmarks

Given two series $\mathbf{x}$ and $\mathbf{y}$, Euclidean distance $d^2(\mathbf{x}, \mathbf{y}) = \sum_i (x_i - y_i)^2$ assumes a perfect alignment of both series as only values with the same time index are compared. If the series are not aligned, x_i might be better compared with some $y_{f(i)}$ where $f : \mathbb{N} \to \mathbb{N}$ is a monotonic index mapping. With dynamic time warping (DTW) [3] the optimal warping path f, that minimizes the Euclidean distance of $\mathbf{x}$ to a warped version of $\mathbf{y}$, is determined. An example warping path f is shown in Fig. 1(left): Two series are shown along the two axes; $\mathbf{x}$ on the left, $\mathbf{y}$ at the bottom. The matrix enclosed by both series encodes the warping path: an index pair (i, j) on the warping path denotes that x_i is mapped to $y_j = y_{f(i)}$. All pairs (i, j) of the warping path, starting at index pair $(0, 0)$ and leading to (m, m), contribute to the overall DTW distance. Despite the fact that DTW is rather old, recent studies [6] still recommend it as the best measure on average over a large range of datasets.

However, value and time are treated differently in time warping approaches: while a monotone but otherwise arbitrary transformation of time is allowed, the values remain untouched during this procedure.[1] This may lead to some surprising results. In Fig. 1(middle) we have two similar series (linearly decreasing, increasing, decreasing segments), depicted in red and green. They are also shown on the x- and y-axis in the leftmost figure, together with the warping path. Both series were standardized, but their range is not identical. If we would

---

[1] In order to get meaningful results with time warping methods, the value range of both series should clearly overlap, as it may be obtained from standardization (to zero mean and unit variance).

ask a human to align both series, an alignment of the local minima and maxima would be natural, revealing the high similarity of both series as they behave identically between the local extrema. The local maximum m of the red curve (near $t = 60$), however, lies below the local maximum of the green curve, so all DTW approaches assign the red maximum to *all points of the green curve above* m. (The assignment is shown in the leftmost figure and by the dotted lines in the middle.) As a consequence, if we shuffle or reorder the green data above m (cf. rightmost subfigure, blue curve), neither the assignment nor the distance changes. This is in contrast to the human perception, who would never consider the blue series being as similar to the red series as the green.

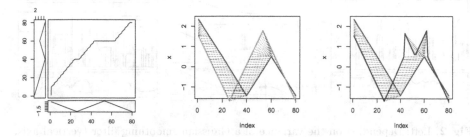

Fig. 1. Behaviour of time warping distances. Left: warping path of two time series (also shown in green and red in the middle). Mid and right: examples series; dotted lines indicate the DTW assignment. Right: Pair of blue and red series; structurally different, but with the same distance as green and red series. (Color figure online)

In this example, a human recognizes the red and green curve as similar because of the similarity of the segments (as suggested by the extrema). This kind of similarity includes time warping to compensate for different segment lengths, but also segmentwise value re-scaling (not done with DTW). The natural segmentation along extrema is also propagated by other authors, e.g. [5] in their landmark model. Landmarks correspond to extrema in the time series and a distance measure is defined on the sequence of landmarks rather than the original series. Landmarks are often employed for time series segmentation, but only seldomly for comparing series directly.

2.2   Interval Tree of Scales

We need not consider all extrema to grasp a time series. In the landmark model of [5], some extrema are skipped based on some a priori defined thresholds. This is typical for smoothing operations, but it is difficult to come up with such a fixed threshold, because different degrees of smoothing may be advisable for different parts of the series. Too much smoothing bears the danger of smearing out important features, too little smoothing may draw off the attention from the relevant features.

This is acknowledged by multiscale methods such as wavelets [4]. Witkin was one of the first who recognized the usefulness of a scale-space representation of time series [7]. The *scale s* denotes the degree of smoothing (variance of Gaussian filter) that is applied to the time series. The scale-space representation of a series depicts the location of extrema (or inflection points) as the scale s increases (cf. Fig. 2(left) for the time series shown at the bottom). The prominence (persistence against smoothing) of an extremum can be evaluated by following it from the original series ($s \approx 0$) to the scale s at which it vanishes (where it gets smoothed away). The scale-space can be considered as a fingerprint of the time series.

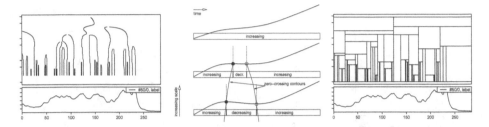

Fig. 2. Left: Depending on the variance of a Gaussian smoothing filter (vertical axis, logarithmic) the number and position of zero crossings in the first derivative varies. Mid: The zero-crossings of the first derivative (extrema in the original series) vanish pairwise. Right: Interval tree of scales obtained from left figure.

Zero-crossings typically vanish pairwise, three consecutive segments (e.g. increasing, decreasing, increasing) turn into a single segment (e.g. increasing), cf. Fig. 2(middle). The scale-space representation can thus be understood as a ternary tree of time series segments where the location of zero-crossings determine the temporal extent of the segment and the (dis-) appearance of zero-crossings limit the (vertical) extent or lifetime of a segment. By tracing the position of an extrema in the scale-space back to the position at $s \approx 0$ we can compensate the displacement caused by smoothing itself. We may thus construct a so-called interval tree of scales [7] (cf. Fig. 2(right)), where the lifetime of a monotone time series segment is represented by a box in the scale-space: its horizontal extent denotes the position of this segment in the series, the vertical extent denotes the stability or resistance against smoothing. Rather than choosing a single smoothing filter beforehand, such a tree represents the time series at multiple scales and allows different views or perspectives on the same series. We consider this to be advantageous for our purpose, because we do not know at which level discriminative features may occur.

3   Visualising Discriminative Features

We have seen that the distance values obtained by established methods such as DTW may be misleading when investigating structural properties of time series. We are looking for a way to visualize discriminative, structural features in sets of labelled series. When addressing the analysts perception, we should adjust the algorithmic approach to the way humans perceive time series. While the temporal alignment of DTW is objective function-driven, humans align time series by aligning landmarks and matching the corresponding segments between the landmarks. The underlying idea of this work is to use the interval tree of scales as the core for the visualisation as it encodes landmarks already (extrema or inflection points). By matching series from the same and/or from different classes we recognize which cases have which landmarks in common. By complementing the interval tree with this information it becomes a tool to not only visualize the structure of a series, but also to distinguish which parts are shared among classes and which help to distinguish classes.

3.1 Graph Representation of the Interval Tree of Scales

We consider the graphical depiction of the interval tree as a tesselation of the time-scale space, which encodes *all possible perceptions* of a time series. We represent a tile in this tesselation that covers the temporal range $[t_1, t_2]$ and the scale range $[s_1, s_2]$ by a quintuple (t_1, t_2, s_1, s_2, o) with $t_1 < t_2$, $s_1 < s_2$ and orientation $o \in \{$increasing, decreasing$\}$. We define a graph representation $G_\mathbf{x} = (V, E)$ of the interval tree as follows: The set of all tiles is denoted as V and makes up the set of nodes in our graph. Two tiles $v = (t_1^v, t_2^v, s_1^v, s_2^v, o^v)$ and $w = (t_1^w, t_2^w, s_1^w, s_2^w, o^w)$ are connected, that is $(v, w) \in E$, if and only if they are adjacent in time

$$t_2^v = t_1^w \tag{1}$$

We define a subset $V_S \subseteq V$ (resp. $V_E \subseteq V$) that contains all start-tiles (resp. end-tiles), that is, tiles which do not have a predecessor (resp. successor) in the graph. A path of n tiles $(v_1, \ldots, v_n)$ in $G_\mathbf{x}$ is called *perception* of series $\mathbf{x}$ if $\forall i : (v_i, v_{i+1}) \subset E$, $v_1 \in V_S$ and $v_n \in V_E$. Such a *perception* represents a segmentation of the time series $\mathbf{x}$ because the time periods of the tiles v_i represent a segmentation of the time range $[1, m]$: Apparently we have $t_1^{v_1} = 1$ (because $v_1 \in V_S$), $t_2^{v_n} = m$ (because $v_n \in V_E$) and subsequent time periods touch due to (1). The sequence of orientations is alternating between increasing and decreasing due to the properties of the original interval tree of scales. The set of all *perceptions* corresponds to all possible paths from a leftmost tile to a rightmost tile, that is, a view of the series $\mathbf{x}$ with a (possibly) different degree of smoothing within each segment.

Cf. Fig. 3: The interval tree of the series on the left (or bottom) consists of 9 (or 7) tiles. The graphs are superimposed on the interval tree. Nodes belonging to V_S (or V_E) are connected to the virtual node S (or E). From these particular graphs we find two perception for the left series $((a_0, a_1, a_2, a_4)$ and $(a_0, a_1, a_2, a_3, a_5, a_6))$ and four for the bottom series (e.g. $(b_0, b_2, b_4, b_7, b_8)$).

3.2 Matching Perceptions

In this section we define a distance measure for two time series $\mathbf{x}$ and $\mathbf{y}$ by means of their respective graphs $G_{\mathbf{x}}$ and $G_{\mathbf{y}}$. To compare time series we have to decide which *parts* of both series correspond and, once the assignment has been made, how well they match. With Euclidean distance or DTW these *parts* correspond to the values at individual time points. Here we assign *time periods* or segments of both series to each other and then compare the subseries of the respective time periods. The segments we consider for this assignment correspond to the tiles in the interval tree (or vertices of each graph). As we want to match the full series we have to cover all points from both series exactly once (with some possible exceptions near the beginning and end of the series) and keep the temporal order of the segments – that is, we have to match perceptions.

Suppose we have two similar series, where one time series $\mathbf{x}$ contains a certain landmark while the other $\mathbf{y}$ does not. When perceiving all the details (path through tiles near the bottom of the interval tree), both series do not match structurally in the number of tiles (because $\mathbf{x}$ has an additional landmark). To perceive them as similar, we have to switch to a coarser scale for $\mathbf{x}$ in the temporal region where the additional landmark resides. That is, a structural comparison of time series corresponds to finding the right path through both graphs $G_{\mathbf{x}}$ and $G_{\mathbf{y}}$ such that the tiles in both sequences correspond to each other. (In the example case of Fig. 3 we have to find one perception (out of 2) for the left series and one perception (out of 4) for the bottom series that correspond best, e.g. (a_0, a_1, a_2, a_4) and (b_1, b_4, b_7, b_8)).

However, given two arbitrary paths $p_{\mathbf{x}} = (v_1^{\mathbf{x}}, \ldots, v_k^{\mathbf{x}})$ from $G_{\mathbf{x}}$ and $p_{\mathbf{y}} = (w_1^{\mathbf{y}}, \ldots, w_l^{\mathbf{y}})$ from $G_{\mathbf{y}}$, the tiles $v_i^{\mathbf{x}}$ and $w_i^{\mathbf{y}}$ may not be directly comparable: as we intend to perform a structural match, segments of different type must not be aligned, e.g., we do not assign increasing $w_1^{\mathbf{y}}$ to decreasing $v_1^{\mathbf{x}}$. (Figure 3: a_0 cannot be associated with b_0.) Either we have to switch to a different level of abstraction again (that is, different perceptions) – or we allow to skip short segments near the beginning and the end of the series. We express this by an alignment $\delta \in \mathbb{N}$ such that tile $w_{1+\delta}^{\mathbf{y}}$ is assigned to $v_1^{\mathbf{x}}$ with $o^{v_1} = o^{w_{1+\delta}}$. (Ex. from Fig. 3: to match $p_{\mathbf{x}} = (a_0, a_1, a_2, a_4)$ to $p_{\mathbf{y}} = (b_0, b_2, b_4, b_7, b_8)$ we need to associate the first node of $p_{\mathbf{x}}$ with the second node of $p_{\mathbf{y}}$, that is, set the offset $\delta = 1$ leading to a comparison of $a_0 \sim b_2, a_1 \sim b_4, \ldots$).

Eventually we compare the shape of the series within a segment. Given a tile $v = (t_1^v, t_2^v, s_1^v, s_2^v, o^v) \in V$ we denote the subseries of $\mathbf{x}$ that the tile refers to by $\mathbf{x}|_v := (x_{t_1^v}, x_{t_1^v+1}, \ldots, x_{t_2^v})$. Given two paths (plus an alignment δ) we evaluate how well the assigned segments match each other by means of a dissimilarity measure d' to compare $\mathbf{x}|_{v_i}$ against $\mathbf{y}|_{w_{i+\delta}}$. As the assigned segments need not be of the same length, the distance $d'(v_i^{\mathbf{x}}, w_{i+\delta}^{\mathbf{y}})$ must cope with segments of different lengths. This might be achieved by stretching one series to the length of the other and apply Euclidean distance afterwards. Other choices will be discussed below.

Thus, among all possible matches of perceptions of time series $\mathbf{x}$ and $\mathbf{y}$, choose the one that minimizes the sum of distances of the corresponding segments:

$$d(\mathbf{x},\mathbf{y}) = \min_{(v_1^{\mathbf{x}},\dots,v_k^{\mathbf{x}})\in G_{\mathbf{x}},(w_1^{\mathbf{y}},\dots,w_l^{\mathbf{y}})\in G_{\mathbf{y}},\delta} \sum_{i=\max\{1,1-\delta\}}^{\min\{k,l+\delta\}} d'(v_i^{\mathbf{x}},w_{i+\delta}^{\mathbf{y}}) \qquad (2)$$

The search for the minimal distance (and thus the best match) can be formalized as a weighted shortest path problem in the graph $\mathcal{G}_{\mathbf{x},\mathbf{y}} = (V_{\mathbf{x},\mathbf{y}}, E_{\mathbf{x},\mathbf{y}})$ with $V_{\mathbf{x},\mathbf{y}} = V_{\mathbf{x}} \times V_{\mathbf{y}}$ and $E_{\mathbf{x},\mathbf{y}} \subseteq V_{\mathbf{x},\mathbf{y}}^2$ where $\forall v,v' \in V_{\mathbf{x}}, \forall w,w' \in V_{\mathbf{y}}$: $((v,w),(v',w')) \in E_{\mathbf{x},\mathbf{y}} \Leftrightarrow (v,v') \in E_{\mathbf{x}} \wedge (w,w') \in E_{\mathbf{y}} \wedge o^v = o^w$ from an arbitrary start node $(v,w) \in V_{\mathbf{x}}^S \times V_{\mathbf{y}}^S$ to an end node $(v',w') \in V_{\mathbf{x}}^E \times V_{\mathbf{y}}^E$. The search for an optimal δ is reformulated by extending the set of start/end tiles by their adjacent tiles. The edge weights are given by the distance d' among the segments. Standard methods such as the Dijkstra algorithm might be used, but a more efficient solution via dynamic programming is advisable and has been implemented for this work. Similar to DTW we have a matrix of assignments (of tiles rather than points) where we seek for the minimal cost path from a start position (bottom left) to end position (top right) as illustrated by the matrix in Fig. 3. This step has complexity $O(n \cdot m)$ with n and m being the number of nodes in the resp. graph (rather than number of points as in DTW).

3.3 Discriminative Features

Determining the distance between two time series includes the identification of the best-matching perceptions. If two series share many low-level features the corresponding tiles are likely to be included in the optimal assignment, whereas series from different classes may have to retract to segments on a coarser scale (with fewer landmarks) as the details of one series have no counterparts in the other. We complement the interval tree of some series $\mathbf{x}$ of class c with this information: We count (when comparing a series to all other) how often segments were involved in the best match of series from the same as well as from different classes. Based on these numbers we highlight tiles that get primarily matched to series of the same class (or other classes). In the visualization, increasing segments will be coloured in blue, decreasing in red, but the opacity is determined by the entropy of the distribution "same class vs. other class" weighted by the total number count (a distribution 5:0 is less relevant than 25:0). Furthermore, we normalize both counts to 100 to avoid a bias towards the *other classes* in multiclass problems (otherwise we expect only a fraction of $\frac{1}{k}$ cases from class c and $\frac{k-1}{k}$ cases from other classes (bias towards 'other class')).

Apart from counting, upon matching all series to one series $\mathbf{x}$, we may collect for each of its tiles all segments from other series that were assigned to it. From these segments we can derive secondary features such as maximal difference (in value), slope, curvature, variance, etc. and construct a classical dataset (with a fixed number of attributes) for each tile and feed it into a standard classifier to identify features that help to distinguish the classes. In this work we consider

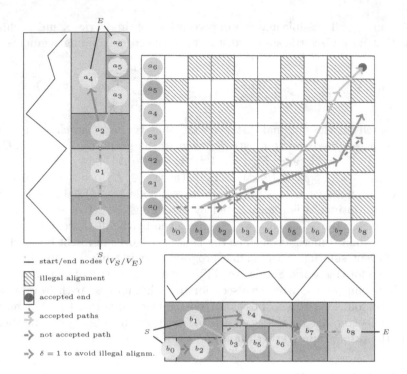

Fig. 3. Two series (left and bottom), represented by their interval tree and respective graph (slope is color-coded). The matrix in the center encodes the assignment of nodes from both graphs, a match of perceptions from both series is thus a path from the bottom left to top right edge of the matrix. (Color figure online)

only the maximal difference (difference between end points of a tile) and report its utility for distinguishing classes by means of the weighted accuracy[2].

Among all possible visualisations (one per time series) we automatically select one per class that offers a perception where the discriminative features are pronounced best and present them to the analyst. Each tile in the interval tree is annotated with the (weighted) class distribution (same vs. other class) and, below, the (weighted) accuracy (only if above 65 %). The bottom left tile in Fig. 5(l) reads as follows: this tile is matched (in the optimal assignement) with 62 % (28 %) of series from the same (other) class(es) and a classifier on the height difference alone yields 66 % (weighted) accuracy.

---

[2] As before, we weigh the cases such that the total weight of series from the same class and series from a different class becomes identical to get accuracies independent of the number of classes.

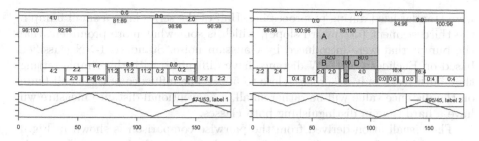

Fig. 4. Series from the right class have a small bump imputed. The tile marked A corresponds to a perception of the series where the bump has been smoothed away; it matches 100 % of the *series from other classes*, while the tiles B–D occur mainly for series of the same class. This turns the tiles A–D into interesting visualization features.

3.4 Segment Distance

For Eq. (2) we employ a distance measure $d'(v_i^\mathbf{x}, w_j^\mathbf{y})$ that is used to compare segments. By scaling the segments and applying Euclidean distance, (2) becomes quite similar to DTW, only that we do not allow arbitrary warping but linear warping between extrema. However, as we have showcased in Sect. 2.1, Euclidean distance and DTW miss a possibility of vertical scaling. We could use Pearson correlation instead, because its built-in normalization compensates for different ranges in both segments. Here, to better focus on structural distances and get most of the interval tree, we treat the value range identical to the temporal range, that is, rescale and shift both series in both dimensions such that their start and end points coincide (in time and value). Then we apply Euclidean distance to the rescaled series (denoted by $d_{ES}(\mathbf{x}, \mathbf{y})$ in (3)) and thereby capture differences in the shape of the segments. But there are more aspects than just shape if we want to adopt to the visual perception: The duration of the segments, the difference in value (range) and in particular the importance, that is, the persistence against smoothing. We therefore penalize d_{ES} by additional factors:

$$\underbrace{d'(v^\mathbf{x}, w^\mathbf{y})}_{\text{segment distance}} = \underbrace{d_{ES}(\mathbf{x}|_v, \mathbf{y}|_w)}_{\text{shape distance}} \cdot \underbrace{f(\Delta t_v, \Delta t_w)}_{\text{delta in duration}} \cdot \underbrace{f(\Delta y_v, \Delta y_w)}_{\text{delta in height}} \cdot \underbrace{f(\Delta s_v, \Delta s_w)}_{\text{delta in importance}}$$

(3)

where for some tile u of series $\mathbf{z} = (z_1, \ldots, z_n)$ we define $\Delta t_u = |t_2^u - t_1^u|$, $\Delta y_u = |z_{t_2^u} - z_{t_1^u}|$, $\Delta s_u = |s_2^u - s_1^u|$ and $f(x, y) = \frac{\max\{x, y\}}{\min\{x, y\}}$. We thus penalize d_{ES} by a factor of 2 if one segment is twice as long, tall, or important (persistent) as the other.

4 Experimental Evaluation

4.1 Sanity Check

We illustrate the approach by using an example similar to that of Fig. 1 in the introduction. All series consist of five linear segments with varying duration that were standardized during preprocessing. Half of the series have a small bump in the third segment (downward slope), which is somewhat more prominent than the bumps that were introduced by Gaussian noise. Standard 1-NN classifiers based on Euclidean or DTW distance have difficulties with this simple setting, they both reach only accuracies close to 50 %. The influence of the small bump on the distance value will be rather small, a conventional distance measure will have a hard time in distinguishing both classes.

The visualisation derived from the pairwise comparison is shown in Fig. 4. While nothing of interest shows up for the class where the bump is absent, the visualisation for the other class clearly reveals the relevant features. Note that the noise introduces local extrema, which vanish quickly as the scale increases, the small bump survives somewhat longer. Some series from the same class match not only the bump but also the noise, but they are small in number. While all of the series from the other class match this segment on a coarse scale (tile A), most of the series from this class subdivide tile A into three subtiles B, C, D. This is easily recognizable from the visualisation, which therefore supports a human in interpreting and understanding the differences of both classes.

4.2   Series from the UCR Repository

In this section we discuss the results on some datasets from the UCR time series repository. All series were standardized in advance. We show a diversive subset from these datasets: motion capture (Gunpoint), shape (Plane, Fish), and mass spectrometry (Coffee). For many other datasets of the same type the visualisation led to comparable results. The visualisation provides less informative results if series from different classes hardly share common properties, but in this case a sophisticated search for discriminative features is not necessary anyway.

Figure 5 shows a series of class 1 from the Coffee dataset. Series from both classes are quite similar in shape, the visualization draws our intention directly to some interesting differences. While 73 % of the series from class 1 exhibited a small, local extremum (tile B), most of the series from class 0 do not have this feature (tile A). A similar observation can be made at tile C and its subnodes. The number 89 in tile C denotes a (weighted) accuracy of 89 % obtained from the absolute difference in value alone (between start and end points of this tile).

Figure 6 displays a series of class 1 from the Gunpoint dataset. For this dataset, we did not use the zero crossings of the first but the second derivative, that is, the interval tree of scales characterizes inflection points rather than extrema. The series record the hand position of subjects drawing a gun, aiming, and returning it to the holster – or performing the same motion without a gun. The visualisation shows the relevance of small features when drawing and returning the gun from/to the holster, which are absent with series from the other class.

Figure 7 depicts a series of class 2 from the Fish dataset. We immediately recognize that the small local extrema in the first large increasing and the first large decreasing segment (in [0, 100] and [120, 220], resp.) are characteristic for the majority of the series from class 2. The upper left tile describes the segment of

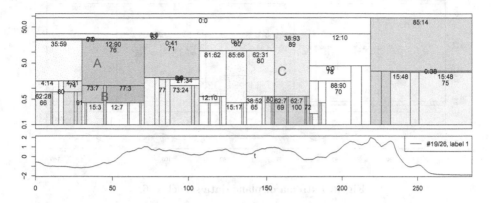

Fig. 5. Extrema of coffee data set, class 1.

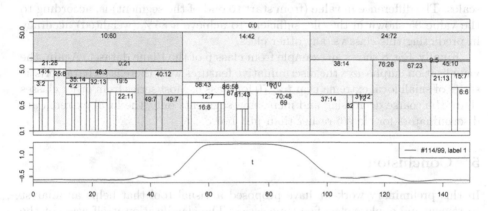

Fig. 6. Inflection points of gunpoint data set, class 1.

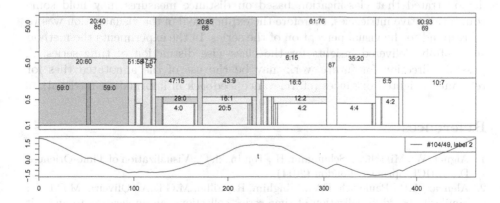

Fig. 7. Extrema of fish data set, class 2.

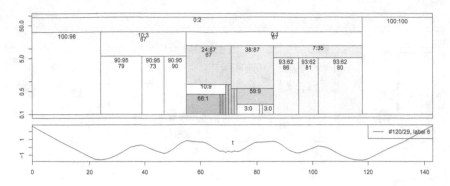

Fig. 8. Extrema of plane data set, class 6.

all series from the beginning to the first minimum that persists over all (shown) scales. The difference in value (from start to end of the segment) is, according to the value 85 shown in the tile, sufficient to achieve a 85 % (weighted) accuracy in predicting this class vs. any other class.

Finally, Fig. 8 shows an example from class 6 of the Plane dataset. Again, the visualisation emphasizes the discriminative features well: there is a characteristic series of small local extrema near $t \approx 70$ for class 6. Most series from other classes share the coarse decreasing and increasing segments only, the local extrema are discriminators for class 6 rather than just noise.

5   Conclusion

In this preliminary work we have proposed a visual tool that helps an analyst to review and explore classified time series. The classification itself was not the focus of this work, but to support the data understanding and the assessment of structural differences to simplify, e.g. subsequent preprocessing steps. We have demonstrated that classification based on distance measures may hold some counterintuitive intricacies, therefore the explicit goal of the visualisation was to correspond to the visual perception of the series. In the experiments the method successfully delivered insights for the class-wise distinction of time series. A possible direction for future work may be the use of the annotated tiles for relevance-weighted distances (cf. relevance feedback in information retrieval).

References

1. Aigner, W., Miksch, S., Schumann, H., Tominski, C.: Visualization of Time-Oriented Data. HCI. Springer, London (2011)
2. Alencar, A.B., Paulovich, F.V., Minghim, R., Filho, M.G.D.A., Oliveira, M.C.F.D.: Similarity-based visualization of time series collections: an application to analysis of streamows. In: International Conference Information Visualisation, pp. 280–286, July 2008

3. Berndt, D.J., Clifford, J.: Finding patterns in time series: a dynamic programming approach. In: Advances in Knowledge Discovery and Data Mining, pp. 229–248. MIT Press (1996)
4. Mallat, S.G.: A Wavelet Tour of Signal Processing. Elsevier Ltd., Amsterdam (2001)
5. Perng, C.-S., Wang, H., Zhang, S.R., Parker, D.S.: Landmarks: a new model for similarity-based pattern querying in time series databases. In: International Conference on Data Engineering (2000)
6. Wang, X., Mueen, A., Ding, H., Trajcevski, G., Scheuermann, P., Keogh, E.: Experimental comparison of representation methods and distance measures for time series data. Data Min. Knowl. Disc. **26**(2), 275–309 (2012)
7. Witkin, A.P.: Scale space filtering. In: Artifical Intelligence (IJCAI), Karlsruhe, Germany, pp. 1019–1022 (1983)

Spotting the Diffusion of New Psychoactive Substances over the Internet

Fabio Del Vigna[1,2][(✉)], Marco Avvenuti[1], Clara Bacciu[2], Paolo Deluca[3],
Marinella Petrocchi[2], Andrea Marchetti[2], and Maurizio Tesconi[2]

[1] Department of Information Engineering, University of Pisa, Pisa, Italy
marco.avvenuti@unipi.it
[2] Institute of Informatics and Telematics (IIT-CNR), Pisa, Italy
{f.delvigna,c.bacciu,m.petrocchi,a.marchetti,m.tesconi}@iit.cnr.it
[3] Institute of Psychiatry, Psychology and Neuroscience,
King's College London, London, UK
paolo.deluca@kcl.ac.uk

Abstract. Online availability and diffusion of New Psychoactive Substances (NPS) represents an emerging threat to healthcare systems. In this work, we analyse drugs forums, online shops, and Twitter. By mining the data from these sources, it is possible to understand the dynamics of drug diffusion and its endorsement, as well as timely detect new substances. We propose a set of visual analytics tools to support analysts in tackling NPS spreading and provide a better insight about drugs market and analysis.

Keywords: NPS data mining · Drugs forums · NPS online shops · Data visualisation and exploration · NPS detection · Visual analytics · Social media analysis

1 Introduction

Noticeably, health departments of European countries are facing a raising issue: the online trade of substances that lay in a grey area of legislation, known as New Psychoactive Substances (NPS). European Union (EU) continuously monitors the market to tackle NPS diffusion, forbid NPS trade and sensitise people to the harmful effects of these drugs[1]. Unfortunately, legislation is typically some steps back and newer NPS quickly replace old generation of substances.

Online shops and marketplaces convey NPS through the Internet [20], without any (or with very few) legal consequences. Quite obviously, this attracts drug consumers, which can legally buy these drugs without risk of prosecution. The risks connected to this phenomenon are high: every year, hundreds of consumers get overdoses of these chemical substances and hospitals have difficulties

---

[1] http://www.emcdda.europa.eu/start/2016/drug-markets#pane2/4; All URLs in the paper have been accessed on July 10, 2016.

ⓒ Springer International Publishing AG 2016
H. Boström et al. (Eds.): IDA 2016, LNCS 9897, pp. 86–97, 2016.
DOI: 10.1007/978-3-319-46349-0_8

to provide effective countermeasures, given the unknown nature of NPS. Furthermore, products sold over the Internet with the same name may contain different substances, as well as possible changes in drug composition over time [5].

Social media and specialised forums offer a fertile stage for questionable organisations to promote NPS as a replacement of well known drugs, whose effects have been known for years and whose trading is strictly forbidden. Furthermore, forums are contact points for people willing to experiment with new substances or looking for alternatives to some chemicals, but also a discussion arena for those at the first experiences with drugs, as well as trying to stop with substance misuse or looking for advice regarding doses, assumption and preparation.

The EU-funded project Cassandra[2] investigates the NPS supply chain, lifecycle, and endorsement, through the analysis of popular social media, drug forums, and online shops. Such analysis is vital to timely detect NPS diffusion: this will support governments and health agencies in confining the progress of substance abuse, prohibiting NPS sales and improving the awareness of citizens towards unhealthy and harmful behaviours.

In this paper, we shed light on the structure and activity of NPS forums and online shops. The main contributions are as follows: (i) we give an insight into two popular forums, Bluelight[3] and Drugsforum[4], hosting drugs discussions since more than one decade; (ii) we map NPS sales (as monitored on online shops) and NPS diffusion and distribution (as monitored on discussion forums); and (iii) we provide automatic support to timely NPS detection.

Overall, we show a successful application of Intelligent Data Analysis techniques to complex systems, such as social networks and hierarchical ones. This eases the human exploration and interpretation of the online universe of drugs, with a support for the interactive visualisation of the data analysis results.

The paper is structured as follows. Next section gives related work in the area. Section 3 gives a panoramic view on our data sources. In Sect. 4, we focus on forums and shops structure and activities, by analysing their data and offering a visualisation of the analysis results. Finally, Sect. 5 concludes the paper and gives directions for future work.

2   Related Work

Recently, academia has started investigating the massive use of social media and online forums to advertise and discuss about psychedelic substances and drugs, and how the preferences of online communities can affect those of consumers. Large forums drew attention, being a primary source of information about NPS and a good sample of consumer tastes [4]. Work in [14] considers the Flashback forum and traces the trend of the discussions, especially in relation with the

---

[2] http://www.projectcassandra.eu.
[3] http://www.bluelight.org.
[4] https://drugs-forum.com.

scheduling of a substance ban. The paper puts in evidence how volumes of discussions drop when a ban is scheduled. In [22], the authors focus on new drugs detection and categorisation by scanning online shops and the dark net. A complete list of the known effects of new drugs, to the publication date, is given in [11,20].

Small subsets of the contents of the Drugsforum and Bluelight forums, which we deeply analyse in the present paper, have been already considered in [21], highlighting how large forums embody a cumulative community knowledge, i.e., a stratified knowledge built over years of forum activities, and showing that drugs effects and dosage are among the most discussed topics.

Other studies explored the abuse of medicines and how these are advertised, e.g., on Twitter, and sold by online pharmacies, with no authorisation [6,13]. Twitter features a rapid spread of contents, especially through small communities of users, which share common interests and tastes. This is the main reason why it has been investigated to mine patterns of drug abuse, also for non-medical purposes, e.g., to improve students performances in study [7,8]. Furthermore, Twitter allows analysts to comprehend rapid disease diffusion and health issues [17], as well as prices and effects of new drugs [16]. Nevertheless, social media play an important role also for contrasting the drugs diffusion [19] and for preventing end users from further consumption [12]. Twitter was also extensively mined to detect geographical diffusion of drug consumers over time [3].

The Web is not the only marketplace where NPS are advertised and sold. Indeed, the TOR network[5] has drawn much attention from drug consumers and resellers, who search for a channel to buy and sell drugs that guarantees their anonymity. This aspect affects trustworthiness of peers, especially when it is not possible to assess users reputation at all. In [9], the authors investigated the impact of reputation in Silk Road, one of the most popular marketplaces for drugs in the dark net. Data analysis often deals with the quality of the results obtained when searching the web. The work in [18] describes the possibility to improve the recall of queries issued to search engines by exploiting all variants and misspelled words.

With respect to related work, this paper addresses a finer-grained, more detailed picture of NPS data sources and NPS data available on the Internet. As an example, the analysis of forums carried on in [21] was limited in time and quantity. In our work, we overcome this limitation, by analysing more than one decade of data, posted by users all over the world. Overall, we dealt with more than 4 million and a half posts and more than 500,000 users. Furthermore, we integrated more than one source, by monitoring two forums, Twitter, and a number of online shops. The results of our analysis are conveniently conveyed to the reader via a set of interactive visual web interfaces, which are being integrated into a dashboard that will help researchers mine the wealth of gathered data. Ultimately, we are aligned with recent advances in data analysis leading to applications in pattern mining of, e.g., medical records and human anatomies [2,10].

---

[5] https://www.torproject.org.

3   Data Sources

This section presents the data sources for our analysis. We collected the data by developing ad-hoc software, which scrapes websites and uses APIs to crawl social media.

3.1   Forums

Bluelight and Drugsforum are two large forums, which host more than a decade of discussion about drugs and addiction. Being particularly rich of information, the two forums provide a historical, worldwide background of drug consumption, comprising that related to NPS. Similar to Google Flu Trends[6] efforts to detect spreading of diseases, the analysis of the forums' content and structure is significant to understand how psychoactive substances have spread out and to study new infoveillance strategies, to timely detect drugs abuse.

The two forums have a hierarchical structure, which enables proper content categorisation. The root of both forums organises content into sub-forums, which can be nested up to several levels of depth. The forums' structures were subject to different content re-organisations over time.

We carried out a Web scraping activity to create a dump of the entire database of discussions from the two forums, following the links between the forums' sections. During the storage phase, we kept track of the forums' hierarchy and structure, maintaining all the tags and metadata associated to each post and thread. Table 1 summarises the amount of data available from the two forums.

3.2   Online Shops

The forums introduced in Sect. 3.1 are a primary source of information about drugs reviews, feelings, effects and preparation, but little information is available about the drugs market, such as prices and bulk quantities. Thus, we focused our attention also on other data sources, dealing with drugs trading.

Online shops sell both legal and illegal substances. Among others, those that sell NPS have grown in popularity, given the relatively low risks in trading such substances. Many online shops accept payments in pounds, euros and dollars. Also, bitcoins are often accepted. This opens up the possibility to track price trends and, indirectly, to estimate the popularity and quality (or purity) of drugs. Furthermore, many of the marketplaces are advertised and mentioned on forums and social media.

We have started an intense scraping activity on a set of online shops to monitor the market availability of different substances. Online shops can be quite easily found through simple queries to search engines (e.g., "legal highs" and "smart drugs"). We set up a battery of scrapers that collect the information that are present on the shops showcases. Data is collected on a weekly basis, and stored in a relational database, to be easy queryable. Table 2 shows the monitored shops.

---

[6] https://www.google.org/flutrends/about/.

Table 1. Drug forums: posts and users

Forum	First post	Last post	Tot posts	Users
Bluelight	22-10-1999	09-02-2016	3,535,378	347,457
Drugsforum	14-01-2003	26-12-2015	1,174,759	220,071

Table 2. Monitored online shops and number of substances they sell

ID	Website	Substances found
1	http://chem-shop.co.uk	7
2	http://researchchemist.co.uk	45
3	http://researchchemistry.co.uk	56
4	http://sciencesuppliesdirect.com	43
5	http://www.bitcoinhighs.co.uk	4
6	http://www.buylegalrc.eu	17
7	http://www.legalhighlabs.com	33
8	http://www.ukhighs.com	51
9	https://www.buyanychem.eu	78
10	https://www.iceheadshop.co.uk	68

3.3 Twitter

Twitter is extensively used by resellers and "pharmacies" to advertise psychoactive substances, and by consumers to discuss their effects and share feelings with others [6,13]. We have collected about 14 million tweets, over the period March 16, 2015–February 2, 2016, using the Streaming API[7], which allows applications to gather tweets in real time fashion. We have used a crawler that fetches data relying on a set of ad-hoc keywords. We have also followed a series of Twitter accounts associated to online shops. In the next section, we will detail the monitored keywords, which we chose among known emerging substances.

4 Data Analysis and Visualization

This section shows the analysis we have carried out over the data sources described in Sect. 3. First, we report on a series of analyses over the two drug forums, with the purpose of figuring out their structural features, how their content is organised, and the geographical distribution of their users. Secondly, we mine the forums textual contents, aiming at looking for new substances mentioned in recent discussions. Finally, we provide a picture on the NPS substances sold on online shops, correlating them with mentions on Twitter and the forums.

---

[7] https://dev.twitter.com/streaming/overview.

4.1 Forums: Structural and Geographical Features

To facilitate the investigation of the forums structural features, we have developed a set of visual interfaces. Figure 1 depicts the screenshot of a zoomable treemap of the two forums. Nested subsections are represented as nested rectangles, the area of which are proportional to the number of posts a subsection contains. Quick visual comparisons of the forums' size and structure may gather meaningful information. For example, compared to Drugsforum, whose structure is quite complex, Bluelight has a shallow organisation. Also, the names of the subsections suggest that the discussion on Drugsforum is mainly focused on drugs and it follows a rigid categorisation, based on the kind of the substance, while the topics on Bluelight are broader and less related specifically to drugs.

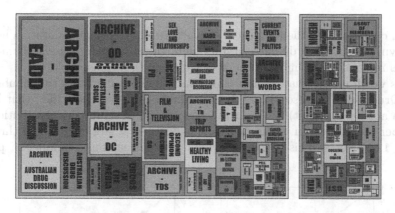

Fig. 1. The structure of Bluelight (left) and Drugsforum (right). Bluelight is about three times bigger.

Figure 2 shows the worldwide distribution of the Drugsforum users. The information has been extracted from the users' profiles (when available). Looking at the figure, we understand that drugs discussions on forums is a wide phenomenon, quite naturally leading to a widespread word of mouth. The colours in the figure are proportional to the density of users. Noticeably, the most involved areas are North America, Australia, UK, and Scandinavia.

We have also investigated some topological aspects of the forum, like the number of posts per user and the number of posts per thread, on both forums (Fig. 4). With the *powerlaw* Python package [1], we have compared the four real data distributions with the exponential, power law, truncated power law and lognormal distributions. The tool measured the $xmin$, no more than 4 for all the cases. Furthermore, with a p-value less than 10^{-8} for all the distributions, the power law distribution results in a better fit than the exponential one, as expected [15]. With regard to the lognormal and truncated power law distributions, the lognormal distribution fits slightly better than the power law one, while the truncated power law distribution fits better than the lognormal one.

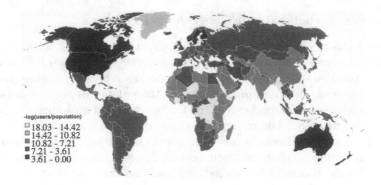

Fig. 2. Geographical distribution of Drugsforum users.

We can conclude that the (truncated) power law distribution assumption holds, as shown in Figs. 3a to d. These results highlight that there is a small amount of users responsible for most of the activity, on both forums.

It is worth noting that, even if Bluelight has about 0.6 times the number of users Drugsforum has (see Table 1), the number of active users (i.e., that have written at least one post) is almost the same for both. As for the distribution of posts per thread, shown in Fig. 5, Bluelight features a large number of threads having 1,000 posts. This is due to a limit on the maximum number of posts

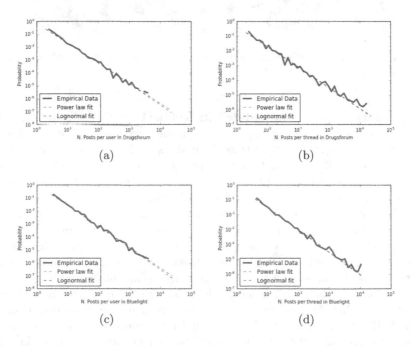

Fig. 3. Probability Density Function of the real data and the different distributions

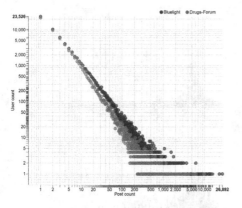

Fig. 4. Posts per user.   **Fig. 5.** Posts per thread.

for certain threads: when exceeding the threshold, the moderators start a new thread for the discussion.

4.2 Content Analysis

A text analysis that is really useful in our scenario is the measurement of volumes of discussion over time, given a term. This investigation helps determining whether some drugs raise in popularity and in which section of the forum this happens, possibly obtaining some clues about the nature of the substance (being it a NPS or not).

Figure 6 shows the frequency of the term "mephedrone" over time, normalised to the whole volume of discussion, for Drugsforum (top) and Bluelight (bottom). Even if not identical, the shapes of the spike are similar, meaning that the substance has gained popularity within both the communities approximately at the same time.

Figure 7 shows a higher level of detail: each line represents a subsection of the forum. As shown in the top-left part of the screenshot, we can choose which forum to analyse. A darker colour indicates a higher frequency of the term, for the corresponding time frame. The search for "mephedrone" in Drugsforum shows a high volume of discussion in the first half of 2010 in a series of subsections, particularly in the one called "Beta-Ketones". This indicates the category of the substance.

As shown in the example of Fig. 8, computing the terms that co-occur with a given one gives interesting insights. Indeed, the generated wordclouds may provide knowledge on substances that are similar, with similar effects and market trends. In the figure, each word occupies an area that is proportional to its frequency. The wordclouds can be generated for both Twitter and the two forums.

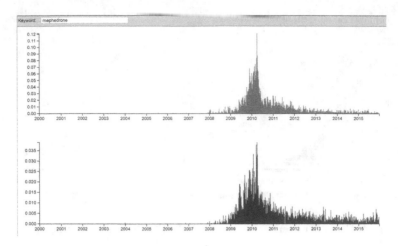

Fig. 6. Frequency of "mephedrone" over time, normalised to the whole volume of discussion, for Drugsforum (top) and Bluelight (bottom).

Fig. 7. Horizon charts showing the frequency of a given term over time, for each subsection of the chosen forum.

Really endorsed drugs are presented and discussed in forums. To timely detect NPS, we have investigated neologisms and terminology on both the forums, to discover new names. As an example, in Fig. 9, we plot the Drugsforum terms that appeared only after 2010. The result clearly indicates a lot of new drugs, appeared on the market from 2010 to 2015. It is possible to notice the name of some new drugs and medicines, such as α-PVP, Diclazepam, Pentedrone, Naphyrone.

4.3   NPS Trading

As a final set of analyses, we have explored the hyperlinks on the forums. Then, we have compared them with a comprehensive list of NPS online shops and with the links in the posts of monitored Twitter accounts. Not surprisingly, they do not overlap, meaning that forums discussions do not link shops. This is mainly due to the specific policies of the forums. We have also tested which are the NPS sold on the shops and also mentioned on forums, finding that almost every substance is mentioned. It is not possible to estimate the trade volume

Fig. 8. Zoomable wordcloud showing the most frequent terms co-occurring with "mephedrone" in the Twitter dataset.

Fig. 9. Zoomable wordcloud showing new terms in Drugsforum after 2010.

Table 3. An excerpt of monitored substances, with no. tweets, posts and shops. Bluelight (BL) and Drugsforum (DF) are the two forums analyzed in this work. The last column highlights the forum where the substance has appeared first.

Drug	Tweets	Post BL	Post DF	Online shops	First seen
MDAI	913	3507	775	1, 3, 4, 9	Bluelight
MDPV	791	11304	3631	9	Drugsforum
Methylone	679	8254	5116	9	Bluelight
AB-CHMINACA	584	16	33	4, 6, 9	Drugsforum
Methiopropamine	515	329	232	2, 3, 7, 8, 9, 10	Bluelight
1P-LSD	483	612	69	1, 2, 3, 4, 9	Bluelight
Etizolam	1592	8629	2630	2, 4, 9	Bluelight
Ethylphenidate	965	2502	1268	2, 7, 9	Bluelight
Synthacaine	217	124	60	3, 4, 9, 10	Drugsforum
Diphenidine	193	779	80	2, 3, 4, 9	Bluelight
Mexedrone	39	113	14	1, 2, 3, 4, 9, 10	Bluelight

of NPS from online shops, but we can try to infer some information about popularity by observing the discussions in forums. Checking the words frequency, we concluded that the very same substances are also advertised through Twitter. Table 3 reports an excerpt of some substances, with a measure of the discussion activity about them on Twitter, on forums and on online shops. In the table, the numbers in the column of online shops are the IDs of the shops, as in Table 2. The meaning is: the drug is mentioned on those shops.

5   Conclusions

Today, New Psychoactive Substances (NPS) lie on a grey area, not precisely addressed by current regulations. NPS rapidly appear on - and suddenly disappear from - the market, with a consistent and continuous introduction of new surrogates, which leaves few margin for intervention by healthcare institutions and governments. This paper has put in evidence some unique features of online NPS forums and shops. Monitoring such websites and elaborating the available data made it possible to explore a large quantity of information, also across platforms, allowing analysts to perform comparisons among them. We also gave a measurement of the relevance of NPS diffusion and advertisement, as well as user engagement. Furthermore, we showed how trading and discussions are correlated, through terms used by both online shops, social media, and forums, despite the prohibition, which hold on forums, to post explicit links to shops. Noticeably, co-occurrences analysis and temporal analysis of neologisms are a valid support for NPS detection.

Currently, the analyses are led by the data scientist, which is assisted by the developed software. The analyses are applicable both to offline datasets and online streaming sources. We aim at fully automating some of the work, e.g., the detection of the psychic and physical effects of NPS on the individual, based on comments by the online users. Finally, we plan to extend the analysis to the dark web marketplaces.

Acknowledgements. This publication arises from the project CASSANDRA (Computer Assisted Solutions for Studying the Availability aNd DistRibution of novel psychoActive substances), which has received funding from the European Union under the ISEC programme: Prevention of and fight against crime [JUST2013/ISEC/DRUGS/AG/6414].

References

1. Alstott, J., Bullmore, E., Plenz, D.: powerlaw: a Python package for analysis of heavy-tailed distributions. PloS One **9**(1), e85777 (2014)
2. de Bono, B., Grenon, P., Helvensteijn, M., Kok, J., Kokash, N.: ApiNATOMY: towards multiscale views of human anatomy. In: Blockeel, H., van Leeuwen, M., Vinciotti, V. (eds.) IDA 2014. LNCS, vol. 8819, pp. 72–83. Springer, Heidelberg (2014)
3. Buntain, C., Golbeck, J.: This is your Twitter on drugs: any questions? In: 24th World Wide Web Conference - Companion Volume, pp. 777–782. ACM (2015)
4. Davey, Z., Schifano, F., Corazza, O., Deluca, P.: e-Psychonauts: conducting research in online drug forum communities. J. Ment. Health **21**(4), 386–394 (2012)
5. Davies, S., et al.: Purchasing legal highs on the Internet - is there consistency in what you get? QJM **103**(7), 489–493 (2010)
6. Freifeld, C.C., Brownstein, J.S., Menone, C.M., Bao, W., Filice, R., Kass-Hout, T., Dasgupta, N.: Digital drug safety surveillance: monitoring pharmaceutical products in Twitter. Drug Saf. **37**(5), 343–350 (2014)

7. Hanson, C.L., Burton, S.H., Giraud-Carrier, C., West, J.H., Barnes, M.D., Hansen, B.: Tweaking and tweeting: exploring Twitter for non medical use of a psychostimulant drug (Adderall) among college students. J. Med. Internet Res. **15**(4), e62 (2013)

8. Hanson, C.L., et al.: An exploration of social circles and prescription drug abuse through Twitter. J. Med. Internet Res. **15**(9), e189 (2013)

9. Hardy, R.A., Norgaard, J.R.: Reputation in the Internet black market: an empirical and theoretical analysis of the Deep Web. J. Inst. Econ. **1**, 1–25 (2015). FirstView Article

10. Hielscher, T., Spiliopoulou, M., Völzke, H., Kühn, J.-P.: Mining longitudinal epidemiological data to understand a reversible disorder. In: Blockeel, H., van Leeuwen, M., Vinciotti, V. (eds.) IDA 2014. LNCS, vol. 8819, pp. 120–130. Springer, Heidelberg (2014)

11. Hillebrand, J., Olszewski, D., Sedefov, R.: Legal highs on the Internet. Subst. Use Misuse **45**(3), 330–340 (2010)

12. Inciardi, J.A., Surratt, H.L., Cicero, T.J., Rosenblum, A., Ahwah, C., Bailey, J.E., Dart, R.C., Burke, J.J.: Prescription drugs purchased through the Internet: who are the end users? Drug Alcohol Depend. **110**(1), 21–29 (2010)

13. Katsuki, T., Mackey, T.K., Cuomo, R.: Establishing a link between prescription drug abuse and illicit online pharmacies: analysis of Twitter data. J. Med. Internet Res. **17**(12), e280 (2015)

14. Ledberg, A.: The interest in eight new psychoactive substances before and after scheduling. Drug Alcohol Depend. **152**, 73–78 (2015)

15. Muchnik, L., Pei, S., Parra, L.C., Reis, S.D., Andrade Jr., J.S., Havlin, S., Makse, H.A.: Origins of power-law degree distribution in the heterogeneity of human activity in social networks. Sci. Rep. **3** (2013)

16. OConnor, K., Pimpalkhute, P., Nikfarjam, A., Ginn, R., Smith, K.L., Gonzalcz, G.: Pharmacovigilance on Twitter? Mining tweets for adverse drug reactions. In: AMIA Annual Symposium, p. 924. American Medical Informatics Association (2014)

17. Paul, M.J., Dredze, M.: You are what you tweet: analyzing Twitter for public health. In: ICWSM, vol. 20, pp. 265–272 (2011)

18. Pimpalkhute, P., Patki, A., Nikfarjam, A., Gonzalez, G.: Phonetic spelling filter for keyword selection in drug mention mining from social media. AMIA Summits Transl. Sci. 90 (2014)

19. Scott, R., Nelson, L., Meisel, Z., Perrone, J.: Opportunities for exploring and reducing prescription drug abuse through social media. J. Addict. Dis. **34**(2–3), 178–184 (2015)

20. Schmidt, M.M., Sharma, A., Schifano, F., Feinmann, C.: Legal highs on the netEvaluation of UK-based websites, products and product information. Forensic Sci. Int. **206**(1), 92–97 (2011)

21. Soussan, C., Kjellgren, A.: Harm reduction and knowledge exchange–a qualitative analysis of drug-related Internet discussion forums. Harm Reduction J. **11**(1), 1–9 (2014)

22. Zawilska, J.B., et al.: Next generation of novel psychoactive substances on the horizon - a complex problem to face. Drug Alcohol Depend. **157**, 1–17 (2015)

Feature Selection Issues in Long-Term Travel Time Prediction

Syed Murtaza Hassan[1], Luis Moreira-Matias[1], Jihed Khiari[1(✉)], and Oded Cats[2]

[1] NEC Laboratories Europe, 69115 Heidelberg, Germany
{syed.hassan,luis.matias,jihed.khiari}@neclab.eu
[2] Department of Transport and Planning, TU Delft, 2600 Delft, Netherlands
o.cats@tudelft.nl

Abstract. Long-term travel time predictions are crucial for tactical and operational public transport planning in schedule design and resource allocation tasks. Similarly to any regression task, its success considerably depend on an adequate feature selection framework. In this paper, we approach the myopia of the State-of-the-Art method RReliefF on mining relevant inter-relationships of the feature space relevant for reducing the entropy around the target variable on regression tasks. A comparative study was conducted using baseline regression methods and LASSO as a valid alternative to RReliefF. Experimental results obtained on a real-world case study uncovered the bias/variance reduction obtained by each approach, pointing out promising ideas on this research line.

Keywords: Travel time prediction · Machine learning · Regression · Feature selection

1 Introduction

One of the most common research problems in transportation is travel time prediction (TTP). The literature on this topic is extensive and covers different application domains such as fleet management, monitoring, control, mass transit and individual navigation [1]. Hereby, we focus on public transport in general and buses in particular. It is possible to distinguish short and long-term travel time prediction problem based on the prediction horizon (e.g. threshold of 2–3 h). Operational tasks (e.g. timetable design) or resource allocation (e.g. vehicle and crew scheduling) requires long-term TTP.

A traditional approach to TTP is regression analysis. It comprises a large number of techniques to estimate the relationship between a set of predictors (i.e. features) and a dependent variable:

$$\hat{f} : x_i, \theta \to \mathbb{R} \text{ such that } \hat{f}(x, \theta) = f(x_i) = y_i, \forall x_i \in X, y_i \in Y \qquad (1)$$

where $f(x_i)$ denotes the true unknown function which is generating the samples' target variable and $\hat{f}(x_i, \theta) = \hat{y}_i$ be an approximation dependent on the feature

© Springer International Publishing AG 2016
H. Boström et al. (Eds.): IDA 2016, LNCS 9897, pp. 98–109, 2016.
DOI: 10.1007/978-3-319-46349-0_9

vector x_i and an unknown parameter vector $\theta \in \mathbb{R}^n$ (given by some induction model M). Notorousily, this approximation will be as good as the adequacy of M to the dependence structure of f as well as the relevancy of the input feature space X. If it has a low number of features, it may not explain the variance of Y, thus leading M to biased models. Coversely, for a large set of features, we may be using features with a low predictive power. In consequence, M may output very complex models which lead to optimal fits on the input dataset (i.e. *local minima*) but a considerably lower ones when tested in any generic inference task. These phenomenons are known as *underfitting* and *overfitting*, respectively.

Automatic Feature Selection [2] is a subfield of study focused on developing algorithms capable of defining adequate feature spaces for supervised learning problems. The idea is to find the feature subset that guarantees solutions (i.e. models) close to the global minima of our generalization error by defining which features to use and which to drop on a particular regression/classification problem. There are mainly two types of feature selection algorithms: (i) *filters*, where the induction model is not take into account to select an adequate feature subset and (ii) *wrappers*, where the feature subset selection process takes into account the induction model (typically through an encapsulated optimization framework). In this paper, we are focused on discussing issues around this topic (i), as well as its impact in the context of long-term TTP tasks.

In transportation science, it is known that the main determinants of bus running times are route length, passenger activity at stops and the number of traffic signals (e.g. [3,4]). Other studies also added driver response to the deviation from the schedule as an explanatory variable [5,6]. However, all of those have estimated linear regression models to identify the impact of potential explanatory variables on bus running times. Consequently, the resulting models often have very limited predictive power.

Attaining better bus travel time predictions can have significant consequences for passenger delays, operator's performance fines and the efficiency of its resource allocation. The inherently complex and uncertain operational environment in which urban bus service operate call for the development of more sophisticated models that can capture non-linear relations between system variables. To the authors' best knowledge, the literature to handle this specific issue is scarce. Mendes-Moreira *et al.* [7] compared Random Forests (RF), Support Vector Machine Regression (SVR) and Progression Pursuit Regression (PPR). On the other hand, the well-known RReliefF [8] was proposed to do an adequate feature selection for each route. As many other methods from the RELIEF*-family, RReliefF is an instance-based learning method which leverages on the concept of neighborhood to define features that can (or cannot) contribute significantly to the entropy reduction on estimating the target variable Y. Consequently, as many other instance-based methods (e.g. k-nearest neighbors), it is highly dependent on an adequate setting of a distance metric that serves this specific purpose (which can easily vary from problem to problem). Moreover, it also has limitations on evaluating inter-relationships among the feature set X which can lead to this effect.

This paper is focused on studying the effects of RReliefF *myopia* to unrealistic distance functions and/or interrelationship on the feature set relevant for predicting the target variable value. To do it so, we propose an the Least Absolute Shrinkage and Selection Operator (LASSO) as a simple and yet valid alternative to RReliefF for this particular domain. The idea is to leverage on the priority that LASSO gives on the bias error reduction - in contrast to RReliefF. Consequently, our contributions are twofold: (1) a practical demonstration on RReliefF limitations through the study of its impact on particular application area; (2) the introduction of LASSO as a valid alternative to this problem due to the high number of relevant interactions among different predictors/features that can reduce bias error. Experimental results of applying the same baseline predictors to a particular real-world case study uncovered the potential of our novel approach.

The remaining of this paper is organized as follows: we start by describing the case study and related data sources. The methodology section presents the feature selection algorithms studied as well as a brief description of the baseline regressors employed. The experimental setup is detailed in Sect. 4, followed by a result report and a comprehensive discussion. We conclude with final remarks and future research directions.

2   Case Study

Our case study is a large urban bus operator in Sweden. We collected data from four high-frequency (maximum planned headway of 11 min between 7:00–19:00) routes A1/A2/B1/B2, i.e. two bus lines A/B. Line A connects residential areas to a public transport interchange hub as well as major shopping areas. B connects the southern parts of the city to the city center, traversing through an interchange, major hospitals as well as a logistic center. The bus operator defines two schedules; a summer schedule taking effect from June 19th till December 14th and a winter schedule taking effect from December 15th till June 18th. Our study covers a period of six months between August 2011 and January 2011 thus including both schedules.

As part of the preprocessing step, a trip pruning was performed by removing trips where more than 80 % of link travel times were missing. In addition, we performed data imputation on the remaining samples by following the interpolation procedure suggested in [9]. The dwell times were also pruned by using the 99 % percentile to remove erroneous measurements. Table 1 presents an overview of the resulting dataset, detailed per route. It contains the (i) total number of trips (NT), (ii) number of stops and (iii) Round Trip Times (RTT).

2.1   Feature Generation

The original features are schedule departure time, daytype and vehicle ID. Unlike RF, SVR and PPR do not support categorical values. Therefore, it is required to

Table 1. Statistics per route. The values are as mean ± s.d. Times in seconds.

	NTrips	Stops	Daily trips	Round trip times
A1	17953	33	134 ± 27	3017 ± 425
A2	16353	33	133 ± 30	2755 ± 480
B1	16280	25	137 ± 23	2607 ± 465
B2	16353	25	134 ± 22	2746 ± 448

generate new features based on the original ones. For the type of day, we use one-hot encoding which generates 7 numerical features corresponding to the day type. Vehicle ids associated with less than 0.5 % of total number of trips were grouped into a single cluster. The remaining vehicle ids were clustered using a clustering technique described in the experiments section. This procedure resulted in four additional features.

Figure 1 illustrates the clustering results for route A2. It illustrates the clustering plot (top-left) and the kernel density estimations for the vehicle ids within each cluster. We used the Bayesian Information Criterion (BIC) to determine the best number of clusters $k = 3$ from the interval $K = [2:20]$. We note that the three clusters are characterized by slightly different p.d.f. This justifies mapping the ids into three distinct features. Since driver rosters are typically assigned

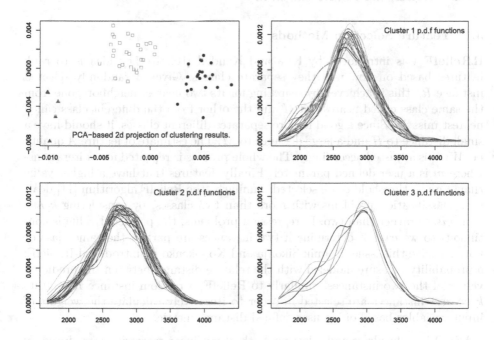

Fig. 1. Clustering results of vehicle ids for route A2.

to individual vehicles throughout their shift, vehicle travel times reflect driving style as well as the propagation of delays from one trip to later ones.

3   Methodology

Feature selection consists in eliminating redundant or non-informative features. Applying feature selection can not only lead to more interpretable models but also attain better results. Redundant features can negatively affect the predictions of models that do not inherently perform such a task. This is also relevant for our TTP framework, where we seek to determine the best set of features for generating predictions. The state-of-the-art method for this domain (proposed by Mendes-Moreira *et al.* [7]) is RreliefF [8]. This instance-based learning algorithm is able to determine features relevance on determining the target variable value. It can handle interdependences on the feature space, missing data and/or different type of functional forms for the dependences. However, its success depends largely on an adequate definition of a distance metric. Moreover, it is focused on reducing variance-type error, neglecting the inter-relationships that can potentially reduce the bias-type one.

Hereby, we compare RReliefF to LASSO as filter feature selection method to highlight why the first is not adequate for this task on long-term TTP problems. This section elaborates formally on the two methods as well on the three used baseline regressors: PPR, RF and SVR.

3.1   Feature Selection Methods

RReliefF was introduced by Kira and Kendell [10]. Its key idea is to rank features based on how well they separate classes. Given a randomly selected instance R_i, this is achieved by searching for its two nearest neighbors, one from the same class called nearest hit H and the other from the different class called nearest miss M. Since a good feature separates different classes, it should have a small distance to H and a large distance to M. The estimate of feature A quality i.e. $W[A]$ is adjusted accordingly. The whole process is repeated for m iterations-where m is a user defined parameter. Finally, features that have a higher value than a given threshold ϕ are selected. Similarly, the ReliefF algorithm [11] deals with classification problems with more than two classes, by considering k *hits* and *misses* rather than two. In regression problems, the predicted value is continuous so we cannot determine if two instances are part of the same class or not. To solve this issue, Robnik-Šikonja and Konokenko [8] introduced RreliefF: a probability measure modeled with the relative distance between the predicted values of the two instances. Similarly to ReliefF, a random instance R_i and its k nearest instances are selected in order to iteratively calculate the weights of input variables based on an user-defined distance metric.

LASSO is a shrinkage and selection method for linear regression introduced by Tibshirani [12]. Similarly to other shrinkage methods, it aims to improve the least-squares estimator by adding constraints on the value of coefficients noted *b.* Given an input data matrix of size $N \times p$ (i.e. N samples defined by $p - 1$ features and a target y), the LASSO estimate is defined by

$$\hat{b}^{lasso} = \underset{b}{\text{argmin}} \sum\nolimits_{i=1}^{N} (y_i - b_0 - \sum\nolimits_{j=1}^{p} x_{i,j} b_j)^2 \qquad (2)$$

subject to

$$\sum\nolimits_{j=1}^{p} |b_j| \leqslant t, t \geqslant 0 \qquad (3)$$

The equivalent *Lagrangian form* is

$$\hat{b}^{lasso} = \underset{b}{\text{argmin}} \left\{ \frac{1}{2} \sum\nolimits_{i=1}^{N} (y_i - b_0 - \sum\nolimits_{j=1}^{p} x_{i,j} b_j)^2 + \lambda \sum\nolimits_{j=1}^{p} |b_j| \right\} \qquad (4)$$

The L_1-norm penalty of LASSO $\sum_{j=1}^{p} |b_j|$ constrains the solution space to go for simpler, low-coefficient models by forcing some of the $n - 1$ features to be shrunk out of the final model. The tuning parameter λ controls the **strength** of the penalty. As it increases, more coefficients are set to zero and hence, less variables are selected. λ is typically set by using a cross-validation search technique over a grid of admissible values.

3.2 Regression Methods

RreliefF and LASSO were tested as filter-type feature selection methods to cope with three baseline regressors: RF, SVR and PPR.

Random Forests is an ensemble method based on classification and regression trees (CART [13]) that was introduced by Leo Breiman in 2001 [14]. The trees are grown by randomly choosing a set of candidate predictors at every node for a sample of the data and then producing the split by choosing the best splitter available. RF combines this with a random selection of samples to train the trees which is referred to as bootstrap aggregating or bagging. RF's hyperparameters are (i) the number of randomly selected predictors to choose from at each split *mtry* and the number of grown trees *ntree*.

Support Vector Machines were introduced by Cortes *et al.* in 1995 [15]. They are primarily binary classifiers that perform their task by constructing hyperplanes in a multidimensional space able to separate instances either linearly on non-linearly. In ϵ-SVM, these hyperplanes are constructed in a way to ensure the largest minimum distance to the training examples. This distance (ϵ) is denominated as *margin*. SVMs can be adapted for regression with a quantitative response by sequentially optimizing an error function where we seek to maximize the geometrical distance between the two hyperplanes $\frac{1}{||w||}$ which is equivalent to minimizing $\frac{1}{2}||w||^2$. To allow examples to be in the margin or to be misclassified, slack variables $\xi_i >= 0$ are introduced. The optimization problem becomes:

$$\underset{w,b}{\text{arg min}} \frac{||w||^2}{2} + C \times \sum_{i=1}^{n} \xi_i \qquad (5)$$

where $C > 0$ is a constant that sets the relative importance of maximizing the margin and minimizing the amount of slack. Kernels are typically used in SVMs to map the data points into higher dimensional feature space, where a linear separation allow a non-linear boundary to be drawn in the original one. Typical kernel include polynomial and radial basis functions. The choice of the kernel depends on the problem and different functions may depend on different hyperparameters.

Projection Pursuit Regression is an additive model that consists of linear combinations of non-linear transformations of linear combinations of explanatory variables (so-called *ridge functions*) [16]. It firstly projects the data matrix of explanatory variables in the optimal direction before applying smoothing functions to those. If *maxterms* (i.e. the number of linear combinations) is sufficiently large, PPR can be considered a universal approximator with considerable similarities to the so-called feed forward neural networks. However and similarly to the latter, complexity constraints need to be formulated to avoid overfitting. The algorithm starts by adding *maxterms* ridge functions. Then, it removes iteratively the least important term until *nterms* terms remain, which is the number of terms in the final model. Both *maxterms* and *nterms* are hyperparameters that need to be tuned beforehand. *optlevel* is a third hyperparameter which controls how thoroughly the models are refitted during this process. To smooth the ridge functions, we use by default Friedman's 'super smoother' *supsmu* which requires to fit the bass/span control.

4   Experiments

The experiments were conducted using the R Software [17]. Data was divided into two sets: a training set and a test set (i-e 70 %/30 %). Statistical independence was assumed to be in place among the routes. Consequently, we ended up having a total of 4 data sets. Vehicle ids were categorized into four groups: one containing all vehicle ids having less than 0.5 % of the total number of trips and 3 obtained through a three-step clustering procedure. First, kernel density estimation was used to generate the p.d.f. for every unique vehicle id. Second, this p.d.f. were clustered by a Gaussian Mixture Model trained using the Expectation-Maximization algorithm. Finally, the Bayesian Information Criterion was used to select the best model.

Package `FSelector` [18] was used for RReliefF. The value used for neighbour.count (the number of nearest examples) in [7] was 10. For robustness reasons, we used neighbour.count $= 50$ with $m = 100$ iterations. For illustrative purposes on this particular issue, we used 0.1 % of total data set as sample size. Similar results were found for a sample size of 0.5 % and 1.0 % of total data set length. A minimum weight threshold was set as $\phi = 0.01$. The default distance metric of `FSelector`'s implementation of RReliefF was used.

We used `glmnet` [19] procedures for fitting LASSO. The best λ was selected using cross validation.

4.1 Hyperparameter Tuning

Package `caret` [20] was utilized for hyperparameter tuning of RF, SVR and PPR. The two methods used in our experiments for hyperparameter optimization are (i) Grid Search (e.g. [7]) and (ii) Random Search [21]. (i) Grid Search exhaustively considers all the parameter combinations specified in a grid of parameter values. Hence, a high computational effort is required for large grids. A valid alternative introduced by Bergstra and Benghio [21] is Random Search. It consists on conducting independent draws from a uniform density using the same configuration space as the one defined by a regular grid. This approach only evaluates a random subsample of grid points - set to 60 in our case - and presents similar results to the grid one on an efficient manner [21].

PPR has five different hyperparameters: nterms, max.terms, optlevel, bass and span (the two latter for *supsmu*). Random Search was used for tuning nterms Package `kernlab` [22] was used for SVR. SVR has six different hyperparameters: kernel, C (for all kernels), epsilon (for all kernels), sigma (for Radial kernel), scale and degree (only for polynomial kernel). Random Search was used for tuning C, sigma, scale and degree. Finally, Package `randomForest` was used for RF. Grid search was used for tuning both hyperparameters, as well as the ones non explicitly mentioned above.

The three abovementioned base learners were evaluated based on the three resulting feature spaces: (1) feature set with all features (12 features), and as well as the ones given by (2) LASSO and (3) RReliefF. The obtained results were compared using two metrics of interest: RMSE and MAE.

4.2 Results and Discussion

The optimal hyperparameter values for the three distinct setups are displayed in Table 2 for RF, PPR and SVR. Figure 3 shows the results of RReliefF for each of the routes. x-axis is the feature set. y-axis is the weight; boxplots. It is evident that only the departure time has a predictive power accordingly with RReliefF. We therefore select departure time as the only feature from RReliefF method for each route. Figure 2 shows the results of LASSO plots for each of the routes. x-axis are different $\log(\lambda)$ values while y-axis are the coefficients. Features after the cut-off are selected to be the most suitable ones.

Finally, the evaluation of SVR, PPR and RF for the three feature sets for each of the routes are presented in Tables 3 and 4 respectively. The tables clearly show that LASSO performs better than RReliefF on this particular task. RF is the algorithm that benefits less of the feature selection process since this task is inherent of its own modelling process. Figure 3 illustrates the myopia of RReliefF on identifying some of the daytypes as relevant for reducing the bias-error around the target variable. As result, underfitted models (using only scheduled departure time) produce bad results - especially for PPR and SVR. These effects are depicted in Fig. 4, where the deficiency of the models output by either PPR and SVR during the peak hours when fed by RReliefF feature subspaces is highlighted. This effect happens because the daytype variables do

Table 2. Optimal hyperparameters setting.

		PPR			SVR		RF	
		nterms	max.terms	optlevel	σ	C	mtry	ntree
LASSO	A1	3	3	1	1179.65	909.04	1	500
	A2	3	3	1	997.91	335.15	1	500
	B1	3	3	1	1369.19	0.509	1	500
	B2	3	3	1	1008.36	72.25	1	700
RrreliefF	A1	7	7	7	65.42	2.84	3	700
	A2	8	8	3	329.67	3.30	3	900
	B1	6	6	3	17.87	22.80	3	900
	B2	7	7	3	71.46	0.076	3	900
ALL	A1	8	8	3	0.15	909.04	6	900
	A2	9	9	3	0.19	318.79	6	500
	B1	11	11	3	0.17	312.95	6	900
	B2	5	5	3	0.24	72.25	6	900

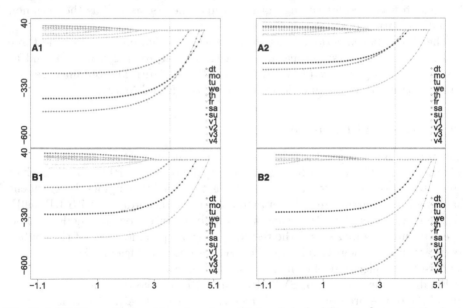

Fig. 2. LASSO results for all routes. A vertical red dashed line is drawn at the best log λ value. This serves as cut.off point. (Color figure online)

not have a particular effect on the variance-error reduction - but mainly only on the bias one. In the authors' opinion, these results illustrate that RReliefF is not the best technique to handle the feature selection task on this particular problem.

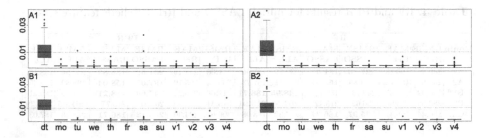

Fig. 3. RRelieF results for all routes. A horizontal red line is drawn at y = 0.01. (Color figure online)

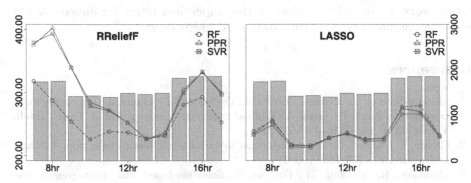

Fig. 4. RRelieF and LASSO comparative analysis (y-axis) using RMSE (scaled on RRelief side) along different scheduled departure times (x-axis). Bars denote the sample size on each timespan (scaled on LASSO side).

Table 3. SVR results for initial, LASSO and RrF-RReliefF feature sets.

Route	RMSE RrF	MAE RrF	RMSE LASSO	MAE LASSO	RMSE ALL	MAE ALL
A1	293.294	224.455	**228.192**	**182.554**	244.888	196.851
A2	260.567	192.483	**196.843**	**154.453**	228.977	180.843
B1	309.361	224.084	**244.650**	**180.188**	281.480	205.387
B2	311.037	231.711	**255.853**	**204.383**	268.029	211.830
ALL	293.564	218.183	**231.384**	**180.394**	255.843	198.477

5   Concluding Remarks

Feature selection is a relevant task in any real-world data mining project. Long-term TTP for public transport planning and/or operational purposes is not an exception. Hereby, we discussed the limitations of RReliefF - the state-of-the-art for this problem. A comprehensive comparison with LASSO was conducted using a real-world case study from a bus operator in Sweden. The obtained results illustrated how dependent RReliefF is on an adequate distance metric that gives different relevance for distinct features - thus leading to a proper normalization of the RReliefF output weights and/or different selection thresholds for each

Table 4. RF and PPR results for initial, LASSO and RrF-RReliefF feature sets.

	RF						PPR					
Route	RMSE RrF	MAE RrF	RMSE LASSO	MAE LASSO	RMSE ALL	MAE ALL	RMSE RrF	MAE RrF	RMSE LASSO	MAE LASSO	RMSE ALL	MAE ALL
A1	240.20	190.18	**227.66**	**181.44**	232.72	187.96	311.66	242.49	**231.52**	**186.83**	232.94	188.16
A2	235.65	180.76	**199.84**	**158.91**	203.11	161.73	263.04	195.19	**197.05**	**158.04**	198.80	159.45
B1	266.32	198.87	249.78	189.70	**248.95**	**188.71**	311.68	226.25	**247.69**	**188.21**	248.33	189.21
B2	283.59	223.68	266.62	215.42	**264.51**	**213.91**	316.31	233.50	**264.21**	213.77	**261.05**	**211.29**
ALL	256.44	198.37	**235.97**	**186.37**	237.32	188.08	300.67	224.36	**235.13**	**186.71**	235.28	187.03

feature accordingly to each one's contribution on model's bias reduction. As future work, we intend to explore further supervised filters for dimensionality reduction purposes on this task - such as Auto-Encoders.

References

1. Moreira-Matias, L., Mendes-Moreira, J., de Sousa, J.F., Gama, J.: Improving mass transit operations by using AVL-based systems: a survey. IEEE Trans. Intell. Transp. Syst. **16**(4), 1636–1653 (2015)
2. Guyon, I., Elisseeff, A.: An introduction to variable and feature selection. J. Mach. Learn. Res. **3**, 1157–1182 (2003)
3. Mishalani, R., McCord, M., Forman, S.: Schedule-based and autoregressive bus running time modeling in the presence of driver-bus heterogeneity. In: Hickman, M., Mirchandani, P., Voß, S. (eds.) Computer-Aided Systems in Public Transport, pp. 301–317. Springer, Heidelberg (2008)
4. Berkow, M., El-Geneidy, A., Bertini, R., Crout, D.: Beyond generating transit performance measures. Transp. Res. Rec. J. Transp. Res. Board **2111**(1), 158–168 (2009)
5. El-Geneidy, A., Horning, J., Krizek, K.: Analyzing transit service reliability using detailed data from automatic vehicular locator systems. J. Adv. Transp. **45**(1), 66–79 (2011)
6. Mazloumi, E., Rose, G., Currie, G., Sarvi, M.: An integrated framework to predict bus travel time and its variability using traffic flow data. J. Intell. Transp. Syst. **15**(2), 75–90 (2011)
7. Mendes-Moreira, J., Jorge, A., de Sousa, J., Soares, C.: Comparing state-of-the-art regression methods for long term travel time prediction. Intell. Data Anal. **16**(3), 427–449 (2012)
8. Robnik-Šikonja, M., Kononenko, I.: An adaptation of relief for attribute estimation in regression. In: Proceedings of the Fourteenth International Conference on Machine Learning, ICML 1997, pp. 296–304 (1997)
9. Mendes-Moreira, J., Moreira-Matias, L., Gama, J., de Sousa, J.: Validating the coverage of bus schedules: a machine learning approach. Inf. Sci. **293**, 299–313 (2015)
10. Kira, K., Rendell, L.A.: A practical approach to feature selection. In: Proceedings of the Ninth International Workshop on Machine Learning, pp. 249–256 (1992)
11. Kononenko, I.: Estimating attributes: analysis and extensions of RELIEF. In: Bergadano, F., Raedt, L. (eds.) ECML 1994. LNCS, vol. 784, pp. 171–182. Springer, Heidelberg (1994). doi:10.1007/3-540-57868-4_57

12. Tibshirani, R.: Regression shrinkage and selection via the LASSO. J. Roy. Stat. Soc. Ser. B (Methodol.) **58**(1), 267–288 (1996)
13. Breiman, L., Friedman, J., Stone, C.J., Olshen, R.A.: Classification and Regression Trees. CRC Press, New York (1984)
14. Breiman, L.: Random forests. Mach. Learn. **45**(1), 5–32 (2001)
15. Cortes, C., Vapnik, V.: Support-vector networks. Mach. Learn. **20**(3), 273–297 (1995)
16. Friedman, J., Stuetzle, W.: Projection pursuit regression. J. Am. Stat. Assoc. **76**(376), 817–823 (1981)
17. R Core Team: R: A Language and Environment for Statistical Computing. R Foundation, Vienna (2012)
18. Romanski, P.: Fselector: selecting attributes. R package version 0.19 (2009)
19. Friedman, J., Hastie, T., Tibshirani, R.: Regularization paths for generalized linear models via coordinate descent. J. Stat. Softw. **33**(1), 1 (2010)
20. Kuhn, M.: Caret package. J. Stat. Softw. **28**(5), 1–26 (2008)
21. Bergstra, J., Bengio, Y.: Random search for hyper-parameter optimization. J. Mach. Learn. Res. **13**(1), 281–305 (2012)
22. Zeileis, A., Hornik, K., Smola, A., Karatzoglou, A.: kernlab-an S4 package for kernel methods in R. J. Stat. Softw. **11**(9), 1–20 (2004)

A Mean-Field Variational Bayesian Approach to Detecting Overlapping Communities with Inner Roles Using Poisson Link Generation

Gianni Costa[✉] and Riccardo Ortale

ICAR-CNR, Via Bucci 41c, 87036 Rende, CS, Italy
{costa,ortale}@icar.cnr.it

Abstract. A novel model-based machine-learning approach is presented for the unsupervised and exploratory analysis of node affiliations to overlapping communities with roles in networks. At the heart of our approach is a new Bayesian probabilistic generative model of directed networks, that treats roles as abstract behavioral classes explaining node linking behavior. A generalized weighted instance of *directed affiliation modeling* rules the strength of node participation in communities with whichever role through *Gamma priors*. Moreover, link establishment between nodes is governed by a *Poisson distribution*. The latter is parameterized so that, the stronger the affiliations of two nodes to common communities with respective roles, the more likely it is the formation of a connection. A coordinate-ascent algorithm is designed to implement mean-field variational inference for affiliation analysis and link prediction. A comparative experimentation on real-world networks demonstrates the superiority of our approach in community compactness, link prediction and scalability.

1 Introduction

Community discovery and role assignment are two complementary tasks. Role assignment attributes within-community interactions to abstract behavioral classes, thus explaining node contributions to community purpose/functionality. Community discovery is an inherent characterization of the behavioral roles. The tight integration of both tasks was pioneered in [4] and further elaborated in [5,6], being of great practical relevance in various domains including (but not limited to) the social, information, ecological, (counter-)intelligence and recommendation [7] ones. Despite their effectiveness in recovering community structures and predicting prospective connections, the generative models developed in [4–6] do not account for a key property of network connectivity in community overlaps. Therein, the probability of a connection between two nodes was found in [18] to increase with the number of common community memberships shared by the two nodes. Besides, scalability with network size in [4–6] is limited.

In this paper, we propose a new approach to the seamless integration of community discovery and role assignment. The devised approach consists in performing variational inference in TOMATOES (*Tie fOrMATion based on cOmmunity*

© Springer International Publishing AG 2016
H. Boström et al. (Eds.): IDA 2016, LNCS 9897, pp. 110–122, 2016.
DOI: 10.1007/978-3-319-46349-0_10

and rolE affiliationS), i.e., a new Bayesian probabilistic generative model of networks. Four are the innovations behind TOMATOES. Firstly, a generalized weighted instance of *directed affiliation modeling* [12] is used for a twofold purpose, namely representing the strength of node affiliations to communities with roles and unveiling realistic community overlaps and nestings [19]. Secondly, link establishment is generalized with respect to [18], by also accounting for the strength of node affiliations. More precisely, the stronger the affiliations of two nodes to common communities with respective roles, the more likely it is the emergence of a connection from one to the other. Thirdly, the Poisson distribution is borrowed from *Poisson factorization* to expedite posterior inference on sparse networks [9]. Fourthly, mean-field variational inference is implemented by a coordinate-ascent algorithm in order to estimate the strength of node affiliations to communities with roles. This allows for the exploratory analysis of node affiliations in networks along with the prediction of prospective links.

An experimental assessment of the devised approach on real-world networks reveals its superiority in community compactness, link prediction and scalability.

This paper is structured as follows. Section 2 presents notation and preliminaries. Section 3 proposes TOMATOES. Section 4 covers the algorithmic implementation of mean-field variational inference. Section 5 discusses latent variable expectation for network exploration and link prediction. Section 6 is devoted to the empirical evaluation of TOMATOES. Section 7 concludes and highlights future research.

2   Preliminaries

The notation used in this paper and some basic concepts are introduced next.

A network is represented as a directed graph $\mathcal{G} = \{N_\mathcal{G}, E_\mathcal{G}\}$, where $N_\mathcal{G} = \{1, \ldots N\}$ is a set of nodes numbered 1 through N and, in addition, $E_\mathcal{G} \subseteq N_\mathcal{G} \times N_\mathcal{G}$ is a set of directed links (or, equivalently, ordered pairs of nodes). By nodes we mean the entities interacting in the network (such as, e.g., individuals, organizations and so forth). Links denote asymmetric interactions between nodes and are summarized into a $N \times N$ binary adjacency matrix L. Let $u \to v$ denote an interaction from node u to node v. L is such that the generic entry $L_{u \to v}$ is 1 iff $u \to v$ is actually a link of $\mathcal{G}$ (i.e., $\langle u, v \rangle \in E_\mathcal{G}$) and 0 otherwise.

Each network $\mathcal{G}$ is inherently characterized by two respective features, i.e., a latent structure C along with an underlying variety R of behavioral roles. The latent structure $C = \{C_1, \ldots, C_K\}$ reflects the organization of nodes into K unobserved communities. Within the individual communities, nodes exhibit connectivity patterns ascribable to abstract behavioral classes. These are formalized as a set $R = \{R_1, \ldots, R_H\}$ of H underlying roles. The generic node can participate in each community with any role, though with a different affiliation strength. In order to accurately capture the affiliations of nodes to communities with roles, we draw inspiration from [19] and distinguish between two types of affiliations, namely *sender* and *receiver*. More precisely, u participates in C_k with role R_h as a sender, whenever it links to other nodes playing any role in C_k. Dually, u participates in C_k with role R_h as a receiver, whenever it is linked to by other nodes playing any role in C_k. Clearly, u can be affiliated to community C_k with role R_h as a sender as well as a receiver. The strength of the sender affiliation of u to community C_k with role R_h is a nonnegative value $\vartheta_{u,k,h}^{(s)}$. $\vartheta_{u,k,h}^{(s)}$ is 0 iff u is not affiliated to C_k with role R_h as a sender. Likewise, the strength of the receiver affiliation of u to community C_k with role R_h is a nonnegative value $\vartheta_{u,k,h}^{(r)}$. $\vartheta_{u,k,h}^{(r)}$ is 0 iff u is not affiliated to C_k with role R_h as a receiver. The strengths of node affiliations are collectively denoted as $\boldsymbol{\Theta} \triangleq \{\vartheta_{u,k,h}^{(s)}, \vartheta_{u,k,h}^{(r)} | u \in \boldsymbol{N}_{\mathcal{G}}, \; C_k \in \mathcal{C} \; R_k \in \boldsymbol{R}\}$. Notice that the dichotomization of node affiliations to communities with roles supplements the dichotomization of node affiliations to communities alone in [19]. Our approach assumes the observation of an input network $\mathcal{G}$. Besides, the strengths of node affiliations in $\boldsymbol{\Theta}$ are treated as random variables, being unknown and not directly measurable.

3   The TOMATOES Model

TOMATOES (*Tie fOrMATion based on cOmmunity and rolE affiliationS*) is an innovative model of directed networks, whose generative process explains the emergence of links between nodes from a Bayesian probabilistic perspective.

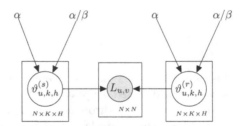

Fig. 1. Directed graphical representation in plate notation of the TOMATOES model

The directed graphical representation of TOMATOES in plate notation is shown in Fig. 1. The latter illustrates the conditional (in)dependencies between the random variables in TOMATOES. The sequence of all interactions among such random variables is the generative process of TOMATOES, whose details are reported in Fig. 2. Essentially, TOMATOES explains the formation of any input network as the result of a two-step process.

At step I, the strength of node affiliations to communities with roles is sampled from Gamma priors, which amounts to implicitly defining a generalized weighted instance of directed affiliation modeling [19]. The rationale behind the choice of Gamma priors is twofold. Firstly, sparseness is fostered in the representation of TOMATOES, which improves its interpretability. Secondly, the affiliation strength for each node does not sum to 1. As a consequence, a strong affiliation of any node to a certain community with some particular role does not

---

I. For each node $u \in N_\mathcal{G}$
- For each community $C_k \in C$ and each role $R_h \in C$
  * draw the affiliation strength $\vartheta_{u,k,h}^{(s)}$ of u to community C_k with role R_h as a sender, i.e., $\vartheta_{u,k,h}^{(s)} \sim Gamma(\vartheta_{u,k,h}^{(s)}|\alpha, \beta)$ where α and β are, respectively, the shape and rate hyperparameters.
  * draw the affiliation strength $\vartheta_{u,k,h}^{(r)}$ of u to community C_k with role R_h as a receiver, i.e., $\vartheta_{u,k,h}^{(r)} \sim Gamma(\vartheta_{u,k,h}^{(r)}|\alpha, \beta)$ where α and β are, respectively, the shape and rate hyperparameters.

II. For each pair of nodes $u, v \in N_\mathcal{G}$
- draw the presence/absence of a link $L_{u,v}$ from u to v, i.e., $L_{u,v} \sim Poisson(L_{u,v}|\lambda_{u,v})$ where $\lambda_{u,v}$ is the rate defined by Eq. 1.

Fig. 2. The probabilistic generative process of the TOMATOES model

diminish the overall strength of all other affiliations of that node. Thus, TOMA-TOES retains the flexibility of mixed-membership modeling, without unnatural assumptions on the structure of community overlaps [18].

The affiliation strengths Θ resulting at the end of step I are then used at step II to rule the establishment of pairwise connections between nodes. This is accomplished by means of a Poisson probability distribution, which is placed over the observed links L as the data likelihood, being beneficial for faster inference on sparse networks [9]. The formation of a link from a node u to a node v is ruled by the Poisson distribution according to the corresponding rate $\lambda_{u,v}$ below

$$\lambda_{u,v} = \sum_{k=1}^{K} \sum_{h=1,h'=1}^{H} \vartheta_{u,k,h}^{(s)} \vartheta_{v,k,h'}^{(r)} \tag{1}$$

Equation 1 allows for an extension of the increasing link probability with the shared community affiliations in [18], that also considers node roles, link direction and affiliation strength. Specifically, the stronger the affiliations of u and v to shared communities with respective roles, the more likely a link from u to v.

4   Approximate Posterior Variational Inference

A posterior distribution is calculated over the latent variables of TOMATOES (i.e., the strengths of node affiliations to communities with respective roles) by means of posterior inference. However, exact posterior inference is intractable under TOMATOES as well as most of the Bayesian models of practical relevance, essentially because of the complexity of the posterior distribution. Therefore, we focus on approximate posterior inference and choose between two widespread methods, i.e., MCMC sampling and variational inference. The former is a probabilistic instance of approximate posterior inference, whereas the latter is a deterministic one. We opt for variational inference, which tends to be faster and more easily scalable on large-scale networks than MCMC sampling [2].

An algorithm implementing mean-field variational inference in TOMATOES is designed to analytically approximate the true posterior distribution over the latent variables given the adjacency matrix L of the observed network $\mathcal{G}$.

The derivation and implementation of mean-field variational inference are simplified, by following the approach in [9]. Accordingly, auxiliary latent variables are added to the original formulation of TOMATOES. Specifically, because of the additive property of Poisson random variables, the generic $L_{u \to v}$ is rewritten as $L_{u \to v} = \sum_{k=1}^{K} \sum_{h,h'=1}^{H} z_{u,v}^{(k,h,h')}$. Here, $z_{u,v}^{(k,h,h')} \sim Poisson(\vartheta_{u,k,h}^{(s)} \vartheta_{v,k,h'}^{(r)})$ is the contribution to $L_{u \to v}$ from the affiliations of u and v to the common community C_k with respective roles R_h and $R_{h'}$. Remarkably, such an application of the auxiliary variables $z_{u,v}^{(k,h,h')}$ preserves the marginal Poisson distribution of $L_{u \to v}$. Let $\boldsymbol{Z} = \{z_{u,v} | u, v \in V_{\mathcal{G}}\}$ denote all auxiliary variables added to TOMATOES, where $\boldsymbol{z}_{u,v} = \{z_{u,v}^{(k,h,h')} | k = 1, \ldots, K$ and $h, h' = 1, \ldots, H\}$. The mean-field family over $\boldsymbol{\Theta}$ and $\boldsymbol{Z}$ has the following factorized form

$$q(\boldsymbol{\Theta}, \boldsymbol{Z} | \boldsymbol{\mu}) = \prod_{u \in N_{\mathcal{G}}, C_k \in C, R_h \in R} q(\vartheta_{u,k,h}^{(s)} | \pi_{u,k,h}) \prod_{v \in N_{\mathcal{G}}, C_k \in C, R_{h'} \in R} q(\vartheta_{v,k,h'}^{(r)} | \xi_{v,k,h'}) \prod_{u,v \in N_{\mathcal{G}}} q(\boldsymbol{z}_{u,v} | \boldsymbol{\gamma}_{u,v})$$

with $\boldsymbol{\mu} \triangleq \{\pi_{u,k,h}, \xi_{v,k,h'}, \boldsymbol{\gamma}_{u,v} | u, v \in N_{\mathcal{G}}, C_k \in C, R_h, R_h' \in R\}$ being the set of all variational parameters. Each such a variational parameter individually conditions a corresponding factor on the right hand side of the above equation. For the class of *conditionally conjugate* models, $\boldsymbol{\mu}$ can be fitted to the observed network $\mathcal{G}$ by means of a simple coordinate-ascent algorithm [2]. It can be proven that TOMATOES (with the addition of the auxiliary latent variables) is conditionally conjugate. We omit such a proof along with the mathematical derivation of the updates used in Algorithm 1 due to space requirements.

The coordinate-ascent variational algorithm operates by iteratively optimizing each variational parameter, while the others remain unchanged, until convergence to a local optimum [1]. Algorithm 1 shows the pseudo-code of such an algorithm for TOMATOES. After a preliminary initialization (line 1), the algorithm enters a loop (lines 2 − 20) to update the individual variational parameters. This loop halts upon convergence, which is met (at line 20) when the difference in the average predictive log likelihood of a validation set $V \subset L$ is smaller than 10^{-6}. Notably, the sums over users involve accounting only for the observed links. This expedites variational posterior inference on sparse adjacency matrices [9].

5   Exploratory and Predictive Tasks

Upon convergence of Algorithm 1, the approximate posterior distribution $q(\boldsymbol{\Theta}, \boldsymbol{Z} | \boldsymbol{\mu})$ under TOMATOES is fit to the input network $\mathcal{G}$ with K latent communities and H underlying roles. This enables the exploratory analysis of $\mathcal{G}$ and the prediction of prospective links. Both tasks involve expectations of specific variables according to standard Bayesian inference as explained in Sects. 5.1 and 5.2.

Algorithm 1 The coordinate-ascent variational algorithm

COORDINATE-ASCENT(L, K, H)
 Input: the adjacency matrix L of an observed network $\mathcal{G}$;
  the numbers K and H of latent communities and underlying behavioral roles, respectively;
 Output: the strengths Θ of node affiliations to communities with respective roles
 1: set the variational parameters $\pi_{u,k,h}$ and $\xi_{v,k,h'}$ equal (except for a random offset) to the prior
   on the corresponding latent variables [9];
 2: **repeat**
 3:   **for all** $u \to v$ such that $L_{u \to v} = 1$ **do**
 4:    **for all** $k = 1, \ldots, K$ **do**
 5:     **for all** $h, h' = 1, \ldots, H$ **do**
 6:      $\gamma_{u,v}^{(k,h,h')} \propto e^{\Psi\left[\pi_{u,k,h}^{(shp)}\right] - log\,\pi_{u,k,h}^{(rate)} + \Psi\left[\xi_{v,k,h'}^{(shp)}\right] - log\,\xi_{v,k,h'}^{(rate)}}$;
 7:     **end for**
 8:    **end for**
 9:   **end for**
10:   **for all** $u \in N_{\mathcal{G}}$ **do**
11:    **for all** $k = 1, \ldots, K$ **do**
12:     **for all** $h = 1, \ldots, H$ **do**
13:      $\pi_{u,k,h}^{(shp)} = \left[\sum_{v \in N_{\mathcal{G}}, R_{h'} \in R} L_{u \to v} \gamma_{u \to v}^{(k,h,h')}\right] + \alpha$;
14:      $\pi_{u,k,h}^{(rate)} = \left[\sum_{v \in N_{\mathcal{G}}, R_{h'} \in R} \frac{\xi_{v,k,h'}^{(shp)}}{\xi_{v,k,h'}^{(rate)}}\right] + \frac{\alpha}{\beta}$;
15:      $\xi_{u,k,h}^{(shp)} = \left[\sum_{v \in N_{\mathcal{G}}, R_{h'} \in R} L_{v \to u} \gamma_{v \to u}^{(k,h',h)}\right] + \alpha$;
16:      $\xi_{u,k,h}^{(rate)} = \left[\sum_{v \in N_{\mathcal{G}}, R_{h'} \in R} \frac{\pi_{v,k,h'}^{(shp)}}{\pi_{v,k,h'}^{(rate)}}\right] + \frac{\alpha}{\beta}$;
17:     **end for**
18:    **end for**
19:   **end for**
20: **until** convergence

5.1 Exploratory Network Analysis

For any node $u \in N_{\mathcal{G}}$, the posterior expectations $E[\vartheta_{u,k,h}^{(s)}]$ and $E[\vartheta_{v,k,h'}^{(r)}]$ denote
the strength of the affiliation of u to the generic community C_k with whichever
role R_h, respectively, as a sender and a receiver. Overall, the affiliation strength
of u to C_k is determined by the role, that maximizes its participation into C_k.
Hence, the affiliation strength $\overline{\vartheta}_{u,k}^{(s)}$ of u to C_k as a sender is defined as $\overline{\vartheta}_{u,k}^{(s)} = max_h E[\vartheta_{u,k,h}^{(s)}]$. Analogously, the affiliation strength $\overline{\vartheta}_{u,k}^{(r)}$ of u to C_k as a receiver
is defined as $\overline{\vartheta}_{v,k}^{(r)} = max_{h'} E[\vartheta_{v,k,h'}^{(r)}]$. In principle, u may not necessarily be
affiliated to C_k. Thus, in order to allow for such a possibility, both $\overline{\vartheta}_{u,k}^{(s)}$ and $\overline{\vartheta}_{u,k}^{(r)}$
are lower-bounded by a threshold ζ. Accordingly, u is affiliated to C_k as a sender
if $\overline{\vartheta}_{u,k}^{(s)} > \zeta$ and as a receiver if $\overline{\vartheta}_{u,k}^{(r)} > \zeta$. The threshold $\zeta = \sqrt{-\frac{1}{H^2} ln(1 - \frac{1}{|V_{\mathcal{G}}|})}$
is estimated through the notion of background link probability in [19].

5.2 Link Prediction

The missing links of $\mathcal{G}$ are ranked by a score, in order to forecast which of such
links will likely emerge in the future. The score $s_{u \to v}$ associated with any missing
link $u \to v \notin E_{\mathcal{G}}$ is defined as the posterior expectation of the corresponding

Poisson distribution rate, i.e.,

$$s_{u \to v} = E \left[\sum_{C_k \in \boldsymbol{C}} \sum_{R_h, R_{h'} \in \boldsymbol{R}} \vartheta_{u,k,h}^{(s)} \vartheta_{v,k,h'}^{(r)} \right]$$

6   Experimental Evaluation

We comparatively investigate TOMATOES on real-world networks. For this reason, four collections of network data are selected from the information, biological and social domains, i.e., *Enron, Neural Network, Twitter* and *Twitter*$_{20\%}$.

The *Enron* corpus (available at http://www.cs.cmu.edu/~enron/) is a large collection of emails generated by 158 employees of the *Enron* Corporation. The cleaned corpus consist of nearly 250,000 messages involving 150 employees, though the number of distinct employees is 148, since two employees are met twice with as many distinct usernames. We focus on the implicit social network arising from the 18,233 emails among the 148 employees.

Neural Network [16] is a directed map of the nervous system of the Nematode Caenorhabditis worm, that consists of 306 neurons and 2,345 connections [17]. We focus on the 297 neurons of the largest connected component.

Twitter data (available at http://snap.stanford.edu/data) was obtained in [13] from 973 ego-networks and consists 81,306 nodes and 1,768,149 directed edges. We additionally consider *Twitter*$_{20\%}$, that is a random sample retaining 20% of the links in the full *Twitter* network.

We conduct an empirical evaluation that is both qualitative and quantitative. The qualitative analysis is focused on *Enron* and its results are summarized by Table 1 and Fig. 3. In particular, Table 1 succinctly indicates the top 5 members with strongest affiliation to each of the 8 communities unveiled by TOMATOES. Besides, Fig. 3 illustrates the frequency of roles in the *Enron* communities.

TOMATOES is quantitatively and comparatively evaluated on the chosen networks hereunder. The selected competitors include all prototypical approaches to the seamless integration of community discovery and role assignment, i.e., the

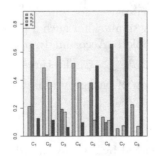

Fig. 3. Occurrence frequency of roles across the 8 *Enron* communities of Table 1

Table 1. An extract from the *Enron* communities unveiled by TOMATOES

Community 1	Community 2	Community 3	Community 4
1 Guzman	1 Zufferli	1 Lewis	1 Taylor
2 Linder	2 Townsend	2 Causholli	2 Staab
3 Donohoe	3 Giron	3 Griffith	3 Ward
4 Williams	4 Whitt	4 Hayslett	4 Hodge
5 Brawner	5 Wolfe	5 Campbell	5 Maggi
Community 5	Community 6	Community 7	Community 8
1 Stepenovitch	1 Saibi	1 Hendrickson	1 Sanchez
2 Meyers	2 Thomas	2 Horton	2 Linder
3 Crandell	3 Love	3 Pimenov	3 Farmer
4 Dean	4 Solberg	4 Quenet	4 Holst
5 Geaccone	5 Rogers	5 Germany	5 Gang

Baysian generative models BLFHM [6], BH-CRM$_{LP}$ [5] and BH-CRM [4]. An additional baseline, namely LDA-G [11], is also considered. LDA-G is a role-unaware approach to community discovery. The comparison against LDA-G is useful to highlight whether node roles actually contribute to improve model performance.

The process of learning a TOMATOES model of the chosen networks involves three steps, i.e., the formation of the training, validation and test sets. We partition the links of each chosen network by random sampling to form such sets. Therein, we include 70 % of the links into the training set, 15 % into the test set and the remaining 15 % into one held-out validation set. Both the test and validation sets have an equal number of present and absent links. More details on the test sets are provided below. Hereinafter, as far as TOMATOES is concerned, we write *Enron, Neural Network, Twitter* and *Twitter*$_{20\%}$ to mean the whole network data without the validation set.

The input parameters K and H are set identically for TOMATOES, BLFHM, BH-CRM and BH-CRM$_{LP}$ on the chosen networks. Specifically, the number K of communities to discover within the *Enron* social network is set to 8, i.e., the number of well connected and topically meaningful communities according to [14]. Also, the number H of roles is set to 4 based on [8], in which the distinct roles of *Enron* employees are grouped into four categories, i.e., *Senior Manager* (*SM*), *Middle Manager* (*MM*), *Trader* (*T*) and *Employee* (*E*). On *Neural Network*, K is set to 5, which is the number of anatomical clusters corresponding to identified functional circuits [15]. Besides, H is fixed to 4 in accordance with the distinct function types of neurons, i.e., *sensory, motor, interneuron, combination* [3]. As far as *Twitter* and *Twitter*$_{20\%}$ are concerned, K is set to 973. Furthermore, H is empirically set to 4. LDA-G unveils on each network as many communities as TOMATOES, BLFHM, BH-CRM and BH-CRM$_{LP}$.

The performance of all competitors on the chosen networks is investigated by looking at community compactness, link prediction and scalability.

Compactness is defined as the average of the shortest distances between nodes of a community. Essentially, it can be interpreted as a network-centric adaptation of a conventional criterion of clustering quality, namely intra-cluster distance, to assess community cohesiveness. As in [20], compactness is calculated by accounting for only a certain number of top-ranked nodes within each community. In particular, we consider the top 10 nodes with highest affiliation strength in each individual community. Table 2 summarizes compactness results. The most compact communities on *Enron*, *Neural Network* and *Twitter*$_{20\%}$ are found by TOMATOES. BLFHM is the runner-up on *Enron*, *Neural Network* and *Twitter*$_{20\%}$. TOMATOES is also the only one competitor with an observed performance on *Twitter*. Indeed, the abbreviation *N.A.* in the entries along the *Twitter* row of Tables 2 and 3 stands for *not available* and, actually, means that we were not able to experiment with BLFHM, BH-CRM, BH-CRM$_{LP}$ and LDA-G on *Twitter* within a reasonable computational time. This finding justifies the comparison over *Twitter*$_{20\%}$ in addition to revealing the meaningful increase in scalability enabled by TOMATOES.

Table 2. Compactness results

Network	TOMATOES	BLFHM	BH-CRM	BH-CRM$_{LP}$	LDA-G
Enron	1.70	1.71	1.98	2.04	2.28
Neural Network	2.14	2.18	2.56	2.68	2.20
Twitter$_{20\%}$	3.62	3.75	3.81	4.05	3.90
Twitter	2.98	*N.A.*	*N.A.*	*N.A.*	*N.A.*

An additional batch of tests is devoted to evaluate the predictive performance of all competitors. Link prediction is the adopted assessment criterion and the design of the tests is borrowed from [10]. Overall, 10 experiments of the predictive performance of the competing models are carried out over the chosen networks. Each experiment consists of two steps. Firstly, the generic input network is separated into a training set and a held-out test set. In particular, the latter is formed by randomly sampling the whole input network to choose a same number of both present and missing links, whose sum amounts to 15 % of the overall number of links in the whole input network. Secondly, the links in the held-out test set are predicted according to the distinct models learnt from the training set. The details on link prediction in TOMATOES are provided in Sect. 5.2, whereas those about BLFHM, BH-CRM, BH-CRM$_{LP}$ and LDA-G appear in [5,6].

The best ROC curves delivered by the competing models across the 10 experiments on *Enron*, *Neural Network* and *Twitter*$_{20\%}$ are reported, respectively, in Fig. 4(a), (b) and (c). Figure 4(d) illustrates the best ROC curve delivered by TOMATOES on *Twitter*. Remarkably, TOMATOES overcomes all other competitors on *Twitter*$_{20\%}$ across the whole range of the false positive rate. On *Enron* and *Neural Network*, TOMATOES overcomes BLFHM, BH-CRM, BH-CRM$_{LP}$ and

LDA-G in very large ranges of the false positive rate. However, on both *Enron* and *Neural Network*, there are still some narrow ranges of the false positive rate where the predictive performance of TOMATOES is not the most distant one from the random-guess diagonal line (in the sense of closeness to the upper left corner). Therefore, in order to more easily compare the overall link-prediction performance of each competitor on *Enron* and *Neural Network*, we reduce the respective ROC curves into the corresponding AUC (Area under the ROC Curve) summary statistics and average across the 10 experiments. The average AUC values on *Twitter*$_{20\%}$ and *Twitter* are computed too.

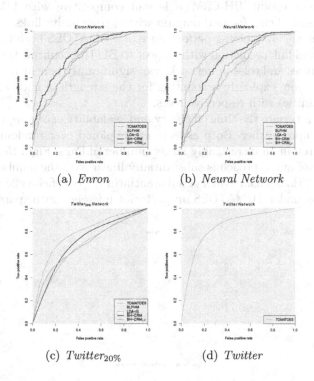

(a) *Enron*   (b) *Neural Network*

(c) *Twitter*$_{20\%}$ (d) *Twitter*

Fig. 4. ROC curves on the chosen networks

Table 3 summarizes the average AUC values for all competitors on the chosen networks. TOMATOES outdoes all other competitors in (average) link prediction. BLFHM is again the runner-up. Notably, according to Table 3, TOMATOES, BLFHM and BH-CRM overcome LDA-G in link prediction. Moreover, based on the results of Table 2, TOMATOES and BLFHM (as well as BH-CRM though with the exception of *Neural Network*) overcome LDA-G in community compactness. Such empirical findings provide substantial evidence in favour of also accounting for node roles in link formation, in order to more accurately capture and predict network connectivity in terms of affiliations to communities and roles.

Table 3. Average AUC results

Network	TOMATOES	BLFHM	BH-CRM	BH-CRM$_{LP}$	LDA-G
Enron	84.38	82.16	80.40	74.81	75.70
Neural Network	82.47	78.27	76.70	67.43	66.52
Twitter$_{20\%}$	80.19	74.28	71.61	69.05	67.24
Twitter	83.64	*N.A.*	*N.A.*	*N.A.*	*N.A.*

From this viewpoint, BH-CRM$_{LP}$ is still competitive with LDA-G in link prediction, despite being focused only on within-community links.

In addition, the overcoming performance of TOMATOES both in community compactness and link prediction with respect to BLFHM confirms that the analysis of communities and roles in networks can significantly benefit from affiliation modeling, i.e., from explicitly accounting for the strength of node affiliations to shared communities with respective roles.

Lastly, we return to the time efficiency and scalability of our approach, which were touched upon earlier. Both aspects are explored over random samples of *Twitter*, that retain an increasingly larger number of links. Figure 5 reveals that the scalability of our approach is substantially linear with the number of links in the input network. Such an evidence substantiates the beneficial effect of Poisson link generation under TOMATOES on posterior inference with sparse networks.

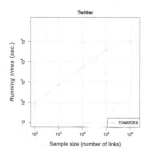

Fig. 5. Scalability over *Twitter*

7   Conclusions

We proposed a new approach to the seamless integration of community discovery and role assignment, that advances research on the simultaneous exploitation of both tasks for a more insightful (unsupervised) exploratory analysis of networks. A comparative evaluation on real-world networks showed the superiority of our approach in community compactness, link prediction and scalability.

Further research is required to study and evaluate the adoption of separate latent interaction factors for nodes and roles.

References

1. Bishop, C.M.: Pattern Recognition and Machine Learning. Springer, New York (2006)
2. Blei, D., Kucukelbir, A., McAuliffe, J.: Variational inference: a review forstatisticians. arXiv:1601.00670 (2016)
3. Chatterjee, N., Sinha, S.: Understanding the mind of a worm: hierarchical network structure underlying nervous system function in C. elegans. In: Banerjee, R., Chakrabarti, B.K. (eds) Progress in Brain Research, pp. 145–153. Elsevier (2008)
4. Costa, G., Ortale, R.: A bayesian hierarchical approach for exploratory analysis of communities and roles in social networks. In: Proceedings of the IEEE/ACM International Conference on Advances in Social Networks Analysis and Mining, pp. 194–201 (2012)
5. Costa, G., Ortale, R.: Probabilistic analysis of communities and inner roles in networks: Bayesian generative models and approximate inference. Soc. Netw. Anal. Min. **3**(4), 1015–1038 (2013)
6. Costa, G., Ortale, R.: A unified generative bayesian model for communitydiscovery and role assignment based upon latent interaction factors. In: IEEE/ACMASONAM, pp. 93–100 (2014)
7. Costa, G., Ortale, R.: Model-based collaborative personalized recommendation on signed social rating networks. ACM Trans. Int. Technol. **16**(3), 20:1–20:21 (2016)
8. Creamer, G., Rowe, R., Hershkop, S., Stolfo, S.J.: Segmentation and automated social hierarchy detection through email network analysis. In: Zhang, H., Spiliopoulou, M., Mobasher, B., Giles, C.L., McCallum, A., Nasraoui, O., Srivastava, J., Yen, J. (eds.) SNAKDD/WebKDD -2007. LNCS (LNAI), vol. 5439, pp. 40–58. Springer, Heidelberg (2009). doi:10.1007/978-3-642-00528-2_3
9. Gopalan, P., Hofman, J., Blei, D.: Scalable recommendation with hierarchical Poisson factorization. In: UAI, pp. 326–335 (2015)
10. Henderson, K., Eliassi-Rad, T., Papadimitriou, S., Faloutsos, C.: HCDF: a hybrid community discovery framework. In: Proceedings of SIAM International Conference on Data Mining, pp. 754–765 (2010)
11. Henderson, K., Eliassi Rad, T.: Applying latent dirichlet allocation to group discovery in large graphs. In: Proceedings of ACM Symposium on Applied Computing, pp. 1456–1461 (2009)
12. Lattanzi, S., Sivakumar, D.: Affiliation networks. In: ACM STOC, pp. 427–434 (2009)
13. McAuley, J., Leskovec, J.: Learning to discover social circles in ego networks. In: NIPS, pp. 548–556 (2012)
14. Pathak, N., Delong, C., Banerjee, A., Erickson, K.: Social topic models for community extraction. In: Proceedings of KDD Workshop on Social Network Mining and Analysis (2008)
15. Sohn, Y., Choi, M.-K., Ahn, Y.-Y., Lee, J., Jeong, J.: Topological cluster analysis reveals the systemic organization of the caenorhabditis elegans connectome. PLoS Comput. Biol. **7**(5), e1001139 (2011)
16. Watts, D.J., Strogatz, S.H.: Collective dynamics of small-world networks. Nature **393**(6684), 440–442 (1998)
17. White, J.G., Southgate, E., Thompson, J.N., Brenner, S.: The structure of the nervous system of the nematode caenorhabditis elegans. Philos. Trans. Royal Soc. B Biol. Sci. **314**(1165), 1–340 (1986)

18. Yang, J., Leskovec, J.: Structure, overlaps of ground-truth communities in networks. ACM Trans. Intell. Syst. Technol. **5**(2), 26:1–26:35 (2014)
19. Yang, J., McAuley, J., Leskovec, J.: Detecting cohesive and 2-mode communities in directed and undirected networks. In: WSDM, pp. 323–332 (2014)
20. Zhang, H., Qiu, B., Giles, C.L., Foley, H.C., Yen, J.: An LDA-based community structure discovery approach for large-scale social networks. In: IEEE ISI, pp. 200–207 (2007)

Online Semi-supervised Learning for Multi-target Regression in Data Streams Using AMRules

Ricardo Sousa[1](✉) and João Gama[1,2]

[1] LIAAD/INESC TEC, Universidade do Porto, Porto, Portugal
rtsousa@inesctec.pt
[2] Faculdade de Economia, Universidade do Porto, Porto, Portugal
jgama@fep.up.pt

Abstract. Most data streams systems that use online Multi-target regression yield vast amounts of data which is not targeted. Targeting this data is usually impossible, time consuming and expensive. Semi-supervised algorithms have been proposed to use this untargeted data (input information only) for model improvement. However, most algorithms are adapted to work on batch mode for classification and require huge computational and memory resources.

Therefore, this paper proposes an semi-supervised algorithm for online processing systems based on AMRules algorithm that handle both targeted and untargeted data and improves the regression model. The proposed method was evaluated through a comparison between a scenario where the untargeted examples are not used on the training and a scenario where some untargeted examples are used. Evaluation results indicate that the use of the untargeted examples improved the target predictions by improving the model.

Keywords: Multi-target regression · Semi-supervised learning · AMRules · Data streams

1 Introduction

Multi-target regression (MTR), also known as Multi-output, Multi-variate or Multi-value, consists of predicting targets of numerical and nominal variables (outputs variables) from a set of other variables (input variables) using a trained relational and functional model [1].

Online data streams systems that use Multi-target regression produce massive amounts of data. Targeting all examples may be impossible, time consuming or expensive. Untargeted examples are abundant due to sensor malfunction, reluctance for sharing sensitive information, high cost of data collection or databases failure [2,3].

As related problems, most of the available methods work on batch mode with high computational and memory requirements [4,5]. Moreover, the literature presents more solutions for classification that cannot be directly applied to regression [6].

© Springer International Publishing AG 2016
H. Boström et al. (Eds.): IDA 2016, LNCS 9897, pp. 123–133, 2016.
DOI: 10.1007/978-3-319-46349-0_11

This regression methods are applied to forecasting and modelling in a wide range of areas such as Engineering Systems (electrical power consumption) [7], Physics (weather forecasting and ecological models) [8], Biology (model of cellular processes) [9] and Economy/Finance (stock price forecasting) [10]. In most of these areas, data from streams is obtained and processed in real time and most data is untargeted [3].

Several authors proposed the use of untargeted data (input variables information only) to improve regression models and targeted examples prediction that lead to the Semi-Supervised Learning (SSL) methods, also called Partially-Supervised Learning or Bootstrapping techniques [3].

Let $\mathcal{D} = \{..., (\mathbf{x}_1, \mathbf{y}_1), (\mathbf{x}_2, \mathbf{y}_2), ..., (\mathbf{x}_i, \mathbf{y}_i), ...\}$ denote an unlimited stream of data examples, where $\mathbf{x}_i = [x_{i,1} \cdots x_{i,j} \cdots x_{i,M}]$ is a vector of data descriptive variables $x_{i,j}$ and $\mathbf{y}_i = [y_{i,1} \cdots y_{i,j} \cdots y_{i,N}]$ is a vector of response output variables $y_{i,j}$ of the ith example (considering one example with the index of reference). The untargeted example is represented with an empty target vector $\mathbf{y}_i = \emptyset$. The objective of SSL consists of using examples $(\mathbf{x}_i, \emptyset)$ to improve the regression model. SSL explores the information for the inference of the targets that is being conveyed by the input variables [6]. These algorithms are very useful if the untargeted data is far more abundant [6]. However, in some scenarios, the algorithm reinforces and propagates the error through the estimations [6].

The objective of this work is to develop an online multi-target algorithm that handle the untarget examples in order to improve the prediction for both target and untargeted examples.

Section 2 briefly reviews semi-supervised learning methods. Section 3 describes the proposed algorithm. Section 4 explains the evaluation method. Finally, the results are presented and discussed in Sect. 5 and the main conclusions are reported in Sect. 6.

2   Related Work

This section briefly reviews some existent SSL algorithms. In the literature, most semi-supervised algorithms work on batch mode where the models are produced with the knowledge of all data [5]. Classification is the more addressed problem in SSL related literature [6]. For the best of our knowledge, no online semi-supervised multi-target regressor was found. Therefore, this review contains essentially descriptions of batch modes algorithms which are fair starting points for online algorithms development.

SSL techniques are essentially divided into five categories: self-training, co-training, generative models and graphs [6]. The self-training is an method that produces an prediction (based on the current model) and a confidence score for the untargeted example [11]. The prediction is used to target the example and the score is used to measure/predict the benefit of this artificially targeted example on the current model. A threshold is used to select the artificially targeted examples that benefit most [11,12]. A self-training batch mode algorithm was proposed by Levatic et al. [2]. This batch algorithm uses ensemble of Predictive

Clustering Trees (PCT) as underlying regressor. The predicted targets mean is used to target the example and a variability measure is used in the example acceptance for training.

The co-training method divides the input variables of the incoming example into two independent/uncorrelated groups and produce two examples with different input variables but with equal targets (targets of the incoming example). Two complementary regressors of the same type yield a prediction for each example which is used to targeting the example of the complementary regressor and for training [6]. COREG is a batch mode algorithm based on co-training that uses k-Nearest Neighbours (kNN) regressors [3]. This algorithm produces a small dataset of examples that are close (according to a predetermined distance metric) to the untargeted example. Each regressor predicts a target for the untargeted example and uses it to re-train the models with all targeted examples. Mean Squared Error (MSE) variation is computed between the scenarios with and without the artificially targeted examples. If MSE is reduced, the artificially targeted example is joined to the targeted examples set. The process stops when none untargeted example is interchanged between target and untargeted sets.

The generative models assume a distribution model of $p(\mathbf{x}_i|\mathbf{y}_j, \theta)$, where θ represent a set of parameters of the model which is identifiable by a Expectation-Maximization or clustering algorithms. The distribution $p(\mathbf{y}_j|\mathbf{x}_i, \theta)$ is computed using the Bayes rule assuming proportionality to $p(\mathbf{x}_i|\mathbf{y}_j, \theta)$ and $p(\mathbf{y}_j)$. The artificial target is computed by maximising $p(\mathbf{y}_j|\mathbf{x}_i, \theta)$ [12]. However, this method is usually applied to classification problems [6].

The graph based methods create models for the association between the inputs and target variables and between output variables themselves using graphs. The examples are the nodes and the weighted links represent their similarity. These models assume that neighbour input nodes tend to produce the similar targets [13]. This model allows to produce $p(\mathbf{y}_j|\mathbf{x}_i)$ (discriminative property) distribution as function of these parametrized associations and founds the target by optimizing it [14,15]. The Continuous Conditional Random Fields (CCRF) explores the relations between input and target variables. A model is created for $p(\mathbf{y}_j|\mathbf{x}_i)$ based on graph parameters and optimised as a function of a target. This approach is an adaptation from classification [14].

3   Semi-supervised Multi-target AMRules

In this section, the Semi-supervised Multi-Target AMRules (SS-AMRules) algorithm and the underlying principles are presented. This SSL algorithm uses the AMRules regressor due to the modularity property that allows the construction of models for particular input variables regions (defined by the rule). The SSL algorithm is based on the assumption that the most likely input variables will benefit the model by reinforcing it even with the artificially targeted examples. This principle approximates to the semi-supervised smoothness assumption which states that if two points (target representation) are close in a very dense region, then their respective response points are also close [11]. If input variables

$\mathbf{x}_i$ are too frequent an artificial target is predicted and the respective example is targeted. The untargeted example occurrence uses the same principle of anomaly detection.

3.1   Rule Learning

Rule Learning is based on implications $R_r = (\mathcal{A}_r \Rightarrow \mathcal{C}_r)$, called rules, where antecedent $\mathcal{A}_r$ is a conjunction of conditions (called literals) that create partitions in the input variables $\mathbf{x}_i$ space and the consequent $\mathcal{C}_r$ is a predicting function (in this context, it is a basic online multi-target regressor). Literals may present the forms $(X_j \leq v)$ and $(X_j > v)$ for numerical data and $(X_j = v)$ and $(X_j \neq v)$ for nominal data, where X_j represents the jth input variable. Rule R_r is said to cover $\mathbf{x}_i$ if $\mathbf{x}_i$ satisfies all the literals in $\mathcal{A}_r$. Support $S(\mathbf{x}_i)$ corresponds to a set of rules that cover $\mathbf{x}_i$ and $\mathcal{C}_r$ returns a prediction $\hat{\mathbf{y}}_i$ if a rule $R_r \in S(\mathbf{x}_i)$.

Data structure $\mathcal{L}_r$ containing the necessary statistics (about the rule and the examples) to the algorithm training and prediction (rule expansion, changes detection and anomalies detection,...) is associated to each rule R_r. In particular, $\mathcal{L}_r$ contains the input variables statistics $\mathcal{I}_r$. Default rule D exists for initial conditions and for the case of none of the current rules covers the example $(S(\mathbf{x}_i) = \emptyset)$. The antecedent of D and is initially empty. Rule set is formed by a set of U learned rules defined as $\mathcal{R} = \{R_1, \cdots, R_r, \cdots, R_U\}$ and a default rule D as depicted in Fig. 1.

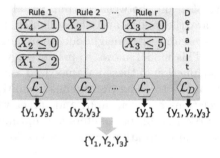

Fig. 1. Rule learning on regression.

3.2   Untargeted Data Handling

The untargeted examples handling essentially performs likelyhood score and artificial target computation for the untargeted examples inputs $\mathbf{x}_i$. The selected likelyhood score was Odd Ratio (OR) measure defined by

$$OR_i = \frac{1}{M} \sum_{j=1}^{M} \log \left(\frac{P(|X_j - \mu_j| \geq |x_{i,j} - \mu_j|)}{1 - P(|X_j - \mu_j| \geq |x_{i,j} - \mu_j|)} \right), \tag{1}$$

where i is the example index, M is the number of input variables, X_j is the jth input variable treated as a random variable and μ_j is the mean of X_j [4]. OR is the mean of probabilities measures ($\frac{P(|X_j-\mu_j|\geq|x_{i,j}-\mu_j|)}{1-P(|X_j-\mu_j|\geq|x_{i,j}-\mu_j|)}$) of each input variable. The higher the OR is the better is the benefit of the artificially targeted example on the current model. For continuous attributes, the Cantelli's inequality [16] is used to estimate $P(|X_j - \mu_j| \geq |x_{i,j} - \mu_j|)$:

$$P(|X_j - \mu_j| \geq k) \leq \begin{cases} \frac{2\sigma_j^2}{\sigma_j^2+k^2} & \text{if } \sigma_j < k \\ 1 & \text{otherwise} \end{cases} \tag{2}$$

where σ_j is the standard deviation of X_j target element and k corresponds to the distance of the actual variable value and its mean $k = |x_{i,j}-\mu_j|$. The artificial target is computed by predicting the target for example using the current model associated to the rules that cover the examples.

3.3   SS-AMRules Training

Algorithm 1 presents the pseudo-code of the SS-AMRules algorithm. On the occurrence of an incoming example $(\mathbf{x}_i, \mathbf{y}_i)$, the SS-AMRules searches for rules R_r of the current rule set that covers the example. Considering a $R_r \in S(\mathbf{x}_i)$, anomaly detection is performed to increase resilience to data outliers (isAnom−aly($\mathcal{I}_r, \mathbf{x}_i$)). In case of anomaly occurrence, the example is simple ignored by the rule otherwise it verifies if the example is untargeted ($\mathbf{y}_i = \emptyset$). For an untargeted example, the OR score is computed from the input variables statistics $\mathcal{I}_r$ and if this score is higher than a predetermined threshold $scoreThreshold$, an artificial target is predicted for the example using the model statistics $\mathcal{L}_r$ and the example become targeted (getPrediction($\mathcal{L}_r, \mathbf{x}_i$)). This prediction is the artificial target of the untargeted example. In case of targeted example identification, the statistics of likelihood score computation $\mathcal{I}_r$ are updated (updateInputStatistics($\mathcal{I}_r$)).

Posteriorly, both target and artificially targeted examples are submitted to the change detector and training process. The Page-Winkle algorithm is used for change detection. The algorithm also performs outputs selection in order to create rules that are specialized on sets of output variables (Compute $\mathcal{O}_c$). The algorithm presents self-training and generative models features. The OR score is the measure of confidence that is estimated from a generative model which describes the input variables space. As consequent ($\mathcal{C}_r$) function, a multi-target perceptron regressor was used as linear predictor due to its models simplicity, low computational cost and low error rates [4].

4   The Evaluation Method

The evaluation consisted of a simulation of a data stream using artificial and real worlds datasets. A d percentage of each dataset examples are used to train a initial model and the remaining examples are used for testing. Iteratively, a

Algorithm 1. Semi-Supervised AMRules

```
 1: $\mathcal{R} \leftarrow \emptyset, D \leftarrow 0$
 2: for all $(\mathbf{x}_i, \mathbf{y}_i) \in \mathcal{D}$ do
 3:     for all $R_r \in S(\mathbf{x}_i)$ do
 4:         if $\neg$isAnomaly($\mathcal{I}_r, \mathbf{x}_i$) then
 5:             if $y_i = \emptyset$ then
 6:                 $ORscore =$ getORScore($\mathcal{I}_r, \mathbf{x}_i$)
 7:                 if $ORscore > scoreThreshold$ then
 8:                     $\mathbf{y}_i =$ getPrediction($\mathcal{L}_r, \mathbf{x}_i$)
 9:                 else
10:                     continue
11:             else
12:                 updateInputStatistics($\mathcal{I}_r$)
13:             if changeDetected($\mathcal{L}_r, \mathbf{x}_i$) then
14:                 $\mathcal{R} \leftarrow \mathcal{R} \setminus \{R_r\}$
15:             else
16:                 $R_c \leftarrow R_r$, update($\mathcal{L}_r$), $expanded \leftarrow$ expand($R_r$)
17:                 if $expanded =$ TRUE then
18:                     Compute $\mathcal{O}'_c$, $\mathcal{O}_c \leftarrow \mathcal{O}'_c$, $\mathcal{R} \leftarrow \mathcal{R} \cup \{R_c\}$
19:
20:     if $S(\mathbf{x}_i) = \emptyset$ then
21:         if $\mathbf{y}_i = \emptyset$ then
22:             $ORscore =$ getORScore($\mathcal{I}_r, \mathbf{x}_i$)
23:             if $ORscore > scoreThreshold$ then
24:                 $\mathbf{y}_i =$ getPrediction($\mathcal{L}_r, \mathbf{x}_i$)
25:         else
26:             updateInputStatistics($\mathcal{I}_r$)
27:         update($\mathcal{L}_D$)
28:         $expanded \leftarrow$ expand($D$)
29:         if $expanded =$ TRUE then
30:             $\mathcal{R} \leftarrow \mathcal{R} \cup \{D\}, D \leftarrow 0$
```

binary random process choose if a test example is untarget with probability p. If the example is imputed as untargeted, the respective target is omitted from the SS-AMRules algorithm perspective. In the experiments, d is 30 % in order to ensure initial model consistency. The chosen p probabilities were 50 %, 80 %, 90 %, 95 % and 99 % which reflect different levels of untargeted examples content on the stream. This evaluation used the prequential mode where the algorithm predicts a target and the error for both targeted and untargeted examples (it uses the hidden targets). Posteriorly, it uses the example for training [17].

The performance measurements used the Error (euclidean norm of the difference between the true and the prediction vector) to measure an example prediction precision (local performance) and the Root Mean Square Error (RMSE) to evaluate global performance. The RMSE benefit was measured by computing the percentage of RMSE reduction between the scenario where no untargeted examples (reference) are used for training and the best scenario (associated to a score threshold) where artificial targeted examples are used. Since the error presented

Table 1. Original datasets description

Dataset	# Examples	# Outputs	# Inputs
Eunite30	8064	5	29
Bicycle	17379	3	12
SCM1d	10103	16	280
SCM20d	9047	16	61

some isolated peaks and noise like aspect in the first experiments, a smoothing median filter was used to better observe the error tendency on the graphs [18]. The window size is 1000 and the window step is 1. Five versions of the Friedman artificial dataset with complex model were generated [19]. These datasets were produced with 128000 examples and each example has 10 inputs and 3 outputs. The models were produced by changing the weights of the complex functions.

Regarding the real world datasets, Eunite03, Bicycle, SCM1d and SCM20d were used [4]. The examples of these datasets were replicated ten times and shuffled, since AMRules uses the Hoeffding bound to determine a sufficient number of examples to produce consistent models.

Table 1 shows the original features of the real world data sets used in these experiments.

5   Results

In this section, the evaluation results are presented and discussed. Figure 2 presents the graphs of error evolution for two cases. The graph at the top reveals a successful error improvement and the graph at the bottom shows an unsuccessful improvement attempt. Each graph presents several curves for several score thresholds that were used to calibrate the algorithm. Curves for score thresholds 0, 1, 3, 3.5, 4 and 5 were selected for clearer plot presentation. The score thresholds 0 and 5 correspond to the scenarios where all and none untargeted examples were used in the training, respectively. The score threshold of 5 is the reference curve. The algorithm starts by training the initial model producing high errors with the first examples.

Referenced by point A on the graph at the top, the algorithm learns the first rules and the error decreases significantly. From point A to B, the algorithm improves the initial model and from point B, the algorithm starts to process untargeted examples. The curves diverge since the score threshold are different and lead to different behaviours on this phase. The graph of the successful case presents error curves (related to artificially targeted examples usage on the training) with lower values than the reference curve (scenario where none untargeted example is used). Moreover, these curves present an error reduction tendency. The score threshold 0 curve presents higher error because some of the accepted untargeted examples damaged the model.

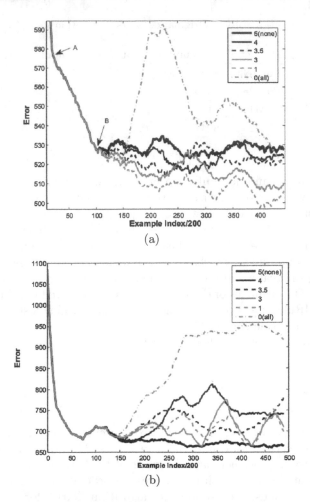

Fig. 2. Error evolution curves (for several score thresholds) of a successful (a) and a unsuccessful (b) cases of improvement attempts. The graph at the top shows the error behaviour of the algorithm for untargeted probability $p = 50\%$ and for SCM20d dataset. The graph at the bottom shows for untargeted probability $p = 99\%$ and for SCM1d dataset.

The graph at the bottom reveals a case where the algorithm produced less accurate models. This fact is due to error propagation through the model that lead to worst predictions in the artificial targeting. This effect leads to a cycle that reinforce the error on each untargeted example processing. In fact, the more untargeted examples arrive the higher is the error.

Figure 3 shows the RMSE as function of score threshold and the untargeted probability for two cases. The graph on the left indicates an approximate optimal score threshold (score threshold = 0.5) that rejects model damaging examples and accepts model reinforcing examples. This threshold is valid for any untargeted probability but it depends on the dataset characteristics (e.g., inputs variables distributions). The $p = 99\%$ scenario is an extreme case where the model is trained essentially with artificially targeted examples and the error propagation can easily occur. The graph on the right presents a dataset where any score threshold produced higher error than the reference scenario. This observation means that most untargeted examples contributed to model damage and the artificial targets conveyed significant errors.

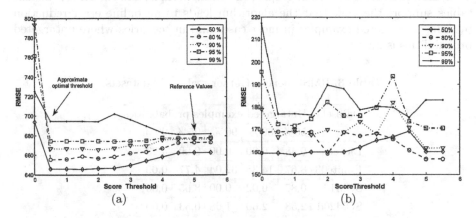

Fig. 3. RMSE as a function of score threshold and untarget example probability. The graph on the left (a) shows a case of successful improvement using the dataset SCM20d and the graph on the right (b) shows a case of unsuccessful improvement attempt using the Bicycle dataset.

Table 2 presents the RMSE benefit for the experiments on artificial datasets for each chosen untargeted examples probabilities. When the value is zero, it means that there was not any score threshold that improved the model.

Table 2. RMSE benefit (%) for artificial datasets.

Datasets	Untargeted examples probabilities				
	50%	80%	90%	95%	99%
FriedModel1	0,13	0,00	0,26	0,00	3,97
FriedModel2	6,83	7,23	6,73	2,72	0,15
FriedModel3	0,01	0,01	0,01	0,00	0,00
FriedModel4	9,11	22,38	21,62	14,65	5,11
FriedModel5	2,31	2,81	1,26	0,00	0,00

RMSE benefit values show that the error decreased in most part of the artificial datasets and for several untargeted examples probabilities. For Fried-Model2, FriedModel4 and FriedModel5 datasets, a significant improvement was achieved. On the other hand, for FriedModel1 and FriedModel3 datasets, the RMSE improvements were very small. Table 3 presents the RMSE benefits for real world datasets in a similar way as the artificial datasets. According to Table 3, the algorithm seem benefit most part of the scenarios. As expected, the more elevated p is the less is the benefit. For Eunite03 and SCM20d datasets, the algorithm produced significant results. But in particular for Bicycle dataset, error reduction did not occur for most untargeted examples probabilities. As general impression, the error evolution graphs, the RMSE graphs and the RMSE benefit tables support the view that the algorithm leads to an online error reduction by using untargeted examples in most cases and in scenarios where untargeted probabilities is high.

Table 3. RMSE benefit (%) for real world datasets.

Datasets	Untargeted examples probabilities				
	50 %	80 %	90 %	95 %	99 %
Bicycle	0,48	0,00	0,00	0,00	4,23
Eunite03	8,67	14,38	32,04	4,77	0,00
SCM1d	0,87	0,02	0,00	8,65	0,00
SCM20d	2,58	2,60	1,68	0,53	0,00

6   Conclusion

In this paper, an online semi-supervised multi-target regression algorithm is addressed. The algorithm reduces the error of prediction on online mode by using untargeted examples in most of evaluation experimental scenarios. However, the error reduction still very small and the score thresholds depend on the dataset. In fact, the algorithm implies an calibration of the score threshold for each data stream.

As future work, this approach can be improved by combining it with Random Rules based algorithms due to multiple prediction diversity feature. Since it is important to know what are the dataset characteristics that lead to error reduction in a semi-supervised scenario, an analytical framework could be constructed. In order to increase the algorithm validity, the evaluation tests will be performed using a higher number of real world datasets with a significant amount of examples.

Acknowledgments. This work was partly supported by the European Commission through MAESTRA (ICT-2013-612944) and the Project TEC4Growth - Pervasive Intelligence, Enhancers and Proofs of Concept with Industrial Impact/NORTE-01-0145-FEDER-000020 is financed by the North Portugal Regional Operational Programme (NORTE 2020), under the PORTUGAL 2020 Partnership Agreement, and through the European Regional Development Fund (ERDF).

References

1. Borchani, H., Varando, G., Bielza, C., Larrañaga, P.: A survey on multi-output regression. Wiley Int. Rev. Data Min. Knowl. Disc. **5**(5), 216–233 (2015)
2. Levatic, J., Ceci, M., Kocev, D., Dzeroski, S.: Semi-supervised learning for multi-target regression. In: Third International Workshop, NFMCP, Held in Conjunction with ECML-PKDD, pp. 3–18 (2014)
3. Zhou, Z.H., Li, M.: Semi-supervised regression with co-training style algorithms. IEEE Trans. Knowl. Data Eng. **19**(11), 1479–1493 (2007)
4. Duarte J., Gama, J.: Multi-target regression from high-speed data streams with adaptive model rules. In: IEEE Conference on Data Science and Advanced Analytics (2015)
5. Goldberg, A.B., Zhu, X., Furger, A., Jun-Ming, X.: OASIS: online active semi-supervised learning. In: Proceedings of the Twenty-Fifth AAAI Conference on Artificial Intelligence, AAAI, San Francisco, California, USA, 7–11 August 2011
6. Kang, P., Kim, D., Cho, S.: Semi-supervised support vector regression based on self-training with label uncertainty: an application to virtual metrology in semiconductor manufacturing. Expert Syst. Appl. **51**, 85–106 (2016)
7. Ozoh, P., Abd-rahman, S., Labadin, J., Apperley, M.: Article: a comparative analysis of techniques for forecasting electricity consumption. Int. J. Comput. Appl. **88**(15), 8–12 (2014)
8. Chalabi, Z., Mangtani, P., Hashizume, M., Imai, C., Armstrong, B.: Article: time series regression model for infectious disease and weather. Int. J. Environ. Res. **142**, 319–327 (2015)
9. Uslana, H.S.V.: Article: quantitative prediction of peptide binding afnity by using hybrid fuzzy support vector regression. Appl. Soft Comput. **43**, 210–221 (2016)
10. Ariyo, A.A., Adewumi, A.O., Ayo, C.K.: Stock price prediction using the arima model. In: Proceedings of the UKSim-AMSS 16th International Conference on Computer Modelling and Simulation, UKSIM 2014, pp. 106–112, Washington, DC, USA. IEEE Computer Society (2014)
11. Chapelle, O., Schlkopf, B., Zien, A.: Semi-Supervised Learning, 1st edn. The MIT Press, Cambridge (2010)
12. Albalate, A., Minker, W.: Semi-supervised and Unsupervised Machine Learning. ISTE/Wiley, London (2011)
13. Verbeek, J.J., Vlassis, N.: Gaussian fields for semi-supervised regression and correspondence learning. Pattern Recogn. **39**(10), 1864–1875 (2006)
14. Radosavljevic, V., Vucetic, S., Obradovic, Z.: Continuous conditionalrandom fields for regression in remote sensing. In: 19th European Conference on Artificial Intelligence, Proceedings of the 2010 Conference on ECAI 2010, pp. 809–814, Amsterdam, The Netherlands. IOS Press (2010)
15. Stojanovic, J., Jovanovic, M., Gligorijevic, D., Obradovic, Z.: Semi-supervised learning for structured regression on partially observed attributed graphs. In: SIAM International Conference on Data Mining (SDM) (2015)
16. Bhattacharyya, B.B.: One sided Chebyshev inequality when the first four moments are known. Commun. Stat. Theor. Methods **16**(9), 2789–2791 (1987)
17. Gama, J., Sebastião, R., Rodrigues, P.P.: On evaluating stream learning algorithms. Mach. Learn. **90**(3), 317–346 (2013)
18. Chen, W.: Passive, Active, and Digital Filters, 3rd edn. CRC Press, Baco Raton (2009)
19. Friedman, J.H.: Multivariate adaptive regression splines. Ann. Stat. **19**(1), 1–67 (1991)

A Toolkit for Analysis of Deep Learning Experiments

Jim O'Donoghue$^{(\boxtimes)}$ and Mark Roantree

Insight Centre for Data Analytics, School of Computing,
Dublin City University, Dublin, Ireland
{jim.odonoghue,mark.roantree}@computing.dcu.ie

Abstract. Learning experiments are complex procedures which generate high volumes of data due to the number of updates which occur
during training and the number of trials necessary for hyper-parameter
selection. Often during runtime, interim result data is purged as the
experiment progresses. This purge makes rolling-back to interim experiments, restarting at a specific point or discovering trends and patterns in
parameters, hyper-parameters or results almost impossible given a large
experiment or experiment set. In this research, we present a data model
which captures all aspects of a deep learning experiment and through an
application programming interface provides a simple means of storing,
retrieving and analysing parameter settings and interim results at any
point in the experiment. This has the further benefit of a high level of
interoperability and sharing across machine learning researchers who can
use the model and its interface for data management.

1 Introduction

In order to tune and optimise machine learning models, a wide range of parameters are required. Finding the best combination of parameters is often complex
and time consuming, as parameter optimisation requires careful monitoring of
each batch of results, which are generated during an update in training. These
results should also be monitored with respect to different combinations of hyper-
parameters. Hyper-parameters (HPs) are those not learned by the model but
instead given as inputs to the algorithm before training. One needs to choose a
set of HPs which allow model parameters to reach a configuration that optimises
a particular performance goal on a dataset for an algorithm during training. Grid
and manual search are the most widely used strategies for HP optimisation and
in both cases, many HP configurations are run to view their effect on algorithm
training in order to determine the best parameters. The main issues in trying to
find *good* parameter settings can be listed as follows:

- Functions require many complete iterations of training to find the optimal
  hyper-parameter configuration - often a manual and lengthy process which
  can lack empirical rigour.

Research funded by Science Foundation Ireland, grant number SFI/12/RC/2289.

H. Boström et al. (Eds.): IDA 2016, LNCS 9897, pp. 134–145, 2016.
DOI: 10.1007/978-3-319-46349-0_12

- In the majority of experiments interim results are stored in-memory and subsequently discarded, save for the final, most accurate learner(s), result(s) and hyper-parameters. This makes backtracking to an earlier parameter set at a point in the experiment, or the analysis of interim learners, impossible.
- There are few languages defined for the exchange of data mining and machine learning (ML) functions and parameters, which provides a barrier to sharing and exchanging the complete set of results captured during the experiment.

There have been a number attempts to address the above problems via frameworks such as CRISP-DM and SEMMA [2]. However, these frameworks are abstract and require the development of more fine-grained methodologies before any benefits can be accrued. To date, a number of ontologies have been created for example, to describe: machine learning experiments [18]; or data mining concepts in general [10]. However, ontologies are expensive to construct, often suited to specific domains and require a significant learning curve for researchers.

Experiment databases, introduced in [4] and expanded upon in [5,16,17] have been defined for similar purposes, but do not focus on the particulars of deep learning experiments. Our approach specifically targets deep learning. The presented JSON and NoSQL solution is inherently more lightweight in its structure to the XML based PMML [13] and better represents the natural tree-like form of deep learning models in comparison to the flat, relational storage paradigm of [17].

1.1 Contribution

In this research, we present the Parameter Optimisation for Learning (POL) data model which captures all aspects of deep learning experiments. The data model formalises the description of a deep learning experiment along with parameters and result data; this enables the design of an application programming interface (API) for the data management of both, facilitating storage and deeper analysis of each trial and learner in the experiment. In specific terms, using a JSON API for our data model provides a platform for historical analyses and comparison across these analyses; a high level of interoperability enabling our results to be shared and evaluated by others; and more efficient learning through the ability to pause experiments, resume from any checkpoint and iterate on results. Our evaluation demonstrates how to achieve a reduced *HP-search-space*, one important requirement in machine learning experiments.

Terminology. For a paper which covers both data modelling and machine learning, it is necessary to clarify the terminology we will use throughout the paper. The conceptual *model* presented in this paper captures all of the *data properties* of a machine learning experiment. For this reason we will use the term **data model** to refer to our representation of these data concepts. When discussing the machine learning aspects of our work, we will use the term **learner** to refer to the model instances learned over the course of training with a particular learning **algorithm**.

Paper Structure. This paper is structured as follows: in the following section, we provide an overview of related research; in Sect. 3, we describe our conceptual model which captures all aspects of deep learning experiments and analysis; in Sect. 4, we present a deployment architecture which uses our conceptual model as an interface layer to deep learning functions, parameter settings and result data which are stored using NoSQL (Mongo) technology; our evaluation is described and discussed in Sect. 5; and finally, in Sect. 6, we provide some conclusions.

2   Related Research

The first model interchange format for predictive data mining functions was PMML [9]. They aimed to provide a mechanism for working with different types of predictive models by defining a convenient language for importing and exporting these models between different systems. Their experience with DM applications had shown the usefulness that a flexible interchange mechanism would provide and they argued that previous interchange formats were proprietary. However, PMML lacks a conceptual abstract model to describe a machine learning experiment, nor does it describe or utilise experiment databases which store interim training results. Instead, its focus is purely on model deployment and interchange and therefore, does not sufficiently describe the hyper-parameter optimisation process, an important function of training deep learning models.

With the Portable Format for Analytics (PFA) [1], the authors provide an abstract description of a machine learning model allowing user-defined algorithms and models. While PFA incorporates JSON for its implementation model, its aim is somewhat different to our own. Similar to our approach, PFA provides a mechanism to export and exchange models where not previously possible. The main difference is that PFA focuses on the deployment of their model to production environments whereas we focus on the analysis of all aspects of a machine learning experiment. This enables the building of more robust learning models and aids in hyper parameter selection. As PFA is a 'mini-language' rather than a data-model, it provides the capability to take in data and score this data according to the algorithm it has learned. As a result, it is a complex process whereas we aimed to define a light-weight, simple format that allows a formal description of a deep learning experiment and model.

In [7], the authors present the MEX vocabulary, a lightweight interchange format for ML experiments, which is an extension to the PROV-O ontology [12]. Their aim is similar to our own, but instead of taking a data-modelling approach, their methodology focuses on a linked-data, semantic web paradigm. Again, similar to the research presented here, [7] aims to provide a means to describe the elements of a learning experiment instead of exhaustively defining all aspects of the knowledge discovery process. However, a physical or implementation model for this ontology is not provided, nor an interface to persistent storage for later evaluation of experiment results. We also believe that the RDF graph-store does not provide as high a level of interoperability offered by JSON.

3   A Conceptual Model for Deep Learning Methods

The goal of our conceptual model is to capture all data properties of the DL *experiment*. There are three broad aspects to our model: model-parameter, hyper-parameter and (interim and final) result data management. The highest level of abstraction is an *experiment* and within that entity are all objects and attributes required to describe parameter and result data. Thus, we refer to our data model as the Parameter Optimisation for Learning (POL) model.

3.1   Model Overview

In Fig. 1, we present a detailed illustration of the POL data model and the 3 levels of data capture required. At the highest level, the **Experiment** class is the entry point to the model and has a 1-to-many relationship with the **Learner** class, meaning an experiment can have multiple occurrences of a learner. As the search space settings remain constant for an experiment, the **Hyper-ParamSearchSpace** is also present at this level. The **Learner** (at one level down from **Experiment**) has a 1-to-many relationship with the **Layer** class, allowing the algorithm to have one or more layers. Within **Learner**, there are three main concepts: *parameters* (weights and biases) which are represented by the **LayerConfiguration**, **Layer** and **Tensor** classes; *hyper-parameters* which are represented by the **HyperParameters** class; and *results* (output from any iteration of the algorithm) which are represented by the **LearnerPerformance**, **Performance**, **ConfusionMatrix** and **Tensor** classes as well as **Indices** which describes the dataset configuration which generated those results. Result data is captured at this layer as the entire Learner is used to produce results. At the lowest level of the data model is the **Layer** class which contains the weights and biases for a Layer in the Learner and as these are multi-dimensional mathematical objects they are represented by the **Tensor** class.

3.2   Model Details

In this section, we provide a detailed description of two of the more important classes of POL data model: `Learner` and `HyperParameters`, as space restrictions prevent a full description of the entire model. `Learner` is described by:

- `learner_type`: Name of the algorithm used in creating the learner, for example: restricted boltzmann machine or recurrent neural network (RNN).
- `learning_type`: Learning task: reinforcement, supervised, unsupervised, etc.
- `optimisation_method`: Optimisation method for the learning function e.g. mini-batch stochastic gradient-descent (MSGD).
- `hyper_parameters`: Input and fixed parameters used by the optimisation process and initialised within the search-space bounds.
- `layers`: A list of `Layer` objects, which transform features into more abstract features or classifications and predictions. A `Layer` contains the model-parameters, weights and biases which make up a `Learner`.

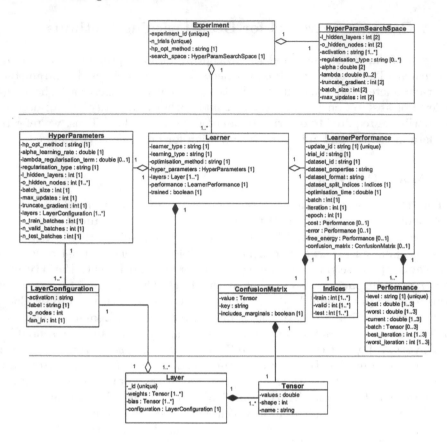

Fig. 1. POL conceptual model

- `performance`: Instance of the `Learner_Performance` object, containing a result snapshot for an update, or the final result if training has finished.
- `trained`: Boolean attribute to indicate if the `Learner`'s model-parameters are optimised. Otherwise, the instance is a snapshot of a particular update.

To represent hyper-parameter optimisation, 2 classes are required in our model: **HyperParamSearchSpace** and **HyperParameters**. The search-space class defines an upper and lower bound for each HP, from which n_trial *hyper-parameters* are instantiated to find the best setting. We now describe **HyperParameters** which details a single configuration generated within the space:

- `hp_opt_method`: The name of the algorithm used to optimise the hyper-parameters.
- `alpha_learning_rate`: Determines the magnitude of parameter updates for one step of gradient descent (GD).

- `lambda_regularisation_term`: Determines the penalty placed on very large or small weights and biases or null if dropout or no regularisation is applied.
- `regularisation_type`: Type of regularisation applied to the model-parameters, for example L1, L2, dropout, dropconnect or none.
- `l_hidden_layers`: Number of hidden layers in a `Learner`; 1 or less is considered shallow whereas anything more is considered deep.
- `o_hidden_nodes`: List where each element describes the number of hidden nodes in each layer.
- `batch_size`: Number of dataset rows to use in MSGD, which affects the algorithm's learning ability. A size of one is synonymous with stochastic GD and a size equal to the number of training samples equates to batch GD.
- `max_updates`: Maximum number of GD updates to apply to a learning function, the bounds depend on the number of model-parameters and rows in the dataset; used as an exit parameter or a patience parameter in early-stopping.
- `truncate_gradient`: Describes how far in the past to pass errors in back-propagation through time.
- `layers`: A list of `LayerConfiguration` objects which detail the setup and label of each layer in the architecture.
- `n_train_batches`, `n_valid_batches`, `n_test_batches`: Number of batches in training, validation and test sets, respectively.

4   Deployment Architecture

We now describe the system architecture where the POL is deployed. It comprises of: Data Storage, Interoperable and Application layers, shown in Fig. 2.

Data Storage Layer. In order to develop an interoperable API to deliver the goals specified in our introduction, the POL data model was implemented in JSON and currently uses MongoDB for storage. This facilitates a direct mapping between the JSON API and the NoSQL database (MongoDB). The efficient storage of parameters and result data, together with the exploitation of key properties of the NoSQL databases to construct the experiment database form part of a future research submission.

Interoperable Layer. The goal of the Interoperable Layer is to facilitate greater flexibility in the learning process but also to facilitate sharing of results for comparison and analysis. The layer has 3 libraries to achieve those goals: the **Setup** library contains all functions to instantiate an experiment, read in the data and configure the database in order for results and snapshots to be processed; the **Evaluation** library contains functions to analyse and rank the performance of different trial-runs in the learning process; and the **Access** library abstracts storage details from the higher level libraries. The Access API is developed using JSON and is a direct implementation of the POL data model. This API contains all of the functionality to write and read attributes before and during a deep learning experiment. The Setup and Evaluation libraries are developed using Python and are currently accessible using Python APIs.

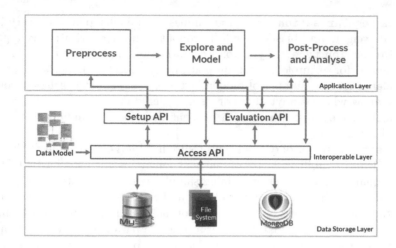

Fig. 2. Operational architecture

These libraries provide higher level functionality for experiment setup and evaluation, both use the Access API to read and write to Mongo.

Application Layer. The major applications which use the toolkit represents different aspects of learning and deep learning experiments: experimental setup (Preprocess); learning (Explore/Model) and evaluation (Post-Process & Analyse). Applications can either interact directly with the Access API to design their own experiments and evaluation functions or use the Python library APIs for easier manipulation of experimental data.

5   Evaluation and Analysis

The aim of our evaluation was *not* to build the most accurate model, but to demonstrate how our interoperable toolkit can be used for the management and analysis of learning experiments. Specifically the aim is HP search space optimisation. The analysis of interim results across all trial-runs was used to fine-tune the full set of hyper-parameter bounds.

The dataset used for evaluation was generated from a series of sensors worn by athletes during Gaelic Football matches and is described elsewhere [14]. Random search was employed as our HP optimisation procedure [3] and 90 *trials* were carried out for 2 *runs* each, giving 180 *trial-runs*. Table 1 shows the *search space* for experiments. Algorithm parameters were randomly initialised according to [8], save for hidden to hidden rectified linear unit (ReLU) weights, initialised according to [11] and optimised with MSGD and Early Stopping [6,15].

5.1   Search Space Reduction: Results and Commentary

We first present summary experiment statistics in Table 2. The experiment consisted of 180 *trial-runs*, during which 40,830 epochs were iterated.

Table 1. Hyper parameters and bounds

Hyper-parameter	Bounds (low, high)	Description
activation	(relu, logistic)	Hidden layer activation
n_hidden_nodes	(1, 10)	Number of hidden layer nodes
truncate_gradient	(5, 100)	Number of time-steps to BP errors
learning_rate_α	(0.0001, 0.9)	Co-efficient for weight updates
max_updates	(10, 10000)	Max possible updates performed
batch_size	(60, 600)	Samples in mini-batch update

The average size of a learner and result *snapshot* was 0.59 MB, leading to *Trial-Runs* being 10.657 MB in total and *Updates* nearly amounting to 2.5 GB. Unlike most machine learning experiments where only the final result is captured, interim results were recorded for every epoch. It is also possible to store results for each batch update within epochs, but this level of granularity was not used in our evaluation due to the obvious cost/benefit in terms of storage and speed.

In our *Evaluation* library, reduce_search_space performs an analysis which uses a set of queries to access results from multiple interim trial-runs. All Learners are evaluated, with the top-k snapshots returned through get_top_k_ids (*Evaluation* library), which then facilitates the retrieval of associated hyper-parameters through get_hyper_parameters in the *Access* library. The coalesce _hyper_parameters analyses the top-k hyper-parameter settings and generates summary statistics before finally, a reduced search space is generated through reduce_search_space.

Table 2. Experiment statistics

Collection	Count	Size (MB)	Avg. object size (MB)
Trial-runs	180	10.657	0.059
Updates	40,089	2,377.193	0.059

Table 3 shows the result of coalesce_hyper_parameters for the Top 20 performing HPs in the 180 *trial-runs*. It calculates the mean, standard deviation, minimum and maximum and the 25th, 50th (median) and 75th percentiles. We have also shown sample HP frequency distribution histograms, which are output from visualise_hp_distribution (Evaluation library) in Figs. 3, 4, 5 and 6. To best understand the beneficial effects of using our system, we now explore the outputs from coalesce_hyper_parameters and visualise_hp_distribution, key functions that allow analysis of interim results from many learners, which is only possible with persistent data management. We have omitted histograms for epochs, updates and batch size due to space restrictions.

Table 3. Hyper-parameter summary statistics

ID	nodes	truncate	alpha	max_updates	steps	epochs	batch_size
mean	4.360	53.980	0.404	3966.870	3520.900	157.900	289.240
std.	2.765	26.251	0.284	2704.307	3569.484	208.755	178.533
min	1	6	0	132	30	2	68
25 %	2	33	0.139	1394	326	8	118
50 %	4	55	0.373	3729	2230	88	242
75 %	6.250	74.250	0.655	6332	7332	155	451.750
max	10	100	0.889	9899	9932	722	619

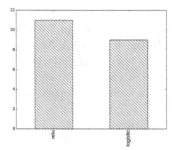
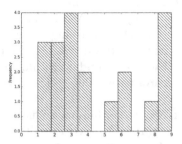

Fig. 3. Activations   **Fig. 4.** Hidden nodes

Activations. We will first consider the activations of the Top 20 *Learners* in our 180 trial runs, shown in Fig. 3. Activations are categorical strings and therefore, require different analyses to numeric HPs. The count of Learners built with ReLUs is close to those with Logistic activations. This result suggests both activations have similar performance but as ReLU outweighs logistic by a ratio of 11:9, reduce_search_space evaluates the ReLU to be the higher performing activation.

Hidden Nodes. The (hidden) nodes summary in Table 3, shows the average value for hidden node count is 4.360 with a median of 4. A median *below* the distribution average (right-skewed distribution), suggests the ideal parameter for this HP would be 4 or less, confirmed in the plot of the frequency distribution in Fig. 4. This means the latent features which describe the input are actually low in cardinality, which shows the dimensionality can likely be reduced.

Truncate_Gradient. Figure 5 shows the number of time-steps recorded for optimising backpropagation through time. The value at the 75th percentile for *truncate gradient* are 74.25 in Table 3, with the mean and median at 53.98 and 55 respectively, indicating a skewed distribution. There are two possibilities for the distribution centering at these values. The first is that time-points near $t_{i-55} : t_i$ have the greatest impact on t_{i+10}, meaning all activity for 55 s before time-point

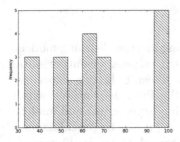

Fig. 5. Gradient truncate

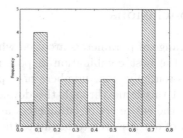

Fig. 6. Learning rate alpha

t_i has the greatest effect on predictions. The second possibility is that for time-points >55 s, the gradient disappears, but this is unlikely as good performance was also demonstrated in the range 90 to 100.

Learning Rate. From Table 3 and Fig. 6, we can see only one configuration in the top 20 had a value below 0.1. The mean was 0.404 and the median was at 0.373, giving a right skew. Both values are quite large for a learning rate and suggest that GD is quite steep.

The above analyses show that the median better represents the central tendency of all parameters. Also, these distributions do not lend themselves to a parametric analysis as shown in the graphs. Therefore, our selection methodology for the bounds of a reduced *search-space* consisted of taking the median and standard deviations for each hyper-parameter resulting from `coalesce_hyper_parameters` (Table 3) and generating a new bound in the range (median − std. dev., median + std. dev.), save for max_updates where we instead use the max and min values, as these parameters had a close to uniform distribution. The more realistic search bounds presented in Table 4 can only be determined using an analysis of the stored history of earlier experiments. This also facilitates augmenting random search with coordinate descent, a process not possible if we simply determine the single best configuration.

Table 4. Reduced hyper-parameter search space

Hyper-parameter	Bounds (low, high)
activation	(relu)
n_hidden_nodes	(1,7)
truncate_gradient	(29, 81)
learning_rate_α	(0.89, 0.657)
max_updates	(30, 9932)
batch_size	(63, 421)

6   Conclusions

A wide range of parameters are used when optimising machine learning models. Finding the best combination of parameters in high volumes of output data, across potentially high numbers of experiments is difficult. In this paper, we addressed this issue through the development of the POL data model, which captures the entire set of parameters used in learning experiments and models. The aim of our research to facilitate the optimisation process by providing data management, analysis and optimisation functions through a standard interface, developed for the POL data model. In effect, a persistent data-store allows us to store *all* models and *all* updates, generating multiple outputs from a single experiment and a direct means of querying interim results. Our evaluation shows how using interim results, distributions can be generated for each hyperparameter of the top 20 performing learners, which can then be analysed and used to determine an empirically reduced search-bounds in which to optimise hyper-parameters. Future work will focus on more robust methods of empirically reducing the search bounds such as the extraction of confidence intervals for each hyper-parameter. Furthermore, we are currently running experiments on two further datasets and intend to compare our approach against other relevant frameworks.

References

1. PFA: Portable format for analytics (version 0.8.1). Technical report, Data Mining Group - PFA Working Group (2015)
2. Azevedo, A., Santos, M.F.: KDD, SEMMA and CRISP-DM: a parallel overview. In: Proceedings of IADIS European Conference on Data Mining 2008, Amsterdam, The Netherlands, 24–26 July 2008, pp. 182–185 (2008)
3. Bergstra, J., Bengio, Y.: Random search for hyper-parameter optimization. J. Mach. Learn. Res. **13**, 281–305 (2012)
4. Blockeel, H.: Experiment databases: a novel methodology for experimental research. In: Bonchi, F., Boulicaut, J.-F. (eds.) KDID 2005. LNCS, vol. 3933, pp. 72–85. Springer, Heidelberg (2006). doi:10.1007/11733492_5
5. Blockeel, H., Vanschoren, J.: Experiment databases: towards an improved experimental methodology in machine learning. In: Kok, J.N., Koronacki, J., Lopez de Mantaras, R., Matwin, S., Mladenič, D., Skowron, A. (eds.) PKDD 2007. LNCS, vol. 4702, pp. 6–17. Springer, Heidelberg (2007). doi:10.1007/978-3-540-74976-9_5
6. Bottou, L.: Stochastic gradient learning in neural networks. In: Proceedings of Neuro-Nîmes. EC2 (1991)
7. Esteves, D., Moussallem, D., Neto, C.B., Soru, T., Usbeck, R., Ackermann, M., Lehmann, J.: MEX vocabulary: a lightweight interchange format for machine learning experiments. In: Proceedings of the 11th International Conference on Semantic Systems, pp. 169–176. ACM (2015)
8. Glorot, X., Bengio, Y.: Understanding the difficulty of training deep feedforward neural networks. In: International Conference on Artificial Intelligence and Statistics, pp. 249–256 (2010)

9. Grossman, R., Bailey, S., Ramu, A., Malhi, B., Hallstrom, P., Pulleyn, I., Qin, X.: The management and mining of multiple predictive models using the predictive modeling markup language. Inf. Softw. Technol. **41**(9), 589–595 (1999)

10. Keet, C., dAmato, C., Khan, Z., Lawrynowicz, A.: Exploring reasoning with the DMOP ontology. In: 3rd Workshop on Ontology Reasoner Evaluation (ORE 2014), vol. 1207, pp. 64–70 (2014)

11. Le, Q.V., Jaitly, N., Hinton, G.E.: A simple way to initialize recurrent networks of rectified linear units. arXiv preprint arXiv:1504.00941 (2015)

12. Lebo, T., Sahoo, S., McGuinness, D., Belhajjame, K., Cheney, J., et al.: PROV-O: the prov ontology. w3c recommendation, 30 April 2013. World Wide Web Consortium (2013)

13. Nurseitov, N., Paulson, M., Reynolds, R., Izurieta, C.: Comparison of JSON and XML data interchange formats: a case study. In: Caine 2009, pp. 157–162 (2009)

14. O'Donoghue, J., Roantree, M., Cullen, B., Moyna, N., Sullivan, C.O., McCarren, A.: Anomaly and event detection for unsupervised athlete performance data. In: Proceedings of the LWA 2015 Workshops, Trier, Germany, 7–9 October 2015, pp. 205–217 (2015)

15. Prechelt, L.: Early stopping — but when? In: Montavon, G., Orr, G.B., Müller, K.-R. (eds.) Neural Networks: Tricks of the Trade. LNCS, vol. 7700, pp. 53–67. Springer, Heidelberg (2012). doi:10.1007/978-3-642-35289-8_5

16. Vanschoren, J., Blockeel, H., Pfahringer, B., Holmes, G.: Experiment databases. Mach. Learn. **87**(2), 127–158 (2012)

17. Vanschoren, J., van Rijn, J.N., Bischl, B.: Taking machine learning research online with OpenML. In: Proceedings of the 4th International Workshop on Big Data, Streams and Heterogeneous Source Mining: Algorithms, Systems, Programming Models and Applications, pp. 1–4 (2015)

18. Vanschoren, J., Soldatova, L.: Exposé: an ontology for data mining experiments. In: International Workshop on Third Generation Data Mining: Towards Service-Oriented Knowledge Discovery (SoKD 2010), pp. 31–46 (2010)

The Optimistic Method for Model Estimation

James Brofos[1]([✉]), Rui Shu[2], and Frank Zhang[1]

[1] The MITRE Corporation, Bedford, MA 01730, USA
jbrofos@mitre.org
[2] Stanford University, Stanford, CA 94305, USA

Abstract. We present the method of *optimistic estimation*, a novel paradigm that seeks to incorporate robustness to errors-in-variables biases directly into the estimation objective function. This approach protects parameter estimates in statistical models from data set corruption. We apply the optimistic paradigm to estimation of linear regression, logistic regression, and Ising graphical models in the presence of noise and demonstrate that more accurate predictions of the model parameters can be obtained.

1 Introduction

In many real-world settings, data is measured with error. Such error invariably affects the maximum likelihood estimates of any parametric model used for learning from the data. For cases where the variables (either explanatory or response variables) in a model have been observed with error, we seek to minimize the error of parameter estimation by introducing an alternative to maximum likelihood.

A popular treatment against data corruption has been to perform robust estimation via the use of an uncertainty set [1,2]. Robust optimization has been used successfully in application to the domains of inventory theory, network flow, option pricing and portfolio management [3,4]. The uncertainty set in robust optimization represents our prior belief on the extent to which the observations have been corrupted. Unlike maximum likelihood, which estimates the true model parameters $\theta_\star$ by choosing θ that maximizes the likelihood function, the goal of robust estimation as presented by [1] is to protect against the worst-case configuration of the data. Formally, let $\mathbf{X}^n$ denote n observations of the random variable $\mathbf{X}$. The data set $\mathbf{X}^n$ contains values that have been incorrectly measured and is thus different from the "true" data set, denoted $\mathbf{X}^n_\star$. We denote the difference as $\boldsymbol{\Delta}_\star = \mathbf{X}^n_\star - \mathbf{X}^n$. Robust estimation seeks to estimate both $\boldsymbol{\Delta}_\star$ and $\theta_\star$ by solving the following optimization problem,

$$\max_\theta \min_{\boldsymbol{\Delta} \in \mathcal{U}} \mathcal{L}\left(\theta; \mathbf{X}^n + \boldsymbol{\Delta}\right), \tag{1}$$

where $\mathcal{L}$ is the likelihood function and $\mathcal{U}$ represents our constraint on the realizations of $\boldsymbol{\Delta}$. While such an approach is shown to minimize the risk of a poor estimate [1], we note that the robust approach reduces the coefficients to zero

H. Boström et al. (Eds.): IDA 2016, LNCS 9897, pp. 146–157, 2016.
DOI: 10.1007/978-3-319-46349-0_13

by assuming a worst-case $\mathbf{X}_*^n$. We therefore believe that the robust optimization paradigm is not well-suited to protecting coefficients against measurement error since in many cases it is not realistic or useful to assume that the true underlying data is as uninformative as possible.

To deal with the randomized corruption of data, which tends to cause underestimation of coefficient magnitudes by maximum likelihood, we present an alternative objective function,

$$\max_{\theta} \max_{\Delta \in \mathcal{U}} \mathcal{L}\left(\theta; \mathbf{X}^n + \Delta\right). \tag{2}$$

We call this approach the "optimistic paradigm." Unlike [1], which considers the minimum over $\Delta \in \mathcal{U}$, the optimistic paradigm considers the maximum. This decision was motivated by previous work in [5], which argued that, by introducing Δ to achieve the best possible value of the likelihood, the optimistic paradigm protects against corruption of the data set by editing the most "suspicious" elements in $\mathbf{X}^n$ to yield the best configuration of $\mathbf{X}^n$ with respect to the likelihood function. We demonstrate the effectiveness of this paradigm in three settings, namely in the optimistic estimation of linear regression, logistic regression and Ising model parameters.

2   Method

2.1   Optimistic Least Squares

2.1.1   Problem Scenario

Given an over-determined system $Ax = b$, where $A \in \mathbb{R}^{m \times n}$ and $b \in \mathbb{R}^m$, an ordinary least squares (OLS) problem is $\min_x \|Ax - b\|$. In [6], the authors account for uncertainty in the data $[A\ b]$ by minimizing the worst-case residual, which they call the robust least squares (RLS) problem,

$$\min_{x} \max_{\|\Delta A \Delta b\|_F \leq \rho} \|(A + \Delta A)x - (b + \Delta b)\|, \tag{3}$$

where $\|*\|_F$ is the Frobenius norm of a matrix, i.e. $\|A\|_F = \sqrt{trace(A^T A)}$.

[6] goes on to prove that the minimization of the worst-case least squares error can be formulated as a Tikhonov regularization procedure. This connection between robustness and regularity appears in other contexts as well; in [7], the authors show that l^1-regularized regression, or Lasso, can also be cast as the following robust optimization problem,

$$\min_{x} \max_{\|\Delta A\|_{\infty,2} \leq \rho} \|(A + \Delta A)x - b\|, \tag{4}$$

where $\|*\|_{\infty,2}$ is the ∞-norm of the 2-norm of the columns.

To introduce the optimistic paradigm, we consider the following problem,

$$\min_{x} \min_{\|\Delta A\|_2 \leq \rho} \|(A + \Delta A)x - b\|, \tag{5}$$

which we cast in its equivalent least squares formulation with explanatory variables $\mathbf{X}^n$, response variable $\mathbf{y}$, and corruption $\boldsymbol{\Delta}_\star = \mathbf{X}_\star^n - \mathbf{X}^n$,

$$\min_{\boldsymbol{\beta}} \min_{\boldsymbol{\Delta} \in \mathcal{U}} \sum_{i=1}^{n} \left(y_i - (\mathbf{X}_i - \boldsymbol{\Delta}_i)^T \boldsymbol{\beta} \right)^2, \tag{6}$$

where the uncertainty set, $\mathcal{U}$, is defined to be,

$$\mathcal{U} = \left\{ \boldsymbol{\Delta} \in \mathbb{R}^{n \times k} : ||\boldsymbol{\Delta}_i||_2^2 \leq \rho \ \forall \ i = 1, \ldots, n \right\}, \tag{7}$$

with ρ being an arbitrary non-negative real number that constrains the Euclidean norm of adjustments made to the explanatory variables.

2.1.2 Optimization

Notice that the objective function is fully separable and that the constraints are independent across rows of $\boldsymbol{\Delta}$. Hence, the optimal solution $\boldsymbol{\Delta}^\star$ is the concatenation of the optimal solutions to row-wise sub-problems. For fixed $\boldsymbol{\beta}$, then, consider the ith sub-problem,

$$\min_{\boldsymbol{\Delta}_i} \left(y_i - (\mathbf{X}_i - \boldsymbol{\Delta}_i)^T \boldsymbol{\beta} \right)^2, \tag{8}$$

$$\text{such that } \boldsymbol{\Delta}_i^T \boldsymbol{\Delta}_i \leq \rho. \tag{9}$$

Observe that the objective function above may be equivalently written,

$$g_i(\omega) = (y_i - \omega)^2, \tag{10}$$

where $\omega = \mathbf{X}_i^T \boldsymbol{\beta}$. Differentiating with respect to ω reveals,

$$\frac{dg_i(\omega)}{d\omega} = \begin{cases} -2(y_i - \omega) > 0 & \text{if } \omega > y_i \quad (\text{``overestimate''}) \\ -2(y_i - \omega) < 0 & \text{if } \omega < y_i \quad (\text{``underestimate''}). \end{cases} \tag{11}$$

Here, we are using the terms underestimate and overestimate to reflect the idea that a prediction is too small or too large relative to the target variable. Logically, the derivatives tell us that if the prediction is larger than the target, then the prediction should be lowered and vice versa if the prediction is smaller than the target. Since ω can only be changed through $\boldsymbol{\Delta}_i$, it suffices to choose $\boldsymbol{\Delta}_i$ to minimize or maximize ω depending on the derivative of $g_i(\omega)$. In particular, since $\boldsymbol{\beta}$ defines a hyperplane on $\mathbb{R}^k$, the direction of maximum increase on the hyperplane is given by $\boldsymbol{\beta}$. Hence, there are several cases,

$$\boldsymbol{\Delta}_i = \begin{cases} -\rho \frac{\boldsymbol{\beta}}{||\boldsymbol{\beta}||} & \text{if } \left(\mathbf{X}_i - \rho \frac{\boldsymbol{\beta}}{||\boldsymbol{\beta}||_2} \right)^T \boldsymbol{\beta} > y_i \ \& \ \omega > y_i \\ \rho \frac{\boldsymbol{\beta}}{||\boldsymbol{\beta}||} & \text{if } \left(\mathbf{X}_i - \rho \frac{\boldsymbol{\beta}}{||\boldsymbol{\beta}||_2} \right)^T \boldsymbol{\beta} < y_i \ \& \ \omega < y_i \\ \frac{-(y_i - \mathbf{X}_i^T \boldsymbol{\beta}) \boldsymbol{\beta}}{||\boldsymbol{\beta}||_2^2} & \text{if } \left(\mathbf{X}_i - \rho \frac{\boldsymbol{\beta}}{||\boldsymbol{\beta}||_2} \right)^T \boldsymbol{\beta} < y_i \ \& \ \omega > y_i \\ \frac{(y_i - \mathbf{X}_i^T \boldsymbol{\beta}) \boldsymbol{\beta}}{||\boldsymbol{\beta}||_2^2} & \text{if } \left(\mathbf{X}_i - \rho \frac{\boldsymbol{\beta}}{||\boldsymbol{\beta}||_2} \right)^T \boldsymbol{\beta} > y_i \ \& \ \omega < y_i. \end{cases} \tag{12}$$

The first case corresponds to the situation where the target is overestimated and the maximal correction still produces an overestimate for the given β. The second case for when the target is underestimated is similar to the first one in that it occurs when a maximal correction still produces an underestimate. For the third and forth cases, the maximal correction causes the overestimate to become an underestimate and vice versa. This suggests that there exists a $0 \leq |\lambda| < |\rho|$ such that $\left(\mathbf{X}_i - \lambda \frac{\beta}{||\beta||_2} \right)^T \beta = y_i$, by which solving for λ yields the optimal correction.

2.2   Optimistic Logistic Regression

2.2.1   Problem Scenario

We consider the case of logistic regression with model parameters β, explanatory variables $\mathbf{X}^n$, and response variable $\mathbf{y}$. Suppose $\mathbf{y}$ is observed with error such that $\mathbf{\Delta}_\star = |\mathbf{y}_\star - \mathbf{y}|$. The optimistic paradigm thus suggests that we consider the following set-up,

$$\max_{\beta} \max_{\mathbf{\Delta} \in \mathcal{U}} \log \mathcal{L} \left(\beta; \mathbf{X}^n, |\mathbf{y} - \mathbf{\Delta}| \right) - \lambda \sum_{i=1}^{n} \Delta_i, \tag{13}$$

where $\log \mathcal{L}$ is the log-likelihood function of a logistic regression, λ is an arbitrary non-negative real number, and $\mathcal{U}$ is defined to be the set,

$$\mathcal{U} = \left\{ \mathbf{\Delta} \in \{0, 1\}^n : \sum_{i=1}^{n} \Delta_i \leq \Gamma \right\}, \tag{14}$$

with Γ being a natural number not exceeding n. An astute observer would note that the second term in Eq. 13 serves as a way of penalizing the use of the uncertainty set, making it a *regularized* optimistic objective function.

2.2.2   Optimization

We develop a method for solving Eq. 13 by following a similar analysis as in [1]. Since $|y_i - \Delta_i| = (-1)^{y_i} \Delta_i + y_i$, the optimal solution to the inner maximization problem,

$$\max_{\mathbf{\Delta} \in \mathcal{U}} \log \mathcal{L} \left(\beta; \mathbf{X}^n, |\mathbf{y} - \mathbf{\Delta}| \right) - \lambda \sum_{i=1}^{n} \Delta_i, \tag{15}$$

is thus equal to the optimal solution of the binary integer program,

$$\max_{\mathbf{\Delta} \in \mathcal{U}} \sum_{i=1}^{n} (-1)^{y_i} \left(\mathbf{X}_i^n \beta \right) \Delta_i - \lambda \sum_{i=1}^{n} \Delta_i. \tag{16}$$

Furthermore, since the polyhedron $\mathcal{U}$ has integer extreme points, the above problem has the same solution as its linear program relaxation. In particular, the relaxation is,

$$\max \quad \sum_{i=1}^{n} \left[(-1)^{y_i} \left(\mathbf{X}_i^n \boldsymbol{\beta} \right) - \lambda \right] \Delta_i \tag{17}$$

$$\text{Such that} \sum_{i=1}^{n} \Delta_i \leq \Gamma \tag{18}$$

$$0 \leq \Delta_i \leq 1 \ \forall \ i = 1, 2, \ldots, n. \tag{19}$$

By strong duality, this linear program has the same objective value as its dual representation,

$$\max \quad -\Gamma p - \sum_{i=1}^{n} q_i \tag{20}$$

$$\text{Such that} -p - q_i \leq - \left[(-1)^{y_i} \left(\mathbf{X}_i^n \boldsymbol{\beta} \right) - \lambda \right] \tag{21}$$

$$p \geq 0 \text{ and } \mathbf{q} \geq \mathbf{0}. \tag{22}$$

Note that the term involving p and $\mathbf{q}$ is trivially concave because it is a linear function. It is also known that the log-likelihood function for a logistic regression is concave when the data matrix $\mathbf{X}^n$ has full rank. It is apparent that the composition of two concave functions is itself concave and, as a result, makes convergence to a maximum easy. We use the following greedy approach in Algorithm 1 for solving $\boldsymbol{\Delta}$ and $\boldsymbol{\beta}$.

Algorithm 1 draws inspiration from the Expectation-Maximization (EM) algorithm, alternating between maximizing the objective with respect to $\boldsymbol{\Delta}$ and $\boldsymbol{\beta}$. This process is continued until convergence is achieved.

Data: The design matrix $\mathbf{X}^n$, the response vector $\mathbf{y}$, a non-negative scalar λ, and a positive integer Γ.
Result: A vector $(p, \mathbf{q}, \boldsymbol{\beta})$ for parameter estimation.
Initialize $\boldsymbol{\beta} = \mathbf{0}$ and $f = -\infty$.
while *True* **do**
    Solve the dual linear program relaxation problem for p and $\mathbf{q}$ using the
    current value of $\boldsymbol{\beta}$. Set $\tilde{\mathbf{y}} = |\mathbf{y} - \boldsymbol{\Delta}|$, where $\boldsymbol{\Delta} = \mathbf{1}\{\mathbf{q} > 0\}$. Compute the $\boldsymbol{\beta}$
    that maximizes the logistic log-likelihood function.
    if $\log \mathcal{L} (\boldsymbol{\beta}; \mathbf{X}^n, \tilde{\mathbf{y}}) + \sum_{i=1}^{n} \Delta_i > f$ **then**
       | Set $f = \log \mathcal{L} (\boldsymbol{\beta}; \mathbf{X}^n, \tilde{\mathbf{y}}) + \sum_{i=1}^{n} \Delta_i$.
    else
       | Break.
    end
end
Output: The current values of $(p, \mathbf{q}, \boldsymbol{\beta})$ that can be used to estimate the model
       parameters.

Algorithm 1: Greedy Parameter Estimation Algorithm

2.3 Optimistic Ising Model

2.3.1 Problem Scenario

We demonstrated in a previous work [5] the application of the optimistic paradigm to Ising model via the approximate method of pseudolikelihood estimation. Here, we demonstrate exact inference using the optimistic paradigm as applied to maximum likelihood. Let $X_1, \ldots, X_k$ be dichotomous random variables taking values in the set $\{-1, +1\}$. Further, let $G = (V, E, \boldsymbol{\theta})$ be a graph with vertex set $V = [k]$, edge set $E \in V \times V$, and edge weights $\boldsymbol{\theta}$. An Ising model associates each of the random variables X_i with a vertex i such that,

$$
\mathbb{P}[X_1 = x_1, \ldots, X_k = x_k]
$$
$$
= \frac{1}{Z(\boldsymbol{\theta})} \exp\left\{ \sum_{(i,j) \in E} \theta_{ij} x_i x_j \right\}. \tag{23}
$$

The Ising model is a special case of a Markov random field and has been well-studied, with important applications in physics and biology [8].

Let $\mathbf{X}^n \in \{-1, +1\}^{n \times k}$ be a set of n random vectors drawn i.i.d. from an Ising model. It is easy to show that a set of sufficient statistics for $\mathbf{X}^n$ is given by,

$$
\left\{ \hat{\mathbb{E}}[X_i X_j] : i < j \wedge i, j \in V \right\}, \tag{24}
$$

which may be regarded as the empirical (with respect to $\mathbf{X}^n$) expected values of products of random variables. Notice that, like the random variables themselves, the sufficient statistics are also bounded to be within the interval $[-1, +1]$. Allowing $\hat{\mu}_{ij}$ to be the empirical expected value of the product of random variables X_i and X_j, the log-likelihood function can be expressed as a function of the sufficient statistics,

$$
\ell(\boldsymbol{\theta}; \mathbf{X}^n) = \boldsymbol{\theta}^T \hat{\boldsymbol{\mu}} - \log Z(\boldsymbol{\theta}). \tag{25}
$$

Although our notation suggests that $\hat{\boldsymbol{\mu}}$ is a matrix, we treat it as a vector and adopt the notation above only to reinforce the idea that each weighted edge in the graph has a corresponding empirical mean value of the product of the two random variables at either end of that edge. If $\mathbf{X}^n$ was observed under error, we can instead consider the likelihood function with respect to $\hat{\boldsymbol{\mu}}$, define the uncertainty set as,

$$
\mathcal{U} = \left\{ \boldsymbol{\Delta} \in \mathbb{R}^k : \|\boldsymbol{\Delta}\|_2^2 \leq \rho \right\}. \tag{26}
$$

and let the optimistic objective function be,

$$
\max_{\boldsymbol{\theta}} \max_{\boldsymbol{\Delta} \in \mathcal{U}} \boldsymbol{\theta}^T (\hat{\boldsymbol{\mu}} + \boldsymbol{\Delta}) - \log Z(\boldsymbol{\theta}) \tag{27}
$$

$$
\text{Such that } 0 \leq \hat{\mu}_{ij} + \Delta_{ij} \leq 1. \tag{28}
$$

By considering the uncertainty set with respect to the sufficient statistics $\hat{\boldsymbol{\mu}}$ instead of directly with respect to $\mathbf{X}$, we avoid the computationally intractable problem of a binary uncertainty set whose cardinality is exponential with respect to $|V|$.

2.3.2 Optimization

Letting $\ell(\boldsymbol{\theta}, \boldsymbol{\Delta}; \mathbf{X}^n) = \boldsymbol{\theta}^T (\hat{\boldsymbol{\mu}} + \boldsymbol{\Delta}) - \log Z (\boldsymbol{\theta})$, we determine the gradient to be,

$$\nabla_{\boldsymbol{\theta}} \ell = \hat{\boldsymbol{\mu}} - \nabla_{\boldsymbol{\theta}} \log Z \tag{29}$$

$$\nabla_{\boldsymbol{\Delta}} \ell = \boldsymbol{\theta}. \tag{30}$$

The authors in [9] further show that $\nabla_{\boldsymbol{\theta}} \log Z$ can be represented as a sum over edges in a graph $\sum_{(i,j) \in E} x_i x_j p(\boldsymbol{x}|\boldsymbol{\theta})$. As it turns out, analysis of the Hessian shows that the objective function is not guaranteed to be concave. We use the interior point algorithm [10] with analytic gradient and numeric Hessian to achieve a local maximum.

3 Numerical Experiments

All experiments were performed on a 2.9 GHz MacBook Pro with 8 GB of memory.

3.1 Optimistic Least Squares

We randomly generated linear systems of five independent variables using parameters drawn from a normal distribution with mean zero and variance 25. The intercepts were drawn from a standard normal distribution. Standard normal noise was added to the target variable. The original (uncorrupted) covariates were drawn from multivariate normal distribution with identity covariance matrix and a uniformly zero mean vector.

In Table 1, we compare the performance of optimistic least squares to OLS on training data in which the covariates are known to have been corrupted with uniformly random noise on a symmetric interval bounded by 'true noise'. Table 1a shows benchmarked mean squared error (MSE) for the coefficient estimates of OLS in the presence of varying amounts of corruption of the covariates (indicated by the column headers) over 110 trials for each configuration of true corruption. In Table 1b, we compare the optimistic and OLS estimates for varying amounts of true and assumed corruption, ρ, (indicated by the column and row headers respectively) in the covariates. There are ten trials for each configuration of assumed and true corruption. In each trial, the same set of covariates and target variables were used to estimate the parameters using the optimistic and OLS methods. The difference in means of the mean squared error of the parameter estimates of the two methods were compared using a paired t-test.

For small bounds on the corruption, OLS outperforms the optimistic approach. However, for larger corruption, we find that the optimistic approach gives, on average, a less biased estimate of the true parameters than does OLS and becomes less sensitive to choice of ρ (Fig. 1).

Table 1. The mean MSE over 110 trials for each configuration of true corruption of parameter estimates obtained using OLS and the mean difference over 10 trials for each configuration of true and assumed corruption of MSE of parameter estimates obtained using OLS versus optimistic least squares on the same data. In (a), standard deviations are included in parentheses. In (b), 95% confidence intervals are included in parentheses. Differences were assigned statistical significance using a paired t-test (*** $p < 0.001$, ** $p < 0.01$, * $p < 0.05$).

(a) Mean β MSE$_{LS}$

true noise	0.0	0.2	0.4	0.6	0.8	1.0
	0.0101	0.0323	0.1147	0.3542	0.8816	1.6111
	(0.0058)	(0.0269)	(0.0759)	(0.3106)	(0.7775)	(1.1975)

(b) β MSE$_{optimistic}$ - β MSE$_{LS}$ for varying configurations of assumed and true corruption

true noise		0.0	0.2	0.4	0.6	0.8	1.0
ρ	0.1	0.3752	0.1436	-0.0244	-0.1431**	-0.3574**	-0.7293***
		(0.0636, 0.6868)	(0.0128, 0.2744)	(-0.0774, 0.0286)	(-0.2388, -0.0475)	(-0.597, -0.1178)	(-1.0483, -0.4103)
	0.2	0.4641	0.3981	0.054	-0.102	-0.5664**	-1.1007***
		(0.196, 0.7321)	(0.2186, 0.5776)	(-0.0042, 0.1122)	(-0.296, 0.0919)	(-0.9269, -0.2058)	(-1.5985, -0.6028)
	0.3	1.1076	1.0202	0.2425	-0.0724	-0.5307**	-0.7439***
		(0.4418, 1.7734)	(0.5581, 1.4823)	(0.113, 0.372)	(-0.1812, 0.0364)	(-0.9308, -0.1305)	(-1.1142, -0.3736)
	0.4	1.4165	1.1357	0.5155	-0.0498	-0.369**	-0.9764***
		(0.9108, 1.9221)	(0.4408, 1.8306)	(0.318, 0.713)	(-0.1636, 0.0639)	(-0.5775, -0.1606)	(-1.4189, -0.5339)
	0.5	1.588	1.1568	1.0962	0.219	-0.2955*	-1.0069*
		(0.9268, 2.2492)	(0.46, 1.8536)	(0.626, 1.5663)	(-0.0467, 0.4847)	(-0.5538, -0.0373)	(-1.8665, -0.1473)
	0.6	1.0358	1.5687	0.9415	0.3539	-0.2579*	-0.9604**
		(0.6722, 1.3994)	(0.7939, 2.3434)	(0.5934, 1.2895)	(0.0859, 0.6219)	(-0.5384, 0.0227)	(-1.6535, -0.2673)
	0.7	2.1971	2.0862	1.1000	0.3869	-0.1209	-1.07**
		(1.066, 3.3283)	(1.1077, 3.0648)	(0.5003, 1.7129)	(0.1594, 0.6144)	(-0.3379, 0.0961)	(-1.7164, -0.4236)
	0.8	1.4402	1.5783	1.3615	0.701	-0.0129	-0.6125*
		(0.8358, 2.0445)	(0.5894, 2.5672)	(0.7902, 1.9328)	(0.0978, 1.3042)	(-0.3516, 0.3258)	(-1.2019, -0.023)
	0.9	2.0876	1.7583	1.6319	0.9936	0.0294	-0.6275***
		(1.0941, 3.0811)	(1.2236, 2.293)	(0.8158, 2.4479)	(0.4477, 1.5394)	(-0.2498, 0.3086)	(-0.948, -0.307)
	1.0	1.9351	1.9594	2.0459	0.585	0.4143	-1.048*
		(0.663, 3.2073)	(1.1926, 2.7262)	(1.0673, 3.0245)	(0.2001, 0.9699)	(-0.2239, 1.0526)	(-2.0247, -0.0713)

3.2   Optimistic Logistic Regression

We generated each logistic regression model coefficient independently at random from a normal distribution with mean zero and standard deviation seven. The matrix of explanatory variables, denoted by $\mathbf{X}^n$, was generated uniformly at random from the unit square. Logistic success probabilities were then generated according to $1/(1 + \exp\{-\mathbf{X}^n\beta\})$, and targets, $\mathbf{y}_\star$, were generated for each observation in $\mathbf{X}^n$ according to these values. Corrupting $\mathbf{y}_\star$ was accomplished first by setting $\Gamma = \lceil \frac{n}{10} \rceil$. Each experiment was repeated one-hundred times. The greedy estimation algorithm was implemented in MATLAB and used the fmincon function with an analytic gradient and Hessian.

We compare the performance of the optimistic estimator to direct maximum likelihood in Fig. 2. In Fig. 2a, measurements of the ℓ_1 and ℓ_2 errors

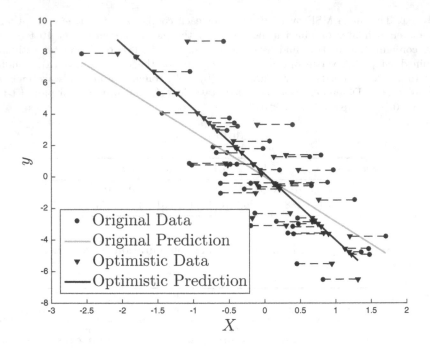

Fig. 1. Simple linear models estimated using optimistic ($\rho = 0.25$) and OLS are shown along with the training data (N = 500) for which the independent variable has been corrupted with uniformly random noise on the interval $[-1, +1]$. Use of the uncertainty set for the optimistic method is denoted by the dashed lines connecting the original data points (denoted by circles) to their modified counterparts (denoted by triangles). For reference, $\beta = (-0.1282427 - 4.44234665)$, $\beta_{\mathrm{LS}} = (0.00825438 - 2.84945464)$, $\beta_{\mathrm{optimistic}} = (0.16238759 - 4.16662121)$.

for optimistic logistic regression models are compared to maximum likelihood under uncertainty in the response variable. The optimistic estimator asymptotically outperforms direct maximum likelihood in estimation accuracy. We set $\lambda \in \{0.1, 2.5, 2.3, 2.2\}$ for each $n \in \{10^1, 10^2, 10^3, 10^4\}$, respectively. Note that λ was not tuned for best performance in these experiments. In Fig. 2b, we evaluate the performance of the optimistic estimator with respect to Γ. These experiments were identical to those described previously, with $\lambda = 2.6$. For large Γ, the optimistic approach with regularization performed identically to direct maximum likelihood. However, for smaller corruption parameters, we find that the optimistic approach gave, on average, a more consistent estimate of the true parameters than does maximum likelihood.

3.3   Optimistic Ising Model

We generated each Ising model coefficient independently at random from a normal distribution with mean zero and standard deviation one. The uncorrupted

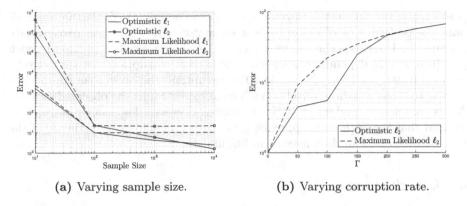

(a) Varying sample size.

(b) Varying corruption rate.

Fig. 2. Comparison of optimistic logistic regression to maximum likelihood logistic regression in terms of error of parameter estimates for different sample sizes and different corruption rates.

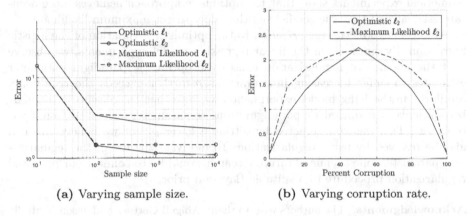

(a) Varying sample size.

(b) Varying corruption rate.

Fig. 3. Comparison of Ising model parameters learned using optimistic estimation versus maximum likelihood in terms of error for different sample sizes and different corruption rates.

data set, $\mathbf{X}_\star$, was generated according to 23 via Gibbs sampling. Corrupting the data was accomplished by setting $\mathbf{X}_{ij} = -(\mathbf{X}_\star)_{ij}$ with some probability p for each $(i, j) \in [n] \times [k]$. Each experiment was repeated three-hundred times. We used the interior point algorithm from MATLAB's fmincon function and provided the analytic gradient.

We compare the performance of the optimistic estimator to direct maximum likelihood in Fig. 3, evaluating estimator performance for a range of sample sizes. In Fig. 3a, measurements of the ℓ_1 and ℓ_2 errors for Ising models are compared to maximum likelihood under uncertainty. Optimistic estimator asymptotically outperforms direct maximum likelihood in estimation accuracy. We set $\rho \in \{0, 0.002, 0.02, 0.2\}$ for each $n \in \{10^1, 10^2, 10^3, 10^4\}$, respectively. Note that

ρ was not tuned for best performance in these experiments. In Fig. 3b, we evaluate the performance of the optimistic estimator with respect to percent data corruption. These experiments were identical to those described previously, with $\rho = \{0, 0.03, 0.1\}$ depending on whether the percent corruption was, respectively, $\{0, 1\}, \{10, 90\}$, or $\{20, \ldots, 80\}$. At 50 % corruption, the optimistic approach is similar to direct maximum likelihood. This is expected, as the observed data set $\mathbf{X}^n$ is completely random at 50 % corruption. However, for smaller corruption, we find that the optimistic approach gave, on average, a better estimate of the true parameters than does maximum likelihood.

4   Conclusions

We developed a new paradigm that protects parameter estimation in statistical models from data set corruption. This approach was applied to estimating a linear regression, logistic regression and Ising model in the presence of noise. Numerical experiments show that the optimistic approach achieves more accurate estimates of the true coefficients than does direct maximum likelihood.

We also introduced *regularization* to the optimistic estimation of the logistic regression. By contrast, in the linear regression model experiments, we observe that the absence of regularization causes the optimistic paradigm parameter estimates to suffer increasing inaccuracy as ρ further exceeds the true noise. Similarly, in the Ising model experiments, we note that the absence of regularization makes the optimistic paradigm prone to parameter overestimation if ρ is too large. This was counteracted by setting a lower ρ, but we believe this can also be resolved by using regularization. To motivate the use of the regularization parameter, one avenue of future research would be to connect our proposed regularization procedure to a suitable Bayesian prior.

Acknowledgments. The authors wish to thank Abigail Gertner and Jason Ventrella of The MITRE Corporation for helpful comments and recommendations. The author's affiliation with The MITRE Corporation is provided for identification purposes only, and is not intended to convey or imply MITRE's concurrence with, or support for, the positions, opinions or viewpoints expressed by the author. Approved for Public Release; Distribution Unlimited. Case Number 16-0621.

References

1. Fertis, A.: A robust optimization approach to statistical estimation problems. Ph.D. thesis, MIT (2009)
2. Bertsimas, D., Sim, M.: The price of robustness. Oper. Res. **52**(1), 35–53 (2004)
3. Bertsimas, D., Brown, D., Caramanis, C.: Theory and applications of robust optimization. SIAM Rev. **53**(3), 464–501 (2011)
4. Gabrel, V., Murat, C., Thiele, A.: Recent advances in robust optimization: an overview. Eur. J. Oper. Res. **235**(3), 471–483 (2014)
5. Brofos, J., Shu, R.: Optimistic and parallel Ising model estimation. Dartmouth CS Technical report. TR2015-766 (2015)

6. Ghaoui, L., Lebret, H.: Robust solutions to least-squares problems with uncertain data. SIAM J. Matrix Anal. Appl. **18**(4), 1035–1064 (1997)
7. Xu, H., Caramanis, C., Mannor, S.: Robust regression and lasso. IEEE Trans. Inform. Theor. **56**, 3561–3574 (2010)
8. Kindermann, R., Snell, J.L.: Markov Random Fields and Their Applications. Contemporary Mathematics, vol. 1. American Mathematical Society, Providence (1980)
9. Wainwright, M.J., Jordan, M.I.: Graphical models, exponential families, and variational inference. Found. Trends Mach. Learn. **1**(1), 62 (2008)
10. Boyd, S., Vandenberghe, L.: Algorithms for Convex Optimization. Cambridge University Press, Cambridge (2009)

Does Feature Selection Improve Classification?
A Large Scale Experiment in OpenML

Martijn J. Post$^{(\boxtimes)}$, Peter van der Putten, and Jan N. van Rijn

Leiden University, Leiden, The Netherlands
m.j.post@umail.leidenuniv.nl,
{p.w.h.van.der.putten,j.n.van.rijn}@liacs.leidenuniv.nl

Abstract. It is often claimed that data pre-processing is an important factor contributing towards the performance of classification algorithms. In this paper we investigate feature selection, a common data pre-processing technique. We conduct a large scale experiment and present results on what algorithms and data sets benefit from this technique. Using meta-learning we can find out for which combinations this is the case. To complement a large set of meta-features, we introduce the Feature Selection Landmarkers, which prove useful for this task. All our experimental results are made publicly available on OpenML.

Keywords: Feature selection · Meta-learning · Open science

1 Introduction

Feature selection can be of value to classification for a variety of reasons. Real world data sets can be rife with irrelevant features, especially if the data was not gathered specifically for the classification task at hand. For instance in many business applications hundreds of customer attributes may have been captured in some central data store, whilst only later is decided what kind of models actually need to be built [14]. Bag of words text classification data will by definition include large numbers of terms that may end up not to be relevant. Micro-array data sets consisting of genetic expression profiles are very wide data sets, whilst the number of instances is typically very small. In general, feature selection may help in terms of making models more interpretable, ensuring that models actually generalize rather than overfit and it will speed up the building of models when costly algorithms are being used. Highly cited surveys exist that provide a more theoretical overview of feature selection [1,6], however classical empirical papers on feature selection are typically based on small numbers of data sets (for example, 3 data sets in [5] and 14 data sets in [10]).

In this paper we investigate the specific question: will feature selection improve binary scoring models for a given data set and algorithm. We base our findings on experiments across a large number of data sets (almost 400) and a range of algorithms, and for repeatability all results have been made available in OpenML, an open science experiment database [20]. This results in a meta-data

© Springer International Publishing AG 2016
H. Boström et al. (Eds.): IDA 2016, LNCS 9897, pp. 158–170, 2016.
DOI: 10.1007/978-3-319-46349-0_14

set that we leverage to learn in what circumstances feature selection may provide better classifications for a given data set algorithm combination. We introduce a number of new meta-features to characterize data sets and algorithms for this purpose.

Our contributions are the following. We conduct two large scale experiments ranging over almost 400 data sets. The first experiment investigates for which algorithms feature selection generally improves predictive performance. This experiment both confirmed well-established conjectures and raised some interesting new findings. The second experiment exploits meta-learning to understand for which data sets feature selection may improve results. We introduce new meta-features, specific to this problem. All our underlying experimental results as well as the meta-data set are made publicly available, for the purposes of verifiability, reproducibility and generalizability.

The remainder of this paper is structured as follows. We will introduce some background in feature selection and meta-learning (Sect. 2) as well some additional meta-features that prove useful (Sect. 3). We will then review the overall experiments and results in terms of when feature selection may add value (Sect. 4), and a meta-learning experiment where we aim to predict whether to use feature selection for a given data set (Sect. 5). Section 6 concludes the paper.

2 Background

In this section we discuss relevant background and related work in feature selection, meta-learning and experiment databases.

2.1 Feature Selection

As discussed in the introduction feature selection can serve a number of purposes, such as improved interpretation, generalization and learning speed. The merits of and methods for feature selection are discussed extensively in a number of classical survey papers, hence we will keep the overview brief here [1,4–6,10]. The goal of feature selection can be to find the optimal set of features that maximizes a given objective, and hence can be seen as a search problem with a given search method, evaluation metric and overall objective, typically some form of predictive power.

Exhaustive search is typically not feasible so different approaches are needed. A simplistic approach would simply select the top features based on predictive power. This is sub optimal, because features may be correlated to features already selected, so not adding much information, or conversely, weak features could jointly actually be predictive, thus subset feature selection rather than rankers are required [6,8]. The evaluation metrics could be so called filter metrics, such as correlation, mutual information or information gain, independent of the classification algorithm used. Alternatively, models could be trained on subsets of features in a so called wrapper approach, which can be valuable if the subsequent learners have very specific biases or limitations [10]. Wrappers do not necessarily perform better than filters [19] so in our work we have focused on a subset filter approach [8]. Feature construction or dimension reduction can be seen as an extension of feature selection, but this is out of scope for this paper. Note that classification methods can also have some embedded element of feature selection built in, but as we will see this is no guarantee that feature selection is no longer required.

2.2 Meta-learning

Meta-learning aims to learn which learning techniques work well on what data. A common task, known as the Algorithm Selection Problem [17], is to determine which classifier performs best on a given data set. We can predict this by training a meta-model on data describing the performance of different methods on different data sets, characterized by *meta-features* [2,11,13]. Meta-features are often categorized as either simple (number of examples, number of attributes), statistical (mean standard deviation of attributes, mean skewness of attributes), information theoretic (class entropy, mean mutual information) or landmarkers [12] (performance evaluations of simple classifiers). Alternatively, performance estimates of algorithms on small subsets of the data set can be used [18].

Experiment databases enable the reproduction of earlier results for verification and reusability purposes, and make much larger studies (covering more classifiers and parameter settings) feasible. Above all, experiment databases allow a variety of studies to be executed by a database look-up, rather than setting up new experiments. An example of such an online experiment database is OpenML [20]. All data sets and experimental results used in this work are made publicly available in OpenML. Similar collaborative platforms exist in the commercial domain, such as Kaggle [3], but these typically lack the ability to store and search low level results in a structured manner.

3 Methods

The field of meta-learning addresses the question what machine learning algorithms work well on what data. The algorithm selection problem, formalised by Rice in [17], is a natural problem from the field of meta-learning. According to the definition of Rice, the problem space P consists of all machine learning tasks from a certain domain, the feature space F contains measurable characteristics calculated upon this data (called meta-features), the algorithm space A is the set of all considered algorithms that can execute these tasks and the performance space Y represents the mapping of these algorithms to a set of performance measures. The task is for any given $x \in P$, to select the algorithm $\alpha \in A$ that maximizes a predefined performance measure $y \in Y$, which is a classification problem. Typically, this problem is addressed by creating a meta-data set. Each example represents an experiment where all algorithms in A are run on a data set from P, the meta-features are measurable characteristics of this data set and the target is the best performing algorithm on this data set. A classifier can then learn to predict for new data sets which algorithm will perform best [22].

In this work we address the following problem. Given a data set and an algorithm, should we use feature selection or not? We aim to solve this in a similar manner. We construct a meta-data set, where each example represents the combination of data set d and algorithm α. The features are measurable characteristics of data set d, and the target is whether the performance of algorithm α is (significantly) better after performing feature selection than without it.

The performance of meta-learning solution typically depends on the quality of the meta-features. Typical meta-features are often categorized as either simple, statistical, information theoretic or landmarkers. The simple meta-features can all be calculated by one single pass over all instances and describe the data set in an aggregated manner. The statistical meta-features are calculated by considering a statistical concept (e.g., standard deviation, skewness or kurtosis), calculate this for all numeric attributes and taking the mean of this. This leads to, e.g., the mean standard deviation of numeric attributes. Likewise, the information theoretic meta-features are calculated by considering a information theoretic concept (e.g., mutual information or attribute entropy), calculate this for all nominal attributes and taking the mean of this. This leads to, e.g., mean mutual information. Landmarkers are performance evaluations of fast classifiers on a data set, characterising the complexity landscape and bias of various learners. Table 1 shows all traditional meta-features used in the experiments.

Table 1. Standard meta-features.

Category	Meta-features
Simple	# Instances, # Attributes, Dimensionality, Default Accuracy, # Observations with Missing Values, # Missing Values, % Observations With Missing Values, % Missing Values, # Numeric Attributes, # Nominal Attributes, # Binary Attributes, Majority Class Size, % Majority Class
Statistical	Mean of Means of Numeric Attributes, Mean Standard Deviation of Numeric Attributes, Mean Kurtosis of Numeric Attributes, Mean Skewness of Numeric Attributes
Information theoretic	Class Entropy, Mean Attribute Entropy, Mean Mutual Information, Equivalent Number Of Attributes, Noise to Signal Ratio
Landmarkers [12]	Accuracy of Decision Stump, Kappa of Decision Stump, Area under the ROC Curve of Decision Stump, Accuracy of Naive Bayes, Kappa of of Naive Bayes, Area under the ROC Curve of Naive Bayes, Accuracy of k-NN, Kappa of k-NN, Area under the ROC Curve of k-NN

Landmarkers are generally considered the most expensive meta-features (in terms of resources), as well as the most useful (in terms of predictive power). Although this might be true for the algorithm selection problem, there are reasons to suspect that this might be different for the task of determining whether or not to perform feature selection. First, many feature selection methods operate on statistical and information theoretical concepts. Second, information about the learning bias of various classifiers seems less relevant, as we try to obtain information about one algorithm at a time.

For this reason, we introduce specific *feature selection landmarkers*. We run a simple (fast) classifier with and without feature selection. By subtracting one from the other, we can see what the effect of feature selection was when using a fast algorithm. Similar to regular landmarkers, we assume that this effect translates to the results of more expensive algorithms as well.

4   Effect of Feature Selection

In this section we will present some explorative results, surveying per algorithm how often feature selection is beneficial and how large the effects are. All data sets, algorithm and experimental results can be obtained from OpenML[1] [20]. Figure 3 also gives some basic insight in the number of features and the dimensionality of the data sets.

4.1   Experiment

All algorithms are evaluated over the data sets using 10-fold cross-validation, with and without feature selection. We measure the difference in Area under the ROC Curve (AUC) for each algorithm with and without feature selection. We prefer AUC over zero one loss accuracy as an evaluation criterion for a variety of reasons. First, if the outcome class distribution is very skewed, a simple majority vote may achieve very high accuracy, whereas in practice this may not be very useful model. Second, false positives and false negative classifications may come at a different cost, but these costs are not known, hence it makes sense to evaluate model performance across the entire model score range.

For feature selection, the Correlation-based Feature Subset Selection (CfsSubsctEval) algorithm is used [8]. We experimented with other feature selection methods as well (i.e., GainRatio and InfoGain) but as the differences in performance were too marginal and subset feature selection is generally considered to be a better approach we stick to CfsSubsetEval.

The data sets that are used in the experiments are all data sets containing between 10 and 200,000 instances. As we are focusing on Area under the ROC Curve, we selected data sets with a binary target. In total 394 data sets from OpenML matched these criteria. Table 2 shows the algorithms that were used and their parameter settings.

---

[1] Full details: http://www.openml.org/s/15.

Table 2. Algorithms used in the experiments. All algorithms are as implemented in Weka 3.7.13 [7] run with default parameter settings, unless stated different.

Algorithm	Model type	Parameter settings
Naive Bayes	Bayesian	
IBk	k-NN	$k = 1$
Stochastic Gradient Descent (SGD)	SVM	
Sequential Minimal Optimization (SMO)	SVM	Polynomial kernel
Logistic	Logistic	ridge = 0.00000001
Multilayer perceptron	Neural network	1 hidden layer
JRip	Rules	
J48	Decision tree	
Hoeffding tree	Decision tree	
REP tree	Decision tree	
RandomForest	Bagging	100 trees
AdaBoost	Boosting	100 iterations

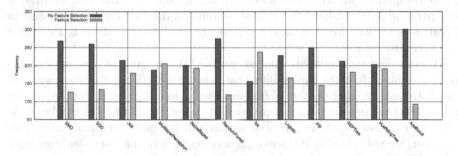

(a) Number of data sets where 'no feature selection' obtained better results (red) and 'feature selection' obtained better results (green)

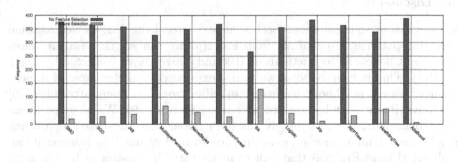

(b) Number of data sets on which 'feature selection' was statistically significant better (green) and not statistically significant better (red)

Fig. 1. Number of data sets on which feature selection improves performance. (Color figure online)

4.2   Results

Figure 1(a) shows for each algorithm in how many cases feature selection yields better results. Figure 1(b) shown for each algorithm in how many cases this difference was also statistically significant (using a double tailed T-test of 0.05).

We can also focus on how big the effect of feature selection per data set is. In Fig. 2 we plotted for some algorithms the difference in performance with and without feature selection. The x-axis represents the various data sets, the y-axes the difference in performance (AUC). The x-axis is sorted on this effect, so we can see the big trends. For every dot above 0, using feature selection yields better results than not using feature selection.

In Fig. 1(a) it is observed that no feature selection is slightly better for every algorithm, except for IBk and Multilayer Perceptron. J48 is noteworthy because it is expected that a tree partitioning algorithm has feature selection embedded. Controversially, the figure shows that feature selection can still add value for many data sets (see also Fig. 2(a)). For the Multilayer Perceptron, Naive Bayes and Hoeffding Tree, in about half of the cases feature selection improves the performance. Figure 1(b) shows that applying feature selection seldom results in a performance gain that is statistically significant. The IBk and Multilayer Perceptron algorithms have the highest amount of data sets where the benefits of feature selection are statistically significant, and behind these is Hoeffding Tree with just over 50 data sets.

Figure 3 shows a univariate analysis on how the amount of features and the dimensionality affect the probability that feature selection improves classification. Although a higher number of features results in a slightly higher percentage of data sets that benefit from feature selection, no clear distinction can be made with just one feature. A similar observation can be made for the dimensionality. Later, we will see that meta-models leveraging multiple meta-features, also highly depend on the number of features in a data set.

4.3   Discussion

The previous experiment shows both some expected behaviour as well as some interesting patterns. First of all, from Fig. 1(a) we can see that feature selection is most beneficial for methods as IBk and Naive Bayes (reflected by Fig. 1 and 2(c)). This is exactly what we would expect: due to the curse of dimensionality, nearest neighbour methods can suffer from too many attributes [16] and Naive Bayes is vulnerable to correlated features [9]. We also see unexpected behaviour. For example, it has been noted that tree-based algorithms such J48 have built-in protection against irrelevant features [15], however it can be observed from Fig. 2(a) that still in many cases it appears to benefit from feature selection. Multilayer Perceptrons are also supposed to learn themselves which features are relevant [21], however Fig. 2(b) shows that in many cases feature selection makes a substantial difference for the better. This delta to the right of the curve is higher than the delta on the left.

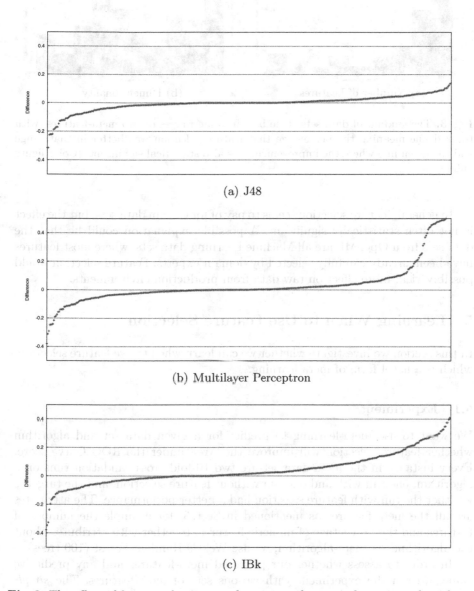

(a) J48

(b) Multilayer Perceptron

(c) IBk

Fig. 2. The effect of feature selection per data set on the x-axis for a given algorithm, sorted by difference in Area under the ROC Curve as the y-axis. When the difference is positive, the algorithm performed better after feature selection.

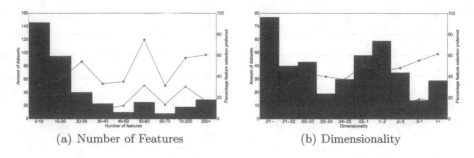

(a) Number of Features   (b) Dimensionality

Fig. 3. The amount of data sets (blue bar) in some ranges of two meta-features, with the red line meaning the percentage that feature selection was better in that range and the green line where the improvement was also statistical significant. (Color figure online)

In general, feature selection seems to pay of for certain data sets, but the effect is not often statistically significant. A possible explanation could be that the data sets from OpenML are all Machine Learning data sets, where most features have been already carefully selected by domain experts. Feature selection would possibly yield more effect on raw data from production environments.

5   Learning When to Use Feature Selection

In this section we investigate whether we can learn when to use feature selection, which is a novel form of meta-learning.

5.1   Experiment

We want to use meta-learning to predict for a given data set and algorithm whether feature selection will improve the Area under the ROC Curve score. Every instance in the meta-data set are two 10-fold cross validation runs on a algorithm, one run with and one run without feature selection, and the target is whether the run with feature selection had a better performance. The attributes are all the meta-features as mentioned in Sect. 3, for example the number of features and the percentage of numeric features, together with attributes about the algorithm. As meta-algorithm, we use Weka's Random Forest (100 trees).

In order to assess whether our proposed meta-features add any predictive value, we run the experiment with various sets of meta-features. The *simple* set contains just the simple meta-features (see Table 1) totalling to 13 features. The *no landmarkers* set contains all simple, statistical and information theoretic meta-features, (i.e., all meta-features from Table 1 except the landmarkers) which are in total 22 features. The *default landmarkers* set contains all meta-features from the no landmarkers set, and the traditionally described landmarkers (i.e., all meta-features from Table 1) which give a total amount of 31 features. The *Feature Selection Landmarkers* set contains all meta-features from the no landmarkers set, and the newly created Feature Selection Landmarkers as described in Sect. 3 thus also 31 features in total. The *All Landmarkers* set is the union of all previous sets totalling up to 40 features.

The data set can also be split in various subsets containing the results of only one algorithm. For example, we can investigate whether we can learn for a given algorithm whether to use feature selection or not.

The main motivation for using meta-learning here is primarily to obtain a further understanding of when feature selection may or may not add value, across multiple dimensions, to complement the analysis in the previous section that mainly focused on the algorithms used. A meta-model could be used in practice to assess beforehand whether performance may be improved in general or for specific algorithms, for example algorithms which are very costly to run. If exhaustive search is possible and reliable (i.e., run all algorithms for all parameters) it may still be preferred over using meta-learning.

5.2   Results

The results are shown in Table 3. Each row represents a partition of the data set, i.e., how well we could predict for each classifier whether we should use feature selection.

Table 3. Area under the ROC Curve scores for various sets of meta-features on different partitions of the meta-data set.

Partition	Simple	No LM	Default LM	FS LM	All LM
J48	0.705	0.703	0.737	0.733	0.731
IBk	0.680	0.700	0.750	0.768	0.783
Multilayer perceptron	0.734	0.704	0.708	0.711	0.710
Logistic	0.623	0.625	0.711	0.676	0.695
SMO	0.642	0.632	0.695	0.713	0.704
SGD	0.705	0.698	0.736	0.746	0.733
Hoeffding tree	0.612	0.617	0.679	0.647	0.670
REP tree	0.593	0.573	0.614	0.591	0.621
Naive Bayes	0.620	0.660	0.714	0.708	0.721
JRip	0.590	0.581	0.595	0.616	0.639
AdaBoost	0.623	0.634	0.638	0.649	0.668
RandomForest	0.712	0.722	0.764	0.774	0.784
Total data set	0.704	0.728	0.765	0.768	0.773

First, from this we conclude that meta-learning can answer the question whether to use feature selection or not. Compared to just predicting majority class (which always has an Area under the ROC Curve of 0.5), we score better on all defined tasks, even with just a set of simple meta-features. Second, we observe that using just the two sets without landmarkers are clearly worse than the sets that use landmarkers. Finally, it appears that the set of default landmarkers and the newly created feature selection landmarkers perform similar. However, putting them together is beneficial.

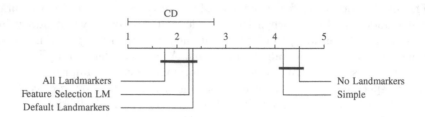

Fig. 4. Results of Nemenyi test. Sets of meta-features are sorted by their average rank (lower is better). Classifiers that are connected by a horizontal line are statistically equivalent.

Figure 4 shows the result of a statistical test. This adds to the empirical evidence that the meta-classifier benefits from the landmarkers. However, there is no statistical evidence that one set is better than another. One interesting observation is that the set of meta-features without landmarkers performs worse than the set of simple meta-features. However, the difference is not statically significant.

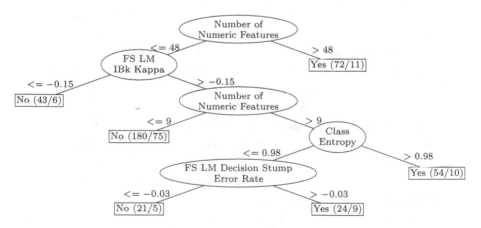

Fig. 5. Decision tree determining whether to use feature selection with a Multilayer Perceptron. Each leaf node contains the amount of correctly classified instances and the amount of misclassified instances.

As an example of deeper inspection of meta-models, Fig. 5 shows a decision tree that determines when to use feature selection in combination with a

Multilayer Perceptron. It splits on meta-features the Number of Attributes, Class Entropy and twice on a Feature Selection Landmarker, suggesting that these are important features. The interplay of these features is interesting. For example if the number of features exceeds 48, feature selection will be useful, if the number of features is smaller than 9 than not, and otherwise it depends on the interplay between the feature selection landmarkers, class entropy and number of features. Real world data sets often have more than 50 features, so this is an indication that even though the OpenML collection of data sets is large, it may be still be skewed towards 'cleaned-up' data sets collected for machine learning and data mining research. By inspecting these meta-models observations like these may surface, and in this case the meta-model will still recommend to apply feature selection for these broader data sets.

6   Conclusion

In this paper we present the results of a large scale experiment on the benefits of using feature selection for classification. We ran 12 algorithms across almost 400 data sets, and created a meta-model to understand when feature selection improves classification accuracy for a given model. Surprisingly, for 41 per cent of algorithm data set combinations feature selection improved the results, but only in 10 per cent of cases this improvement was statistically significant. A possible explanation for this low percentage could be that the data sets from OpenML consist mostly of features that have already been carefully selected by domain experts. The experimental setting would possibly yield other results on raw data from production environments, which would be an interesting direction for future work. Major deciding factors are the number of attributes in the data set, the relative difficulty of the task as measured by landmarkers and the algorithm type. Across algorithms, nearest neighbor benefits most often, but also algorithms that have feature selection built in (such as decision trees) may still benefit.

   Future work will focus on extending the set of Feature Selection landmarkers, aiming to perform even better on the meta-leaning task. Having a publicly available meta-data set enables the community to actively participate in this process.

References

1. Blum, A.L., Langley, P.: Selection of relevant features and examples in machine learning. Artif. Intell. **97**(1–2), 245–271 (1997)
2. Brazdil, P., Gama, J., Henery, B.: Characterizing the applicability of classification algorithms using meta-level learning. In: Bergadano, F., Raedt, L. (eds.) ECML 1994. LNCS, vol. 784, pp. 83–102. Springer, Heidelberg (1994). doi:10. 1007/3-540-57868-4_52
3. Carpenter, J.: May the best analyst win. Science **331**(6018), 698–699 (2011)
4. Chandrashekar, G., Sahin, F.: A survey on feature selection methods. Comput. Electr. Eng. **40**(1), 16–28 (2014)

5. Dash, M., Liu, H.: Feature selection for classification. Intell. Data Anal. **1**(3), 131–156 (1997)
6. Guyon, I., Elisseeff, A.: An introduction to variable and feature selection. J. Mach. Learn. Res. **3**, 1157–1182 (2003)
7. Hall, M., Frank, E., Holmes, G., Pfahringer, B., Reutemann, P., Witten, I.H.: The WEKA data mining software: an update. ACM SIGKDD Explor. Newsl. **11**(1), 10–18 (2009)
8. Hall, M.A.: Correlation-based feature subset selection for machine learning. Ph.D. thesis, University of Waikato, Hamilton, New Zealand (1998)
9. John, G.H., Langley, P.: Estimating continuous distributions in Bayesian classifiers. In: Proceedings of the Eleventh Conference on Uncertainty in Artificial Intelligence, pp. 338–345. Morgan Kaufmann Publishers Inc. (1995)
10. Kohavi, R., John, G.H.: Wrappers for feature subset selection. Artif. Intell. **97**(1), 273–324 (1997)
11. Peng, Y., Flach, P.A., Soares, C., Brazdil, P.: Improved dataset characterisation for meta-learning. In: Lange, S., Satoh, K., Smith, C.H. (eds.) DS 2002. LNCS, vol. 2534, pp. 141–152. Springer, Heidelberg (2002). doi:10.1007/3-540-36182-0_14
12. Pfahringer, B., Bensusan, H., Giraud-Carrier, C.: Tell me who can learn you and I can tell you who you are: landmarking various learning algorithms. In: Proceedings of the 17th International Conference on Machine Learning, pp. 743–750 (2000)
13. Pinto, F., Soares, C., Mendes-Moreira, J.: Towards automatic generation of metafeatures. In: Bailey, J., Khan, L., Washio, T., Dobbie, G., Huang, J.Z., Wang, R. (eds.) PAKDD 2016. LNCS (LNAI), vol. 9651, pp. 215–226. Springer, Heidelberg (2016). doi:10.1007/978-3-319-31753-3_18
14. van der Putten, P., van Someren, M.: A bias-variance analysis of a real world learning problem: the coil challenge 2000. Mach. Learn. **57**(1), 177–195 (2004)
15. Quinlan, J.R.: Induction of decision trees. Mach. Learn. **1**(1), 81–106 (1986)
16. Radovanović, M., Nanopoulos, A., Ivanović, M.: Hubs in space: popular nearest neighbors in high-dimensional data. JMLR **11**, 2487–2531 (2010)
17. Rice, J.R.: The algorithm selection problem. Adv. Comput. **15**, 65–118 (1976)
18. van Rijn, J.N., Abdulrahman, S.M., Brazdil, P., Vanschoren, J.: Fast algorithm selection using learning curves. In: Fromont, E., Bie, T., Leeuwen, M. (eds.) IDA 2015. LNCS, vol. 9385, pp. 298–309. Springer, Heidelberg (2015). doi:10.1007/978-3-319-24465-5_26
19. Tsamardinos, I., Aliferis, C.: Towards principled feature selection: relevancy, filters and wrappers. In: Proceedings of the Ninth International Workshop on Artificial Intelligence and Statistics (2003)
20. Vanschoren, J., van Rijn, J.N., Bischl, B., Torgo, L.: OpenML: networked science in machine learning. ACM SIGKDD Explor. Newsl. **15**(2), 49–60 (2014)
21. Verikas, A., Bacauskiene, M.: Feature selection with neural networks. Pattern Recogn. Lett. **23**(11), 1323–1335 (2002)
22. Vilalta, R., Giraud-Carrier, C.G., Brazdil, P., Soares, C.: Using meta-learning to support data mining. IJCSA **1**(1), 31–45 (2004)

Learning from the News: Predicting Entity Popularity on Twitter

Pedro Saleiro[1,2](✉) and Carlos Soares[1,3]

[1] DEI-FEUP, University of Porto, Rua Dr. Roberto Frias, s/n, Porto, Portugal
{pssc,csoares}@fe.up.pt
[2] LIACC, University of Porto, Rua Dr. Roberto Frias, s/n, Porto, Portugal
[3] INESC-TEC, University of Porto, Rua Dr. Roberto Frias, s/n, Porto, Portugal

Abstract. In this work, we tackle the problem of predicting entity popularity on Twitter based on the news cycle. We apply a supervised learning approach and extract four types of features: (i) signal, (ii) textual, (iii) sentiment and (iv) semantic, which we use to predict whether the popularity of a given entity will be high or low in the following hours. We run several experiments on six different entities in a dataset of over 150M tweets and 5M news and obtained F1 scores over 0.70. Error analysis indicates that news perform better on predicting entity popularity on Twitter when they are the primary information source of the event, in opposition to events such as live TV broadcasts, political debates or football matches.

Keywords: Prediction · News · Social media · Online reputation monitoring

1 Introduction

Online publication of news articles has become a standard behavior of news outlets, while the public joined the movement either using desktop or mobile terminals. The resulting setup consists of a cooperative dialog between news outlets and the public at large. Latest events are covered and commented by both parties in a continuous basis through the social media, such as Twitter. When sharing or commenting news on social media, users tend to mention the most predominant entities mentioned in the news story [1]. Therefore, entities, such as personalities, organizations, companies or geographic locations, can act as latent interlinks between online news and social media.

Online Reputation Monitoring (ORM) focuses on continuously tracking what is being said about entities (e.g. politicians) on social media and online news. Automatic collection and processing of comments and opinions on social media is now crucial to understand the reputation of individual personalities and organizations and therefore to manage their public relations. However, ORM systems would be even more useful if they would be able to know in advance if social media users will talk a lot about the target entities or not. For instance, on April

H. Boström et al. (Eds.): IDA 2016, LNCS 9897, pp. 171–182, 2016.
DOI: 10.1007/978-3-319-46349-0_15

4th 2016, the UK Prime-minister, David Cameron, was mentioned on the news regarding the Panama Papers story. He didn't acknowledge the story in detail on that day. However, the news cycle kept mentioning him about this topic in the following days and his mentions on social media kept very high. He had to publicly address the issue on April 9th, when his reputation had already been severely damaged, blaming himself for not providing further details earlier.

We hypothesize that for entities that are frequently mentioned on the news (e.g. politicians) it is possible to establish a predictive link between online news and popularity on social media. We cast the problem as a supervised learning classification approach: to decide whether popularity will be high or low based on features extracted from the news cycle. We define four set of features: signal, textual, sentiment and semantic. We aim to respond to the following research questions: **RQ1:** Are online news valuable as source of information to effectively predict entity popularity on Twitter? **RQ2:** Do online news carry different predictive power based on the nature of the entity under study? **RQ3:** How different thresholds for defining high and low popularity affect the effectiveness of our approach? **RQ4:** Do performance remains stable for different time of predictions? **RQ5:** What is the most important feature set for predicting entity popularity on Twitter based on the news cycle? **RQ6:** Do individual set of features represent different importance for different entities?

2   Related Work

In recent years, a number of research works have studied the relationship and predictive behavior of user response to the publication of online media items, such as, commenting news articles, playing Youtube videos, sharing URLs or retweeting patterns [2–5]. The first attempt to predict the volume of user comments for online news articles used both metadata from the news articles and linguistic features [4]. The prediction was divided in two binary classification problems: if an article would get any comments and if it would be high or low number of comments. Similarly other works, found that shallow linguistic features (e.g. TF-IDF or sentiment) and named entities have good predictive power [6,7].

Research work more in line with ours, tries to predict the popularity of news articles shares (url sharing) on Twitter based on content features [2]. Authors considered the news source, the article's category, the article's author, the subjectivity of the language in the article, and number of named entities in the article as features. Recently, there was a large study of the life cycle of news articles in terms of distribution of visits, tweets and shares over time across different sections of the publisher [8]. Their work was able to improve, for some content type, the prediction of web visits using data from social media after ten to twenty minutes of publication.

Other line of work, focused on temporal patterns of user activities and have consistently identified broad classes of temporal patterns based on the presence of a clear peak of activity [3,9–11]. Classes differentiate by the specific amount and duration of activity before and after the peak. Crane and Sornette [9] define endogenous or exogenous origin of events based on being triggered by internal aspects of the social network or external, respectively. They find that hashtag popularity is mostly influenced by exogenous factors instead of epidemic spreading. Other work [10] extends these classes by creating distinct clusters of activity based on the distributions in different periods (before, during and after the peak) that can be interpreted based on semantics of hashtags. Consequently, the authors applied text mining techniques to semantically describe hashtag classes. Yang and Leskovec [3] propose a new measure of time series similarity and clustering. Authors obtain six classes of temporal shapes of popularity of a given phrase (meme) associated with a recent event, as well as the ordering of media sources contribution to its popularity.

Recently, Tsytsarau et al. [12] studied the time series of news events and their relation to changes of sentiment time series expressed on related topics on social media. Authors proposed a novel framework using time series convolution between the importance of events and media response function, specific to media and event type. Their framework is able to predict time and duration of events as well as shape through time.

Compared to related work, we focus on a different problem - online reputation monitoring - where it is necessary to track what is being said about an entity on social media on a continuous basis. Therefore, our problem consists on assessing the impact of the news cycle on the time series of popularity of a target entity on social media.

3   Approach

The starting point of our hypothesis is that for entities that are frequently mentioned on the news (e.g. politicians) it is possible to predict popularity on social media using signals extracted from the news cycle. The first step towards a solution requires the definition of entity popularity on social media.

3.1   Entity Popularity

There are different ways of expressing the notion of popularity on social media. For example, the classical way of defining it is through the number of followers of a Twitter account or likes in a Facebook page. Another notion of popularity, associated with entities, consists on the number of retweets or replies on Twitter and post likes and comments on Facebook. We define entity popularity based on named entity mentions in social media messages. Mentions consist of specific surface forms of an entity name. For example, "Cristiano Ronaldo" might be mentioned also using just "Ronaldo" or "#CR7".

Given an set of entities $E = \{e_1, e_2, ..., e_i, ...\}$, a daily stream of social media messages $S = \{s_1, s_2, ..., s_i, ...\}$ and a daily stream of online news articles $N = \{n_1, n_2, ..., n_i, ...\}$ we are interested in monitoring the mentions of an entity e_i on the social media stream S, i.e. the discrete function $f_m(e_i, S)$. Let T be a daily time frame $T = [t_p, t_{p+h}]$, where the time t_p is the time of prediction and t_{p+h}

is the prediction horizon time. We want to learn a target popularity function f_p on social media stream S as a function of the given entity e_i, the online news stream N and the time frame T:

$$f_p(e_i, N, T) = \sum_{t=t_p}^{t=t_{p+h}} f_m(e_i, S)$$

which corresponds to integrating $f_m(e_i, S)$ over T.

Given a day d_i, a time of prediction t_p, we extract features from the news stream N until tp and predict f_p until the prediction horizon $tp + h$. We measure popularity on a daily basis, and consequently, we adopted t_{p+h} as 23:59:59 everyday. For example, if t_p equals to 8 a.m., we extract features from N until 07:59:59 and predict f_p in the interval 08:00–23:59:59 on day d_i. In the case of t_p equals to midnight, we extract features from N on the 24 h of previous day d_{i-1} to predict f_p for the 24 h of d_i.

We cast the prediction of $f_p(e_i, N, T)$ as a supervised learning classification problem, in which we want to infer the target variable $\hat{f}_p(e_i, N, T) \in \{0, 1\}$ defined as:

$$\hat{f}_p = \begin{cases} 0(low), & \text{if } P(f_p(e_i, N, T) \leq \delta) = k \\ 1(high), & \text{if } P(f_p(e_i, N, T) > \delta) = 1 - k \end{cases}$$

where δ is the inverse of cumulative distribution function at k of $f_p(e_i, N, T)$ as measured in the training set, a similar approach to [4]. For instance, $k = 0.5$ corresponds to the median of $f_p(e_i, N, T)$ in the training set and higher values of k mean that $f_p(e_i, N, T)$ has to be higher than k examples on the training set to consider $\hat{f}_p = 1$, resulting in a reduced number of training examples of the positive class *high*.

3.2 News Features

Previous work has focused on the influence of characteristics of the social media stream S in the adoption and popularity of memes and hashtags [11]. In opposition, the main goal of this work is to investigate the predictive power of the online news stream N. Therefore we extract four types of features from N: (i) *signal*, (ii) *textual*, (iii) *sentiment* and (iv) *semantic*, as depicted in Table 1. One important issue is how can we filter relevant news items to e_i. There is no consensus on how to link a news stream N with a social media stream S. Some works use URLs from N, shared on S, to filter simultaneously relevant news articles and social media messages [2]. As our work is entity oriented, we select news articles with mentions of e_i as our relevant N.

Signal Features - This type of features depict the "signal" of the news cycle mentioning e_i and we include a set of counting variables as features, focusing on the total number of news mentioning e_i in specific time intervals, mentions on news titles, the average length of news articles, the different number of news outlets that published news mentioning e_i as well as, features specific to the

Table 1. Summary of the four type of features we consider: (i) signal, (ii) textual, (iii) sentiment and (iv) semantic, 21 in total

Number	Feature	Description	Type
Signal			
1	*news*	number of news mentions of e_i in $[0, t_p]$ in d_i	Int
2	*news d_{i-1}*	number of news mentions of e_i in $[0, t_p]$ in d_{i-1}	Int
3	*news total d_{i-1}*	number of news mentions of e_i in $[0, 24[$ in d_{i-1}	Int
4	*news titles*	number of title mentions in news of e_i in $[0, t_p]$ in d_i	Int
5	*avg content*	average content length of news of e_i in $[0, t_p]$ in d_i	Float
6	*sources*	number of different news sources of e_i in $[0, t_p]$ in d_i	Int
7	*weekday*	day of week	Categ
8	*is weekend*	true if weekend, false otherwise	Bool
Textual			
9–18	*tfidf titles*	TF-IDF of news titles $[0, t_p]$ in d_i	Float
19–28	*LDA titles*	LDA-10 of news titles $[0, t_p]$ in d_i	Float
Sentiment			
29	*pos*	number of positive words in news titles $[0, t_p]$ in d_i	Int
30	*neg*	number of negative words in news titles $[0, t_p]$ in d_i	Int
31	*neu*	number of neutral words in news titles $[0, t_p]$ in d_i	Int
32	*ratio*	*positive/negative*	Float
33	*diff*	*positive − negative*	Int
34	*subjectivity*	$(positive + negative + neutral)/\sum words$	Float
35–44	*tfidf subj*	TF-IDF of subjective words (pos, neg and neu)	Float
Semantic			
45	*entities*	number of entities in news $[0, t_p]$ in d_i	Int
46	*tags*	number of tags in news $[0, t_p]$ in d_i	Int
47–56	*tfidf entities*	TF-IDF of entities in news $[0, t_p]$ in d_i	Float
57–66	*tfidf tags*	TF-IDF of news tags $[0, t_p]$ in d_i	Float

day of the week to capture any seasonal trend on the popularity. The idea is to capture the dynamics of news events, for instance, if e_i has a sudden peak of mentions on N, a relevant event might have happened which may influence f_p.

Textual features - To collect textual features we build a daily profile of the news cycle by aggregating all titles of online news articles mentioning e_i for the daily time frame $[0, t_p]$ in d_i. We select the top 10000 most frequent terms (unigrams and bi-grams) in the training set and create a document-term matrix R. Two distinct methods were applied to capture textual features.

The first method is to apply TF-IDF weighting to R. We employ singular value decomposition (SVD) to capture similarity between terms and reduce dimensionality. It computes a low-dimensional linear approximation σ. The final set of features for training and testing is the TF-IDF weighted term-document matrix R combined with σR which produces 10 real valued latent features.

When testing, the system uses the same 10000 terms from the training data and calculates TF-IDF using the IDF from the training data, as well as, σ for applying SVD on test data.

The second method consists in applying Latent Dirichlet allocation (LDA) to generate a topic model of 10 topics (features). The system learns a topic-document distribution θ and a word distribution over topics φ using the training data for a given entity e_i. When testing, the system extracts the word distribution of the news title vector r on a test day d'_i. Then, by using φ learned on training data, it calculates the probability of r belonging to one of the 10 topics learned before. The objective of extracting this set of features is to create a characterization of the news stream that mentions e_i, namely, which are the most salient terms and phrases on each day d_i as well as the latent topics associated with e_i. By learning our classifier we hope to obtain correlations between certain terms and topics and f_p.

Sentiment features - We include several types of word level sentiment features. The assumption here is that subjective words on the news will result in more reactions on social media, as exposed in [13]. Once again we extract features from the titles of news mentioning e_i for the daily time frame $[0, t_p]$. We use a sentiment lexicon as *SentiWordNet* to extract subjective terms from the titles daily profile and label them as positive, neutral or negative polarity. We compute count features for number of positive, negative, neutral terms as well as difference and ratio of positive and negatives terms. Similar to textual features we create a TFIDF weighted term-document matrix R using the subjective terms from the title and apply SVD to compute 10 real valued sentiment latent features.

Semantic features - We use the number of different named entities recognized in N on day d_i until tp, as well as, the number of distinct news category tags extracted from the news feeds metadata. These tags, common in news articles, consist of author annotated terms and phrases that describe a sort of semantic hierarchy of news categories, topics and news stories (e.g. "european debt crisis"). We create a TF-IDF weighted entity-document and TF-IDF tag-document matrices and applied SVD to each of them to reduce dimensionality to 10. The idea is to capture interesting entity co-occurrences as well as, news stories that are less transient in time and might be able to trigger popularity on Twitter.

3.3 Learning Framework

Let x be the feature vector extracted from the online news stream N on day d_i until tp. We want to learn the probability $P(\hat{f}_p = 1 | X = x)$. This can be done using the inner product between x and a weighting parameter vector $w \in \mathbb{R}$, $\mathbf{w}^\top \mathbf{x}$.

Using logistic regression and for binary classification one can unify the definition of $p(\hat{f}_p = 1 | x)$ and $p(\hat{f}_p = 0 | x)$ with

$$p(\hat{f}_p | x) = \frac{1}{1 + e^{-\hat{f}_p w^\top x}}$$

Given a set of z instance-label pairs $(x_i, \hat{f}_{p_i})$, with $i = 1, ..., z$ and $\hat{f}_{p_i} \in \{0, 1\}$ we solve the binary class L2 penalized logistic regression optimization problem, where $C > 0$

$$\min_w \frac{1}{2} w^\top w + C \sum_{i=1}^{n} \log(1 + e^{-\hat{f}_{p_i} w^\top x_i})$$

We apply this approach following an entity specific basis, i.e. we train an individual model for each entity. Given a set of entities E to which we want to apply our approach and a training set of example days $D = \{d_1, d_2, ..., d_i, ...\}$, we extract a feature vector x_i for each entity e_i on each training day d_i. Therefore, we are able to learn a model of w for each e_i. The assumption is that popularity on social media f_p is dependent of the entity e_i and consequently we extract entity specific features from the news stream N. For instance, the top 10000 words of the news titles mentioning e_i are not the same for e_j.

4   Experimental Setup

This work uses Portuguese news feeds and tweets collected from January 1, 2013 to January 1, 2016, consisting of over 150 million tweets and 5 million online news articles[1]. To collect and process raw Twitter data, we use a crawler, which recognizes and disambiguate named entities on Twitter [14–16]. News data is provided by a Portuguese online news aggregator. This service handles online

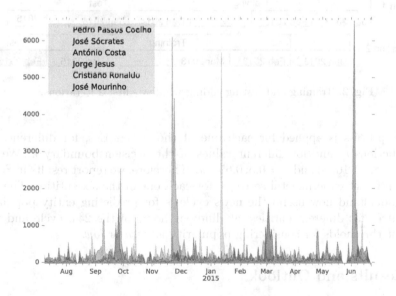

Fig. 1. Daily popularity on Twitter of entities under study.

[1] Dataset is available for research purposes. Access requests via e-mail.

news from over 60 Portuguese news outlets and it is able to recognize entities mentioned on the news.

We choose the two most common news categories: politics and football and select the 3 entities with highest number of mentions on the news for both categories. The politicians are two former Prime-ministers, José Sócrates and Pedro Passos Coelho and the incumbent, António Costa. The football entities are two coaches, Jorge Jesus and José Mourinho, and the most famous Portuguese football player, Cristiano Ronaldo.

Figure 1 depicts the behavior of daily popularity of the six entities on the selected community stream of Twitter users for each day from July 2014 until July 2015. As expected, it is easily observable that in some days the popularity on Twitter exhibits bursty patterns. For instance, when José Sócrates was arrested in November 21st 2014 or when Cristiano Ronaldo won the FIFA Ballon d'Or in January 12th 2015.

We defined the years of 2013 and 2014 as training set and the whole year of 2015 as test set. We applied a monthly sliding window setting in which we start by predicting entity popularity for every day of January 2015 (i.e. the test set) using a model trained on the previous 24 months, 730 days (i.e. the training set). Then, we use February 2015 as the test set, using a new model trained on the previous 24 months. Then March and so on, as depicted in Fig. 2. We perform this evaluation process, rolling the training and test set until December 2015, resulting in 365 days under evaluation.

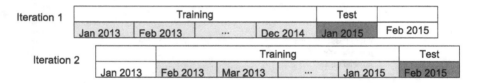

Fig. 2. Training and testing sliding window - first 2 iterations.

The process is applied for each one of the six entities, for different time of predictions t_p and for different values of the decision boundary k. We test $tp = 0, 4, 8, 12, 16, 20$ and $k = 0.5, 0.65, 0.8$. Therefore, we report results in Sect. 5 for 18 different experimental settings, for each one of the six entities. The goal is to understand how useful the news cycle is for predicting entity popularity on Twitter for different entities, at different hours of the 24 h cycle and with different thresholds for considering popularity as *high* or *low*.

5   Results and Outlook

Results are depicted in Table 2. We report F1 on positive class since in online reputation monitoring is more valuable to be able to predict *high* popularity than *low*. Nevertheless, we also calculated overall Accuracy results, which were better than the F1 reported here. Consequently, this means that our system is fairly capable of predicting *low* popularity. We organize this section based on the research questions we presented in the Introduction (Sect. 1).

Table 2. F1 score of popularity *high* as function of t_p and k equal to 0.5, 0.65 and 0.8 respectively.

Entity\t_p(hour)	0	4	8	12	16	20
$k = 0.50$						
António Costa	0,76	0,67	0,74	0,77	0,75	0,72
José Sócrates	0,77	0,66	0,73	0,75	0,75	0,75
Pedro Passos Coelho	0,72	0,63	0,70	0,70	0,74	0,71
Cristiano Ronaldo	0,35	0,41	0,45	0,37	0,35	0,32
Jorge Jesus	0,73	0,68	0,69	0,68	0,69	0,70
José Mourinho	0,62	0,46	0,51	0,56	0,55	0,45
$k = 0.65$						
António Costa	0,61	0,60	0,66	0,64	0,60	0,60
José Sócrates	0,63	0,57	0,62	0,66	0,64	0,62
Pedro Passos Coelho	0,58	0,57	0,65	0,67	0,67	0,65
Cristiano Ronaldo	0,29	0,35	0,42	0,41	0,36	0,30
Jorge Jesus	0,63	0,61	0,63	0,59	0,62	0,64
José Mourinho	0,56	0,39	0,48	0,56	0,47	0,38
$k = 0.80$						
António Costa	0,48	0,51	0,55	0,53	0,44	0,49
José Sócrates	0,48	0,42	0,47	0,53	0,47	0,35
Pedro Passos Coelho	0,47	0,46	0,56	0,56	0,52	0,54
Cristiano Ronaldo	0,14	0,29	0,31	0,26	0,20	0,21
Jorge Jesus	0,50	0,48	0,51	0,48	0,57	0,56
José Mourinho	0,32	0,32	0,36	0,41	0,41	0,36

RQ1 and **RQ2:** Results show that performance varies with each target entity e_i. In general, results are better in the case of predicting popularity of politicians. In the case of football personalities, Jorge Jesus exhibits similar results with the three politicians but José Mourinho and specially Cristiano Ronaldo represent the worst results in our setting. For instance, when Cristiano Ronaldo scores three goals in a match, the burst on popularity is almost immediate and not possible to predict in advance.

Further analysis showed that online news failed to be informative of popularity in the case of live events covered by other media, such as TV. Interviews and debates on one hand, and live football games on the other, consist of events with unpredictable effects on popularity. Cristiano Ronaldo can be considered a special case in our experiments. He is by far the most famous entity in our experiments and in addition, he is also an active Twitter user with more than 40M followers. This work focus on assessing the predictive power of online news and its limitations. We assume that for Cristiano Ronaldo, endogenous features from the Twitter itself would be necessary to obtain better results.

RQ3: Our system exhibits top performance with $k = 0.5$, which corresponds to balanced training sets, with the same number of *high* and *low* popularity examples on each training set. Political entities exhibit F1 scores above 0.70 with $k = 0.5$. On the other hand, as we increase k, performance deteriorates. We observe that for $k = 0.8$, the system predicts a very high number of false positives. It is very difficult to predict extreme values of popularity on social media before they happen. We plan to tackle this problem in the future by also including features about the target variable in the current and previous hours, i.e., time-series auto-regressive components.

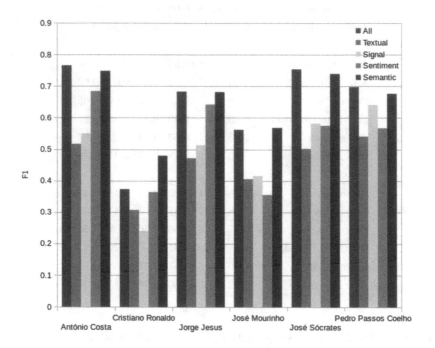

Fig. 3. Individual feature type F1 score for $t_p = 12$ at $k = 0.5$.

RQ4: Results show that time of prediction affects the performance of the system, specially for the political entities. In their case, F1 is higher when time of prediction is noon and 4 p.m. which is an evidence that in politics, most of the news events that trigger popularity on social media are broadcast by news outlets in the morning. It is very interesting to compare results for midnight and 4 a.m. The former use the news articles from the previous day, as explained in Sect. 3, while the latter use news articles from the first 4/8 h of the day under prediction. In some examples, Twitter popularity was triggered by events depicted on the news from the previous day and not from the current day.

RQ5 and **RQ6:** Figure 3 tries to answer these two questions. The first observation is that the combination of all groups of features does not lead to substantial improvements. Semantic features alone achieve almost the same F1 score as the combination of all features. However in the case of Mourinho and Ronaldo, the combination of all features lead to worse F1 results than the semantic set alone.

Sentiment features are the second most important for all entities except José Mourinho. Signal and Textual features are less important and this was somehow a surprise. Signal features represent the surface behavior of news articles, such as the volume of news mentions of e_i before t_p and we were expecting an higher importance. Regarding Textual features, we believe that news articles often refer to terms and phrases that explain past events in order to contextualize a news article.

In future work, we consider alternative approaches for predicting future popularity of entities that do not occur everyday on the news, but do have social media public accounts, such as musicians or actors. In opposition, entities that occur often on the news, such as economics ministers and the like, but do not often occur in the social media pose also a different problem.

References

1. Saleiro, P., Teixeira, J., Soares, C., Oliveira, E.: TimeMachine: entity-centric search and visualization of news archives. In: Ferro, N., Crestani, F., Moens, M.-F., Mothe, J., Silvestri, F., Nunzio, G.M., Hauff, C., Silvello, G. (eds.) ECIR 2016. LNCS, vol. 9626, pp. 845–848. Springer, Heidelberg (2016). doi:10.1007/978-3-319-30671-1_78
2. Asur, S., Bandari, R., Huberman, B.: The pulse of news in social media: forecasting popularity. In: ICWSM 2012 (2012)
3. Yang, J., Leskovec, J.: Patterns of temporal variation in online media. In: Proceedings of the Fourth ACM International Conference on Web Search and Data Mining, pp. 177–186. ACM (2011)
4. Weerkamp, W., Tsagkias, M., De Rijke, M.: Predicting the volume of comments on online news stories. In: CIKM 2009, pp. 1765–1768. ACM (2009)
5. He, X., Gao, M., Kan, M.-Y., Liu, Y., Sugiyama, K.: Predicting the popularity of web 2.0 items based on user comments. In: Proceedings of the 37th International ACM SIGIR Conference on Research & Development in Information Retrieval, pp. 233–242. ACM (2014)
6. Gottipati, S., Jiang, J.: Finding thoughtful comments from social media. In: COLING, pp. 995–1010 (2012)
7. Louis, A., Nenkova, A.: What makes writing great? First experiments on article quality prediction in the science journalism domain. Trans. Assoc. Comput. Linguist. 1, 341–352 (2013)
8. Castillo, C., El-Haddad, M., Pfeffer, J., Stempeck, M.: Characterizing the life cycle of online news stories using social media reactions. In: Proceedings of the 17th ACM Conference on Computer Supported Cooperative Work & Social Computing, pp. 211–223. ACM (2014)

9. Crane, R., Sornette, D.: Robust dynamic classes revealed by measuring the response function of a social system. Proc. Nat. Acad. Sci. **105**(41), 15649–15653 (2008)
10. Lehmann, J., Gonçalves, B., Ramasco, J.J., Cattuto, C.: Dynamical classes of collective attention in Twitter. In: Proceedings of the 21st International Conference on World Wide Web, pp. 251–260. ACM (2012)
11. Romero, D.M., Meeder, B., Kleinberg, J.: Differences in the mechanics of information diffusion across topics: idioms, political hashtags, and complex contagion on Twitter. In: Proceedings of the 20th International Conference on World Wide Web, pp. 695–704. ACM (2011)
12. Tsytsarau, M., Palpanas, T., Castellanos, M.: Dynamics of news events and social media reaction. In: Proceedings of the 20th ACM SIGKDD International Conference on Knowledge Discovery and Data Mining, pp. 901–910. ACM (2014)
13. Reis, J., Olmo, P., Benevenuto, F., Kwak, H., Prates, R., An, J.: Breaking the news: first impressions matter on online news. In: ICWSM 2015 (2015)
14. Boanjak, M., Oliveira, E., Martins, J., Rodrigues, E.M., Sarmento, L.: TwitterEcho: a distributed focused crawler to support open research with twitter data. In: WWW 2012, pp. 1233–1240. ACM (2012)
15. Saleiro, P., Rei, L., Pasquali, A., Soares, C.: Popstar at replab 2013: name ambiguity resolution on Twitter. In: CLEF 2013 Eval. Labs and Workshop Online Working Notes (2013)
16. Saleiro, P., Amir, S., Silva, M., Soares, C.: Popmine: tracking political opinion on the web. In: 2015 IEEE International Conference on Computer and Information Technology; Ubiquitous Computing and Communications; Dependable, Autonomic and Secure Computing; Pervasive Intelligence and Computing (CIT/IUCC/DASC/PICOM), pp. 1521–1526. IEEE (2015)

Multi-scale Kernel PCA and Its Application to Curvelet-Based Feature Extraction for Mammographic Mass Characterization

Sami Dhahbi[✉], Walid Barhoumi, and Ezzeddine Zagrouba

Research Team on Intelligent Systems in Imaging and Artificial Vision (SIIVA)
RIADI, Laboratory, ISI, Ariana, Tunisia
sami.dhahbi@laposte.net, walid.barhoumi@enicarthage.rnu.tn,
ezzeddine.zagrouba@fsm.rnu.tn

Abstract. Accurate characterization of mammographic masses plays a key role in effective mammogram classification and retrieval. Because of their high performance in multi-resolution texture analysis, several curvelet-based features have been proposed to describe mammograms, but without satisfactory results in distinguishing between malignant and benign masses. This paper tackles the problem of extracting a reduced set of discriminative curvelet texture features for mammographic mass characterization. The contribution of this paper is twofold. First, to overcome the weakness of PCA to cope with the nonlinearity of curvelet coefficient distributions, we investigate the use of kernel principal components analysis (KPCA) with a Gaussian kernel over curvelet coefficients for mammogram characterization. Second, a new multi-scale Gaussian kernel is introduced to overcome the shortcoming of single Gaussian kernels. Indeed, giving that faraway points may contain useful information for mammogram characterization, the kernel must emphasis neighbor points without neglecting faraway ones. Gaussian kernels either fail to emphasis neighborhood (high sigma values) or ignore faraway points (low sigma values). To emphasis neighborhood without neglecting faraway points, we propose to use a linear combination of Gaussian kernels with several sigma values, as a kernel in KPCA. Experiments performed on the DDSM database showed that KPCA outperforms state-of-the-art curvelet-based methods including PCA and moments and that the multi-scale gaussian kernel outperforms single gaussian kernels.

Keywords: Mammography · Kernel PCA · Curvelet transform · Multi-scale gaussian kernel

1 Introduction

Despite their effectiveness for early breast cancer detection, a major concern with nationwide mammogram screening programs is the high false-positive and false-negative rates. According to worldwide statistics in 2012 [1], this disease is the most frequent female cancer (25 % of all cancers) and the leading cause of cancer

© Springer International Publishing AG 2016
H. Boström et al. (Eds.): IDA 2016, LNCS 9897, pp. 183–191, 2016.
DOI: 10.1007/978-3-319-46349-0_16

deaths among women (14.71 % of total cancer deaths). Screening mammography is the most cost-effective tool to detect breast cancer at early stages, that increases treatment options and thus decreases mortality rates [2,3]. However, due to the difficulty of mammogram interpretation, radiologists may miss cancers and/or misinterpret non-cancerous lesions. Furthermore, the huge amount of mammograms to be analyzed and the limited number of qualified radiologists restrict double reading, even though it is clinically effective [4]. Hence, Computer-aided diagnosis (CAD) systems provide a more efficient solution by replacing the second reader and assist radiologists in their clinical routines [5]. Recently, content based image retrieval (CBIR) systems have emerged as a promising alternative to traditional CAD systems. On the one hand, radiologists' doubt about the output of black-box CAD systems, which analyze an input mammogram and return a decision about its malignancy, impede their widespread. On the other hand, by displaying similar mammograms with their corresponding ground truth, CBIR can gather radiologists' confidence. Many frameworks on mammogram retrieval have been proposed, dealing with relevance feedback [6], similarity measure [7], scalability [8], multi-view information fusion [9] and feature extraction [10,11]. In particular, feature extraction for accurate characterization of mammographic masses, a key step in effective mammogram retrieval, is both interesting and challenging. Because of their high performance in multi-resolution texture analysis, several curvelet-based features have been proposed in recent years to describe mammograms [10,12], but without satisfactory results in distinguishing between malignant and benign masses. This is mainly due to the difficulty of mammogram interpretation. For instance, Gedik et al. [12] performed a PCA analysis on curvelet sub-bands and used the first principal components to describe mammograms. By using the PCA analysis, the authors assumed a normal distribution of curvelet coefficients. However, the distributions of curvelet coefficients computed from mammogram images are non gaussian and therefore PCA analysis is not valid in this case [10]. To handle the non-gaussianity of curvelet coefficients, Dhahbi et al. [10] used higher order curvelet moments. Even though curvelet moments outperformed PCA-based curvelet analysis for mammogram description, the obtained results were not satisfactory and far from being usable in clinical application.

In this paper we deal with the problem of extracting a discriminative set of curvelet texture features for mammogram retrieval. For this, we propose to use kernel principal component analysis (KPCA). The gaussian KPCA overcomes the shortcoming of PCA to cope with the non-gaussianity of curvelet coefficient distributions. Furthermore, in the context of mammogram retrieval, faraway points are not as important as neighbor ones but they may contain valuable information. In this way, the kernel designed for mammogram retrieval must satisfy the two following conditions: (i) emphasize neighbor points, (ii) take into consideration faraway points. However, single gaussian kernels can satisfy either the first condition or the second one but not both. Indeed, high (*resp.* low) sigma values fail to emphasize neighborhood (*resp.* ignore faraway points). To emphasize neighborhood without neglecting faraway points, we propose a multi-scale gaussian kernel which is the combination of gaussian kernels with sigma values obtained through the successive division of a given sigma value. Experiments performed on the DDSM database showed that KPCA outperform state-of-the-art curvelet-based methods including PCA. Besides, the proposed multi-scale gaussian kernel outperforms single gaussian kernels. Indeed, the main contributions of this work can be summarized as follows. Firstly, we investigate the use of KPCA to extract a reduced set of discriminative features from curvelet coefficients for mammographic masses retrieval. Unlike traditional PCA, KPCA can effectively handle the non-gaussianity of curvelet coefficient distributions. Curvelet-based KPCA features outperform state-of-the-art curvelet-based features, including linear methods (curvelet-based PCA features) and nonlinear methods (curvelet moments). In addition, we propose a multi-scale gaussian kernel that emphasize neighborhood without neglecting faraway point, and thus it is more adopted for image retrieval. The proposed kernel is a mixture of several gaussian kernels with higher and lower sigma values.

The rest of this paper is organized as follows. Proposed methods are presented in Sect. 2. Section 3 is devoted for experimental results. Conclusion and future works are given in the last section.

2 Proposed Method

The proposed method is described here while presenting the curvelet transform, the kernel PCA and the multi-scale gaussian kernel.

2.1 Curvelet Transform

Curvelet is a multi-scale and multidirectional geometric transform that overcome the inherent limitations of wavelet-like transforms [13]. Compared with wavelet, curvelet exhibits desirable properties of directionality, anisotropy, efficient representation of smooth objects with discontinuities along curves, and optimally sparse representation. Indeed, several studies have shown the superiority of curvelet-based methods over wavelet-based ones for image denoising [14] and feature extraction [15]. The discrete curvelet coefficient of an image $f(x, y)$ is defined as follows:

$$c(j, l, k) = \sum_{x=0}^{N} \sum_{y=0}^{N} f(x, y) \overline{\varphi_{j,l,k}(x, y)} \tag{1}$$

where $\varphi_{j,l,k}(x, y)$ is the discrete mother curvelet; and j, l, and k are the parameters of scale, orientation and translation, respectively. To perform fast discrete curvelet transform, two implementations that differ in the way curvelets at a given scale and angle are translated with respect to each other, can be used. The first implementation is based on unequally spaced fast fourier transforms (USFFT), whereas the second one is based on the wrapping of specially selected fourier samples. In this study, we used the USFFT implementation since it is used in similar studies on curvelet-based mammogram analysis [10,12,15].

2.2 Kernel PCA

PCA is the most widely used tool for feature reduction that has been applied successfully in several areas. This method was used in [12] to extract a reduced set of features for mammogram analysis. However, PCA-based curvelet feature extraction does not yield good results in the case of mammogram analysis [10]. This is due to the fact that the distribution of curvelet coefficients is non-gaussian [10], whereas PCA allows only linear dimension reduction. To solve this problem, we propose to use KPCA to cope with the non-linearity of curvelet coefficient distributions. The performance of the KPCA depends on the appropriate choice of the kernel functions and their parameters. Commonly used kernels include polynomial kernels, gaussian kernels and sigmoid kernels. In particular, gaussian kernel (Eq. 2) exhibits several good properties (isotropy, stationary, ...) and outperforms all the other kernels. This kernel is by far the most widely used kernel and therefore is used here.

$$k_G^\sigma(x, y) = exp(-\frac{\|x - y\|^2}{2\sigma^2}), \tag{2}$$

where σ (sigma) is the width parameter. A large width allows to take into account faraway points, whereas a low width emphasizes neighbor points. The choice of the optimal value of σ is application dependent, and it is typically selected using cross-validation approaches.

2.3 Multiscale Gaussian Kernel PCA

In the previous section, we have argued why kernel PCA is more appropriate than linear PCA for curvelet coefficients reduction. In this section, we propose a kernel that is more suitable for mammogram retrieval than (single) gaussian kernel. A large kernel width will not emphasize neighborhood, whereas a low one will omit faraway points. To tackle this problem, we propose a multi-scale gaussian kernel. In fact, in image retrieval, the kernel function must emphasize the neighborhood without neglecting faraway points. Indeed, the neighbors of a point are more likely to be similar to this point. Therefore, the kernel function must emphasize the neighborhood. For faraway points, even though they are most likely to be dissimilar to the point in question, they could contain useful information about this point. Thus, they must not be neglected. Our idea is that weighting data points is better than simply discarding some of them. For this, a good kernel must satisfy two properties: (i) emphasize neighbor points, (ii) consider the all points. In gaussian kernel, the value of sigma defines how the kernel will handle neighbor and faraway points. A low value of sigma emphasizes the neighborhood and discards faraway points, whereas a high value of sigma takes into account all points but does not emphasize neighborhood (Fig. 1). Moderate values of sigma make a trade-off between satisfying property (i) or (ii), but not both at the same time (Fig. 1). Therefore, we propose a multi-scale gaussian kernel defined as follows:

$$k_{MSG}^\sigma(x, y) = \sum_{i=1}^{N} \lambda_i . k_G^{2^{-i}\sigma}(x, y), \tag{3}$$

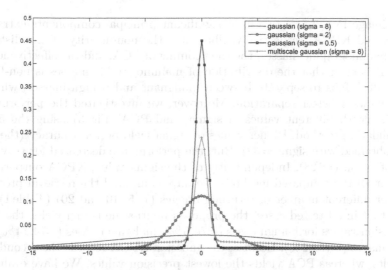

Fig. 1. Illustrations of single and multi-scale gaussian kernels.

where $k_G^{2^{-i}\sigma}$ is a gaussian kernel of width $2^{-i}\sigma$, and λ_i is a weighting parameter. The proposed kernel is a mixture of different gaussian kernels with higher, intermediate and lower sigma values. Thus, the multi-scale gaussian kernel is a positive definite kernel (a linear combination of positive definite kernels). In addition, the proposed kernel can simultaneously handle all points while emphasizing neighbor points (Fig. 1). Another advantage of the proposed multi-scale gaussian kernel is that it allows to mix several kernels while handling one parameter.

3   Experimental Results

In the experiments, we used the Digital Database for Screening Mammography (DDSM), which is currently the largest public mammogram database [16]. The dataset includes 2604 cases, and each case is composed of 4 mammograms. Screening mammography typically involves 2 views of the breast: Cranial-Caudal view (CC) and Medio-Lateral-Oblique view (MLO). Hence when we take into consideration left and right breast, we have 4 images for every case. The database includes the ground truth of each mammogram, mainly its diagnostic result (benign or malignant) and the location of lesions. Since the whole mammographic image comprises the pectoral muscle and the background with a lot of noise, features were computed on a limited ROI that contains the prospective abnormality, while removing the unwanted parts. This is a crucial step to extract and focus on the appropriate part of the mammography. In fact, we performed a manual crop-ping based on the information provided in the ground truth. Thus, the dataset consists of 1914 ROIs (862 benign and 1052 malignant). Figure 2 illustrates the first contribution of this paper, which is the use of KPCA as a more effective alternative than PCA for curvelet-based mammographic masses characterization. This figure, which visualized the projection of each image in the first two most significant principal components extracted by PCA and KPCA, respectively; illustrates the nonlinearity of the distribution of mammographic masses, the shortcoming of PCA, and the effectiveness of KPCA. It is clear that the distribution of mammographic masses is non-linear and the PCA fails to separate between malignant and benign masses, whereas KPCA allows a better separation. Moreover, we investigated the performance of KPCA (with different values of sigma) and PCA (Fig. 3) using the 5-fold cross validation method. Higher values of sigma yield better accuracy (the best results obtained with sigma $= 32$). But, the performance decreased for very high values of sigma ($=128$). Independently of the sigma value, KPCA outperforms PCA. For all the compared methods, we have computed the retrieval precision values for different number of returned images (1, 5, 10 and 20) (Table 1). We can see that in all tested cases, the proposed multi-scale kernel yields the highest precision values for benign cases ($=76.5\%$), malignant cases ($=74.14\%$) and overall images ($=75.82\%$). Besides, the gaussian KPCA records the second better results, whereas PCA yields the lowest precision values. We have evaluated also the classification performance via the accuracy, false positive and false negative values. Again, the multi-scale gaussian kernel achieves the highest accuracy ($=78.75\%$) and the lowest false-positive ($=16.37\%$) and false-negative values ($=24.07\%$). The obtained results show that KPCA (gaussian and multi-scale) permit to obtain better results than curvelet moments. This shows that KPCA is more appropriate than moment theory to handle non linear dimension reduction. The poor results obtained with the linear PCA confirm the incapability of this method to cope with non linear dimension reduction problems. Finally, the best results obtained with MS-KPCA and its superiority over gaussian KPCA confirms our assumption that a kernel must emphasize neighbor points without neglecting faraway ones (Table 2).

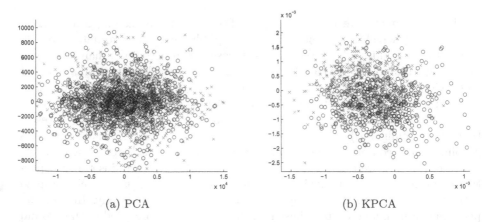

(a) PCA   (b) KPCA

Fig. 2. Distribution of malignant (red points) and benign (blue points) mammographic samples. (Color figure online)

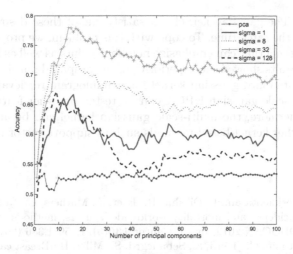

Fig. 3. Performance of gaussian KPCA with different values of sigma against the number of principal components. Accuracy computed using a KNN classifier (K = 10).

Table 1. Retrieval precision at different values of K.

	Benign				Malignant				Overall			
K	1	5	10	20	1	5	10	20	1	5	10	20
Moments	61.17	64.12	61.23	61.11	63.12	62.76	62.14	60.99	62.15	63.44	61.69	61.05
PCA	52.29	52.47	52.04	52.16	48.7	47.52	48.36	48.69	50.5	50.0	50.42	50.43
G-KPCA	69.87	69.54	67.84	64.29	73.67	70.76	67.76	62.6	71.77	70.15	67.8	63.45
MS-KPCA	76.5	74.65	73.64	72.25	75.14	72.77	71.6	68.4	75.82	73.71	72.62	70.32

Table 2. Classification accuracy at different values of K.

	False negatives				False positives				Accuracy			
K	1	5	10	20	1	5	10	20	1	5	10	20
Moments	38.83	34.87	.35.79	38.61	36.88	34.89	37.63	39.11	62.15	65.12	63.29	61.14
PCA	47.71	45.92	43.14	41.88	51.3	54.24	52.88	51.31	50.5	49.92	51.99	53.4
G-KPCA	30.13	27.71	27.8	26.82	26.33	25.54	25.2	28.14	71.77	73.38	73.5	72.52
MS-KPCA	23.5	20.81	16.37	16.86	24.86	24.18	25.2	25.65	75.82	77.51	78.28	78.75

4   Conclusion

In this paper, we proposed new feature extraction methods for mammographic masses retrieval. The proposed method used the multi-resolution texture analysis based on curvelet transform. To overcome the inherited limitation of linear PCA while handling non-linear dimension reduction, KPCA was used to extract a reduced set of discriminative features form curvelet coefficients. The obtained results showed that gaussian PCA-based feature extraction outperforms state-of-the-art methods that used PCA [12] and moment theory [10]. However, in mammogram retrieval, we must emphasize neighbor points without neglecting faraway points. The gaussian kernel can satisfy one of these desired properties but not both at the same time. To cope with this problem, we proposed a multi-scale gaussian kernel that can emphasize neighbor points as well as faraway ones. Experimental results showed the superiority of the proposed multi-scale gaussian kernel over the standard gaussian kernel for mammogram retrieval. As a future work, the multi-scale gaussian KPCA can be tested on other feature reduction problems. Furthermore, the multi-scale gaussian kernel can be an appropriate candidate for other kernel-based methods such as support vector machine.

References

1. Ferlay, J., Soerjomataram, I., Dikshit, R., Eser, S., Mathers, C., Rebelo, M., Bray, F.: Cancer incidence and mortality worldwide: sources, methods and major patterns in GLOBOCAN 2012. Int. J. Cancer **136**(5), E359–E386 (2015)
2. Hofvind, S., Ursin, G., Tretli, S., Sebudegrd, S., Mller, B.: Breast cancer mortality in participants of the Norwegian breast cancer screening program. Cancer **119**(17), 3106–3112 (2013)
3. Nelson, H.D., Tyne, K., Naik, A., Bougatsos, C., Chan, B.K., Humphrey, L.: Screening for breast cancer: an update for the US preventive services task force. Ann. Intern. Med. **151**(10), 727–737 (2009)
4. Dinnes, J., Moss, S., Melia, J., Blanks, R., Song, F., Kleijnen, J.: Effectiveness and cost-effectiveness of double reading of mammograms in breast cancer screening: findings of a systematic review. Breast J. **10**(6), 455–463 (2001)
5. Astley, S.M., Gilbert, F.J.: Computer-aided detection in mammography. Clin. Radiol. **59**(5), 390–399 (2004)
6. Wei, C.H., Li, C.T.: Calcification descriptor and relevance feedback learning algorithms for content-based mammogram retrieval. In: Astley, S.M., Brady, M., Rose, C., Zwiggelaar, R. (eds.) IWDM 2006. LNCS, vol. 4046, pp. 307–314. Springer, Heidelberg (2006)
7. Bedo, M., dos Santos, D., Ponciano, M., de Azevedo, P., Traina, C.: Endowing a content-based medical image retrieval system with perceptual similarity using ensemble strategy. J. Digit. Imaging **29**(1), 22–37 (2016)
8. Jiang, M., Zhang, S., Li, H., Metaxas, D.N.: Computer-aided diagnosis of mammographic masses using scalable image retrieval. IEEE Trans. Biomed. Eng. **62**(2), 783–792 (2015)
9. Dhahbi, S., Barhoumi, W., Zagrouba, E.: Multi-view score fusion for content-based mammogram retrieval. In: 8th International Conference on Machine Vision (2015)
10. Dhahbi, S., Barhoumi, W., Zagrouba, E.: Breast cancer diagnosis in digitized mammograms using curvelet moments. Comp. Biol. Med. **64**, 79–90 (2015)
11. Gardezi, S., Faye, I., Eltoukhy, M.: Analysis of mammogram images based on texture features of curvelet sub-bands. In: 5th International Conference on Graphic and Image Processing (2014)
12. Gedik, N., Atasoy, A.: A computer-aided diagnosis system for breast cancer detection by using a curvelet transform. Turk. J. Electr. Eng. Comput. Sci. **21**(4), 1002–1014 (2013)
13. Candes, E.J., Donoho, D.L.: Curvelets: a surprisingly effective nonadaptive representation for objects with edges. Stanford Univ Ca Dept of Statistics (2000)
14. Starck, J.L., Cands, E.J., Donoho, D.L.: The curvelet transform for image denoising. IEEE Trans. Image Process. **11**(6), 670–684 (2002)

15. Eltoukhy, M.M., Faye, I., Samir, B.B.: A comparison of wavelet and curvelet for breast cancer diagnosis in digital mammogram. Comput. Biol. Med. **40**(4), 384–391 (2010)
16. Heath, M., Bowyer, K., Kopans, D., Kegelmeyer, W., Moore, R., Chang, K., Munishku-maran, S.: Current status of the digital database for screening mammography. In: Karssemeijer, N., Thijssen, M., Hendriks, J., van Erning, L. (eds.) Digital Mammography. Computational Imaging and Vision, vol. 13, pp. 457–460. Springer, Amsterdam (1998)

Weakly-Supervised Symptom Recognition
for Rare Diseases in Biomedical Text

Pierre Holat[1](✉), Nadi Tomeh[1], Thierry Charnois[1], Delphine Battistelli[2],
Marie-Christine Jaulent[3], and Jean-Philippe Métivier[4]

[1] LIPN, University of Paris 13, Sorbonne Paris Cité, Paris, France
pierre.holat@lipn.univ-paris13.fr
[2] MoDyCo, University of Paris Ouest Nanterre La Défense, Paris, France
[3] Inserm, Paris, France
[4] GREYC, University of Caen Basse-Normandie, Caen, France

Abstract. In this paper, we tackle the issue of symptom recognition for
rare diseases in biomedical texts. Symptoms typically have more complex and ambiguous structure than other biomedical named entities. Furthermore, existing resources are scarce and incomplete. Therefore, we
propose a weakly-supervised framework based on a combination of two
approaches: sequential pattern mining under constraints and sequence
labeling. We use unannotated biomedical paper abstracts with dictionaries of rare diseases and symptoms to create our training data. Our
experiments show that both approaches outperform simple projection
of the dictionaries on text, and their combination is beneficial. We also
introduce a novel pattern mining constraint based on semantic similarity
between words inside patterns.

Keywords: Information extraction · Pattern mining · CRF · Symptoms
recognition · Biomedical texts

1 Introduction

Orphanet encyclopedia is the reference portal for information on rare diseases
(RD) and orphan drugs. A rare disease is a disease that affects less than 1 over
2,000 people. There are between 6,000 and 8,000 rare diseases and 30 million
people are concerned in Europe. The Orphanet initiative aims to improve the
diagnosis, care and treatment of patients with such diseases. Among its activities, Orphanet maintains a rare disease database containing expert-authored and
peer-reviewed syntheses describing current knowledge about each disease. The
syntheses are produced by human specialists following a manual, time-consuming
monitoring of the medical literature. The aim of our work is to automatically
acquire new knowledge related to rare diseases; we focus on the task of *symptom
recognition* in medical publication abstracts.

We use the term *symptom* to refer to features of a disease, as noticed and
described by a patient (functional sign), or as observed by a healthcare profes-
sional (clinical sign) without distinction. The linguistic structure of symptoms is

© Springer International Publishing AG 2016
H. Boström et al. (Eds.): IDA 2016, LNCS 9897, pp. 192–203, 2016.
DOI: 10.1007/978-3-319-46349-0_17

typically more complex than other biomedical named entities [3] for various reasons as discussed in [10]. They manifest a considerably larger variability in forms, ranging from simple nouns to whole sentences, and a larger number of syntactic and semantic ambiguities. In the following examples, symptoms as identified by an expert are shown in bold:

- With disease progression patients additionally develop **weakness** and **wasting of the limb and bulbar muscles**.
- Diagnosis is based on clinical presentation, and **glycemia and lactacidemia levels after a meal (hyperglycemia and hypolactacidemia), and after three to four hour fasting (hypoglycemia and hyperlactacidemia)**.

Furthermore, few works have focused on symptom recognition and therefore existing resources are limited and incomplete: to our knowledge, no dataset that is fully annotated with symptoms is available to allow for supervised learning.

To address these issues, we propose a weakly-supervised approach to symptom recognition that combines three independent lexical resources (Sect. 2.1): a corpus of unannotated medical paper abstracts and two dictionaries, one for rare diseases and another for symptoms.

We project the dictionaries on the abstracts to create an annotated dataset to train subsequent models. Since the dictionaries are not exhaustive, the annotation is only partial (weak). Given the annotated dataset, we formalize the problem of symptom recognition in two complementary ways: (a) as supervised sequence labeling for which we use Conditional Random Fields (CRF) [8] (Sect. 2.3); and (b) as sequential pattern mining under constraints [2] (Sect. 2.4). We combine these approaches in a pipeline architecture (Sect. 2).

Our contribution is threefold. First, we show experimentally (Sect. 3) that sequence labeling and pattern mining are both adequate formalizations of the task, for they outperform a simple projection of the dictionaries on the text; Furthermore, we show that their combination is beneficial since CRFs allow for rich representation of words while pattern mining privileges modeling their context. Second, we introduce a novel pattern mining constraint based on distributional similarities between words (Sect. 2.4). Third, we created a gold standard for evaluation (Sect. 2.1), manually annotated with symptoms by human experts.

2   A Pipeline Architecture for Symptom Recognition

In this section we describe our iterative pipeline approach to symptom recognition, in the spirit of [11]. Figure 1 depicts the overall architecture of our system.

Input data to our approach contain a collection of unannotated article abstracts and two dictionaries, one for diseases and one for symptoms. First, the dictionaries are projected on the abstracts to obtain partial annotations; the resulting annotated data is used to train a CRF sequence labeler and as an input to a sequential pattern mining algorithm; the learned CRF model and the extracted patterns are used to extract symptoms from the test data, separately or in combination; these symptoms are then compared to manually annotated

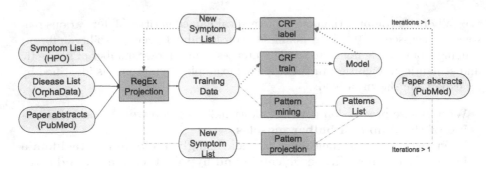

Fig. 1. Overall architecture of the system.

gold standard for evaluation using F-measure. It should be noted that the learned models can be applied on the training data (using cross-validation) to discover new symptoms to be added to the dictionary, and the whole process can be iterated.

2.1 Datasets and Evaluation

The input to our system is composed of three independent online resources:

- The first dataset is a corpus of 10,000 article abstracts extracted from the biomedical literature available on PubMed.[1] To build it, we extracted 100 biomedical paper abstracts for each one of 100 rare diseases selected from OrphaData in advance by an expert;
- The second dataset is a dictionary of 17,469 distinct phenotype anomalies provided by the Human Phenotype Ontology (HPO).[2] A phenotype is all the observable characteristics of a person, such as their morphology, biochemical or physiological properties. It results from the interactions between a genotype (expression of an organism's genes) and its environment. Since many rare diseases are genetic, we follow [11] and consider the above anomalies to be symptoms. This dictionary is not exhaustive;
- The third dataset is a dictionary of 16,576 distinct names of rare disease and their aliases, provided by OrphaData,[3] a comprehensive, high-quality resource related to rare diseases and orphan drugs. This dictionary is not exhaustive.

The testing data are made of 50 biomedical paper abstracts with an average of 184 word by abstract. A first automatic annotation was made on the data, then we ask two medical experts to review the generated symptoms and add missing ones to build a gold standard. Our experts have labeled 407 symptoms and their position in the testing data, so an average of 8,1 symptom by abstract.

---

[1] http://www.ncbi.nlm.nih.gov/pubmed.
[2] http://human-phenotype-ontology.github.io.
[3] http://www.orphadata.org.

Performance is evaluated using the standard precision, recall and F-measure.[4]

2.2   Weak Annotation by Projection

Given the input datasets, the first step in our workflow is to project the dictionaries of symptoms and diseases on the abstracts contained in our trainings set. The projection step produce only weak (partial) annotation since Orpha-Data and HPO lists are not exhaustive; they do not contain all symptoms and diseases, nor the various linguistic forms they can take.

The corpus and dictionaries are preprocessed using TreeTagger:[5] texts are tokenized, and each token is lemmatized and part-of-speech (POS) tagged. Each term in the dictionaries (possibly composed of several tokens) is matched against the corpus by comparing using regular expressions. Terms coming from HPO are often generic (e.g. "weakness") and may be supplemented in medical texts with adjectives or object complements (e.g. "severe weakness of the tongue"). Thus, once a term matches, it can be expanded to its nominal phrase using the POS tags assigned to surrounding terms.

The partially annotated corpus resulting from the projection is used as a training set for the subsequent models.

2.3   Symptom Recognition as Sequence Labeling

We formalize the problem of symptom recognition as a supervised sequence labeling problem with BIO notation, for which we use Conditional Random Fields (CRF) [8]. An abstract text is seen as a sequence of words, each of which is labeled with one of three possible labels: B (beginning of a symptom), I (inside a symptom), and O (outside a symptom). Thus, a symptom is a word segment corresponding to a label B potentially followed by consecutive I labels, as shown in Fig. 2.

O   B   I   O   O   B   I   I   O
clinically **silent tumors** often demonstrate **subclinical hormonal activity** .

Fig. 2. Symptom recognition as BIO sequence labeling. Symptoms are bolded.

The main advantage of CRFs is their conditional nature which allows for rich representations of words in a sequence. It is possible to incorporate multiple information sources in the form of feature functions, without having to model their interactions explicitly. On the contrary, applying sequential pattern mining on such rich representations is prohibitively intractable.

We use the same set of features used for named entity recognition in [5] as implemented in Stanford NER recognizer.[6] They include the current word, the

---

[4] Using the script provided by http://www.cnts.ua.ac.be/conll2000/chunking/output. html which take the same input data format (BIO) as our data.

[5] http://www.cis.uni-muenchen.de/~schmid/tools/TreeTagger.

[6] http://nlp.stanford.edu/software/CRF-NER.shtml.

previous and next one, as well as all the words in window of a given size (n-gram features); orthographic features characterizing the form of the words; prefixes and suffixes; and several conjunctions thereof.

2.4 Symptom Recognition as Sequential Pattern Mining

While the sequence labeling CRF approach presented above can easily employ rich representation of words, it cannot efficiently capture rich context. For instance, considering all possible sub-sequences as context for the current word is computationally intractable because their number grows exponentially in the size of the sequence.

To address this issue, we propose to use a sequential pattern mining approach with an emphasis on the context more than the words themselves. Sequential pattern mining allows to take into account the language sequentiality and is one of the most studied and challenging task in data mining. Since its introduction by Agrawal and Srikant [1], the problem has been well formalized:

Let $\mathcal{I} = \{i_1, i_2 \ldots i_m\}$ be the finite set of items. An itemset is a non-empty set of items. A sequence S over $\mathcal{I}$ is an ordered list $\langle it_1, \ldots, it_k \rangle$, with it_j an itemset over $\mathcal{I}$, $j = 1 \ldots k$. A k-sequence is a sequence of k items (i.e., of length k), $|S|$ denotes the length of sequence S. $\mathbb{T}(\mathcal{I})$ will denote the (infinite) set of all possible sequences over $\mathcal{I}$. A *sequence database* $\mathcal{D}$ over $\mathcal{I}$ is a finite set of doubles (SID, T), called transactions, with $SID \in \{1, 2, \ldots\}$ an identifier and $T \in \mathbb{T}(\mathcal{I})$ a sequence over $\mathcal{I}$.

Definition 1 (*Inclusion*). *A sequence $S' = \langle is'_1\ is'_2\ \ldots\ is'_n \rangle$ is a subsequence of another sequence $S = \langle is_1\ is_2\ \ldots\ is_m \rangle$, denoted $S' \preceq S$, if there exist $i_1 < i_2 < \ldots i_j \ldots < i_n$ such that $is'_1 \subseteq is_{i_1}$, $is'_2 \subseteq is_{i_2} \ldots is'_n \subseteq is_{i_n}$.*

Definition 2 (*Support*). *The support of a sequence S in a transaction database $\mathcal{D}$, denoted $Support(S, \mathcal{D})$, is defined as: $Support(S, \mathcal{D}) = |\{(SID, T) \in \mathcal{D} | S \preceq T\}|$. The frequency of S in $\mathcal{D}$, denoted $freq_S^{\mathcal{D}}$, is $freq_S^{\mathcal{D}} = \frac{Support(S, \mathcal{D})}{|\mathcal{D}|}$.*

Given a user-defined minimal frequency threshold σ, the problem of sequential pattern mining is the extraction of all the sequences S in $\mathcal{D}$ such that $freq_S^{\mathcal{D}} \geq \sigma$. The set of all frequent sequences for a threshold σ in a database $\mathcal{D}$ is denoted $FSeqs(\mathcal{D}, \sigma)$,

$$FSeqs(\mathcal{D}, \sigma) = \{S \mid freq_S^{\mathcal{D}} \geq \sigma\} \tag{1}$$

Our data contains the lemma and the POS of each word. So in our context, let $\mathcal{I}$ be the finite set of all words and part-of-speech tag. An itemset is a non-empty set of the lemma and the part-of-speech of a word. We also add a special item (#symptom#) in $\mathcal{I}$. This item will be used as a placeholder for each annotated symptom in the training data, as shown in the following example:

$< \{we, PP\}\{find, VBD\}\{that, IN\}\{clinically, RB\}\{\text{#symptom#}\}$
$\{often, RB\}\{demonstrate, VBP\}\{\text{#symptom#}\}\{., SENT\} >$

Using sequential pattern mining on such sequences allows us to extract linguistic patterns covering symbolic symptoms. However, using only the user-defined minimal frequency threshold σ as a constraint, pattern mining typically yields an exponential number of patterns. Pattern mining under constraints [17] is a powerful paradigm to target relevant patterns [14]. Therefore, we used a pattern mining algorithm under the most used constraints in the literature in addition to σ. The *minimal and maximal gap* constraint imposes a limit on the number of words separating items of a pattern. The *minimal and maximal length* constraint limits the number of items in a pattern. We also used a *belonging constraint* specific to our task, a pattern must contain our specific item #symptom#.

Semantic Similarity Constraint. In addition to the above-mentioned constraints, we introduce a new *semantic similarity constraint* based on the distributional properties of the words, estimated from a large unannotated corpus.

We observed during initial experiments a high level of redundancy in extracted patterns, such as a succession of conjunctions or prepositions for instance. We designed a constraint with a limit on the similarity of two adjacent items of a pattern. Therefore, this constraint is designed to discard redundant, uninformative patterns.

To be able to quantify the level of redundancy in the pattern, we used the distributional hypothesis [6]: words that occur in the same context tend to have similar meanings, this hypothesis is the basis for models like *Word2Vec* [12,13], which learns a low-dimensional continuous vector representation of words from large amount of text. To train the *Word2Vec* model, we extracted 7,031,643 biomedical paper abstract from PubMed, that's 8.7 GB of input data for a final model of 1.10 GB containing 1,373,138 words in a biomedical context. The learned *Word2Vec* model is loaded by the data mining algorithm, each item i will have an associated vector V_i allowing to measure the cosine distance $D_c(V_i, V_j)$ between two consecutive items.

Definition 3 (Semantic Similarity Constraint). *Given a user-defined maximal similarity threshold ζ. Let i_d the last item of a sequence S, an extension of S by a new item i_n is possible only if :*

$$D_c(V_{i_d}, V_{i_n}) \leqslant \zeta \tag{2}$$

The semantic similarity constraint is anti-monotonic, the Proposition 1 allows an efficient pruning of the search space.

*Property 1 (**Effect of the anti-monotonic semantic similarity constraint**). Let S be a sequence. If S does not respect the semantic similarity constraint, it does not exist a sequence S' with $S \preceq S'$ which will respect the constraint. Therefore, the search space of all the extensions of S can be pruned.*

3   Experiments and Results

In this section we evaluate the performance of our system on the test set. We evaluate each component individually, as long as their combination. We also evaluate the contribution of the new similarity-based constraint.

3.1   Individual Module Results

In this section we compare symptom recognition results, shown in Table 1, for each of the recognition modules independently of the other.

Table 1. Details of the best results with tuned parameters for each module

Module	Parameters	Precision	Recall	F-measure
Dictionary		**57.58**	14.00	22.53
CRF	Bag of words	**56.31**	14.25	22.75
CRF	ngrams, ngramLength=6	**56.14**	15.72	24.57
Pattern	σ=0.05 %, Gap=0, ζ=0.4	23.12	**38.57**	**28.91**

Dictionary Projection Results. We first created a baseline by projection of the 17,469 symptoms that we gathered in our dictionary on the testing data using regular expressions. Of all our results, this baseline have the best precision but the worst recall.

CRF Results. The CRF module, successfully learned to generalize from the symptoms of the dictionary as indicated by the 12 % point increase in recall and 2.5 % in increase in precision, which resulted in 9 % increase in F-measure.

Pattern Mining Results. Since we replaced each symptom by the item #symptom# in the training data, we only represent the context. A pattern like "(such,JJ) (as,IN) (#symptom#)" can be applied on the testing data and discover symptoms that were not in the training data. Hence, the Pattern mining module has an increase of 175 % of recall, but a 60 % decline in precision, for a final amelioration of the F-measure by 28 %. It was expected that the precision would drop, but in our context of helping human expert process data, the recall is more important to maximize: missing a potential new symptom is more harmful than producing a large number of false-positives. Figure 3 lists some of the extracted patterns.

Table 2 shows the difference between the annotation of each module and the annotation of the human expert on two sentences.

```
(treatment,NN) (IN) (#symptom#) : 62
(development,NN) (of,IN) (#symptom#) : 43
(patient,NNS) (with,IN) (#symptom#) : 295
(diagnosis,NN) (IN) (#symptom#) : 98
(patient,NNS) (IN) (#symptom#) : 306
(case,NN) (of,IN) (#symptom#) : 48
(such,JJ) (as,IN) (#symptom#) : 91
(IN) (patient,NNS) (IN) (#symptom#) : 163
(NNS) (such,JJ) (as,IN) (#symptom#) : 46
(in,IN) (patient,NNS) (IN) (#symptom#) : 89
(in,IN) (patient,NNS) (with,IN) (#symptom#) : 88
(IN) (patient,NNS) (with,IN) (#symptom#) : 161
```

Fig. 3. Examples of extracted patterns with their support.

Table 2. Examples of each module annotations.

	diffuse	palmoplantar	keratoderma	and	precocious
Expert Annotation	B	I	I	O	B
Pattern Annotation	B	I	I	I	I
CRF Annotation	B	I	I	O	O

	primary	immunodef.	disorders	with	residual	cell-mediated	immunity
Expert	B	I	I	O	O	O	O
Pattern	B	I	I	I	I	I	I
CRF	O	B	O	O	O	O	O

3.2   Impact of Training Data Size

Figure 4 shows the variation in precision, recall and F-measure for our modules with increasing size of training data. From 1.000 to 10.000 biomedical abstracts, there is no visible impact on the CRF, even if the best score, in term of F-Measure, is on the maximum size data. Pattern Mining improves its recall with more data, but the precision tends to drop. The best score is on the middle size data.

3.3   Model Combination Analysis

We clearly see on the Fig. 4 the difference in the precision/recall ratio of the numerical and the symbolic modules even if they are very close in terms of F-measure. Because the CRF maximizes the precision and the pattern mining maximizes the recall, we tried to do a combination of the best CRF module result and the best pattern mining module. If the two models made the same decision we keep it, if not we promote the choice of "B". This combination gives better results (cf. Table 3) than each modules separately. In the same logic,

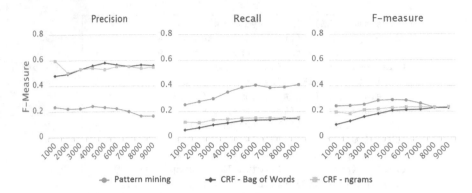

Fig. 4. Impact of the number of abstracts used for training

Table 3. Details of the best results for each module and their combination.

Module	Precision	Recall	F-measure
Dictionary (D)	**57.58**	14.00	22.53
CRF - ngram	**56.14**	15.72	24.57
Pattern mining	23.12	**38.57**	28.91
Combination	23.46	**39.31**	**29.38**
CRF - ngram + D	**56.90**	16.22	25.24
Pattern mining + D	23.35	**39.07**	29.23
Combination	23.46	**39.31**	**29.38**

we also combine our bests results with the baseline using the projection of the symptom dictionary. If that last enhancement has improved each module result individually, it did not improve the combination of the best CRF and pattern mining result.

3.4 Impact of Semantic Similarity Constraint

The purpose of the semantic similarity constraint is to reduce the number of patterns extracted without jeopardizing the classification accuracy, and like most data mining constraint the results are threshold dependent. Figure 5 shows the success of our constraint, a 30 % reduction of extracted patterns without losses in F-Measure. It also shows that a too strong threshold for the constraint lower the performance. With a maximal semantic similarity below 0.4 the constraint tends to produce very few patterns. In terms of cosine distance, a threshold of 0.2 is so low that semantically divergent words would be considered similar.

3.5 Iterative Learning Analysis

In this experiment, unlike the next results, the whole annotation process is iterated a number of times in an attempt to discover more symptoms. For the

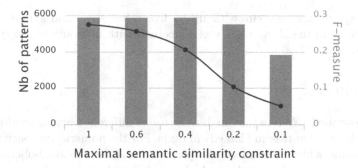

Fig. 5. Impact of the semantic similarity constraint (abscissa) on the number of extracted patterns (black line) and scoring (blue column). (Color figure online)

pattern mining module, there is no amelioration because after the first iteration, symptom placeholders cover more parts of the sequence, and the new symptoms are noisy. We almost never find a new symptom after the first iteration. An interesting perspective would be to try this iterative learning with different data set each time, by splitting the data into subsets. It is different for the CRF module, it learns more and more new symptoms at each iterations, but unfortunately it is also mainly noise because the precision tends to decrease.

4   Related Work

To our knowledge, there is no annotated dataset which can be used to train a supervised model specific for symptom recognition. Most of the studies are based on clinical reports or narrative corpora without symptom annotation and therefore can not be used in our context for symptom monitoring. Such corpora include the Mayo Clinic corpus [15] and the 2010i2b2/VA Challenge corpus [18]. Other existing biomedical datasets annotate only diseases; they include the NCBI disease corpus [4] which consists of 793 PubMed abstracts with 6,892 disease mentions and 790 unique disease concepts mapped to the Medical Subject Headings (MeSH),[7] and the Arizona Disease Corpus (AZDC) [9] which contains 2,784 sentences from MEDLINE abstracts annotated with disease mentions and mapped to the Unified Medical Language System (UMLS)[8].

Symptom recognition [10] is a relatively new task, often included in more general categories such as *clinical concepts* [19], *medical problems* [18] or *phenotypic information* [16]. Even on these categories, few studies take advantage of considering the linguistic context in which symptoms appear, and they are more focused on the linguistic analysis. In [15], the authors notice that most of the symptoms are given in relation with an anatomic location, while in [7] the authors state, after annotating their corpus with MeSH, that 75 % of the

---

[7] https://www.nlm.nih.gov/mesh/meshhome.html.
[8] https://www.nlm.nih.gov/research/umls/.

signs and symptoms co-occur with up to five other signs and symptoms in a sentence. None of the above work was concerned with fully automatic symptom recognition.

5   Conclusion

We have described a system that enable the use of different learning modules for a symptom recognition task in biomedical texts. For the numeric approach we used a CRF module which maximized the precision and for the symbolic approach we used a pattern mining module which maximize the recall. We introduced a new semantic constraint for the pattern mining process which remove redundant patterns without decline in the scoring. Both approach (symbolic and numerical) have been combined to further enhanced the results. A first future direction will be to enhance the combination of the modules. An idea is to use the patterns extracted as features for the CRF. A second future direction is to enhance our similarity constraint to take into account more distant redundancy in a pattern, or to apply the constraint differently in function of the words part-of-speech.

Acknowledgments. This work is supported by the French National Research Agency (ANR) as part of the project Hybride ANR-11-BS02-002 and the "Investissements d'Avenir" program (reference: ANR-10-LABX-0083).

References

1. Agrawal, R., Srikant, R.: Mining sequential patterns. In: Proceedings of the Eleventh International Conference on Data Engineering, pp. 3–14 (1995)
2. Béchet, N., Cellier, P., Charnois, T., Crémilleux, B.: Sequence mining under multiple constraints. In: Proceedings of the 30th Annual ACM Symposium on Applied Computing, pp. 908–914 (2015)
3. Cohen, K.B.: BioNLP: biomedical text mining. In: Handbook of Natural Language Processing, 2nd edn. (2010)
4. Doğan, R.I., Leaman, R., Lu, Z.: NCBI disease corpus: a resource for disease name recognition and concept normalization. J. Biomed. Inf. **47**, 1–10 (2014)
5. Finkel, J.R., Grenager, T., Manning, C.: Incorporating non-local information into information extraction systems by gibbs sampling. In: Proceedings of the 43rd Annual Meeting on Association for Computational Linguistics, pp. 363–370 (2005)
6. Harris, Z.S.: Distributional structure. Word **10**(2–3), 146–162 (1954)
7. Kokkinakis, D.: Developing resources for swedish bio-medical text mining. In: Proceedings of the 2nd International Symposium on Semantic Mining in Biomedicine (SMBM) (2006)
8. Lafferty, J., McCallum, A., Pereira, F.C.: Conditional random fields: probabilistic models for segmenting and labeling sequence data (2001)
9. Leaman, R., Miller, C., Gonzalez, G.: Enabling recognition of diseases in biomedical text with machine learning: corpus and benchmark. In: Proceedings of the 2009 Symposium on Languages in Biology and Medicine, vol. 82(9) (2009)
10. Martin, L., Battistelli, D., Charnois, T.: Symptom extraction issue. In: Proceedings of BioNLP 2014, pp. 107–111 (2014)

11. Métivier, J.P., Serrano, L., Charnois, T., Cuissart, B., Widlöcher, A.: Automatic symptom extraction from texts to enhance knowledge discovery on rare diseases. In: Holmes, J.H., Bellazzi, R., Sacchi, L., Peek, N. (eds.) Artificial Intelligence in Medicine. LNCS, vol. 9105, pp. 249–254. Springer, Heidelberg (2015). doi:10.1007/978-3-319-19551-3_33

12. Mikolov, T., Chen, K., Corrado, G., Dean, J.: Efficient estimation of word representations in vector space. arXiv preprint arXiv:1301.3781 (2013)

13. Mikolov, T., Sutskever, I., Chen, K., Corrado, G.S., Dean, J.: Distributed representations of words and phrases and their compositionality. In: Advances in Neural Information Processing Systems, pp. 3111–3119 (2013)

14. Pei, J., Han, J., Wang, W.: Constraint-based sequential pattern mining: the pattern-growth methods. J. Intell. Inf. Syst. **28**(2), 133–160 (2007)

15. Savova, G.K., Masanz, J.J., Ogren, P.V., Zheng, J., Sohn, S., Kipper-Schuler, K.C., Chute, C.G.: Mayo clinical text analysis and knowledge extraction system (ctakes): architecture, component evaluation and applications. J. Am. Med. Inf. Assoc. **17**(5), 507–513 (2010)

16. South, B.R., Shen, S., Jones, M., Garvin, J., Samore, M.H., Chapman, W.W., Gundlapalli, A.V.: Developing a manually annotated clinical document corpus to identify phenotypic information for inflammatory bowel disease. BMC Bioinform. **10**(9), 1 (2009)

17. Srikant, R., Agrawal, R.: Mining sequential patterns: generalizations and performance improvements. In: Proceedings of the 5th International Conference on Extending Database Technology: Advances in Database Technology, pp. 3–17 (1996)

18. Uzuner, Ö., South, B.R., Shen, S., DuVall, S.L.: 2010 i2b2/va challenge on concepts, assertions, and relations in clinical text. J. Am. Med. Inf. Assoc. **18**(5), 552–556 (2011)

19. Wagholikar, K.B., Torii, M., Jonnalagadda, S.R., Liu, H.: Pooling annotated corpora for clinical concept extraction. J. Biomed. Semant. **4**(1), 1–10 (2013)

Estimating Sequence Similarity from Read Sets for Clustering Sequencing Data

Petr Ryšavý[(✉)] and Filip Železný

Department of Computer Science, Faculty of Electrical Engineering,
Czech Technical University in Prague, Prague, Czech Republic
{rysavpe1,zelezny}@fel.cvut.cz

Abstract. Clustering biological sequences is a central task in bioinformatics. The typical result of new-generation sequencers is a set of short substrings ("reads") of a target sequence, rather than the sequence itself. To cluster sequences given only their read-set representations, one may try to reconstruct each one from the corresponding read set, and then employ conventional (dis)similarity measures such as the edit distance on the assembled sequences. This approach is however problematic and we propose instead to estimate the similarities directly from the read sets. Our approach is based on an adaptation of the Monge-Elkan similarity known from the field of databases. It avoids the NP-hard problem of sequence assembly and in empirical experiments it results in a better approximation of the true sequence similarities and consequently in better clustering, in comparison to the first-assemble-then-cluster approach.

Keywords: Read sets · Similarity · Hierarchical clustering

1 Introduction

Sequencing means reading the sequence of elements that constitute a polymer, such as the DNA. The *human genome project* [5] completed in 2003 was a prime example of sequencing, resulting in the identification of almost the entire genomic sequence (over 3 billion symbols) of a single human. Sequencing becomes technologically difficult as the length of the read sequence grows. The common principle of *new-generation sequencing* (NGS) is that only very short substrings (10's to 100's of symbols) are read at random positions of the sequence of interest. It is usually required that the number of such read substrings, called *reads*, is such that with high probability each position in the sequence is contained in multiple reads; the number of such reads is termed *coverage*. The complete sequence is determined by combinatorial assembly of the substrings guided by their suffix-prefix overlaps. For example, one possible assembly of reads {AGGC, TGGA, GCT} is AGGCTGGA. Short reads imply low cost of wet-lab sequencing traded off with high computational cost of assembly. Indeed the assembly task can be posed as searching the Hamiltonian path in a graph of mutual overlaps.

One of the central tasks in computational biology is to infer phylogenetic trees, which typically amounts to hierarchical clustering of genomes. When they

© Springer International Publishing AG 2016
H. Boström et al. (Eds.): IDA 2016, LNCS 9897, pp. 204–214, 2016.
DOI: 10.1007/978-3-319-46349-0_18

are represented only through sequences read-sets, the bioinformatician is forced to reconstruct the sequences from the read-sets prior to clustering. This of course entails the solution of the NP-hard assembly problem for each data instance with little guarantees regarding the quality of the resulting putative sequence. This motivates the question whether the assembly step could be entirely avoided. We address this question here by proposing a similarity function computable directly on the read sets, that should approximate the true similarities on the original sequences.

Related work includes studies on clustering NGS data (e.g. [1,7]). They however deal with clustering *reads* and we are not aware of a previous attempt to cluster *read-sets*. The paper [16] proposes a similarity measure for NGS data, but again it operates on the level of reads. The previous work [4,13] also aims at avoiding the assembly step in learning from NGS data but these studies concern supervised classification learning and they do not elaborate on read-set similarity.

In the following section, we design the similarity (or, reversely) distance function. Then we provide a brief theoretical analysis of it. In Sect. 4 we compare it to the conventional approach on genomic data and then we conclude the paper.

2   Distance Function Design

The functor $|.|$ will denote the absolute value, cardinality and length (respectively) for a number, set, and string argument. Let $\text{dist}(A, B)$ denote the Levenshtein distance [6] between strings A and B. The function measures the minimum number of edits (insertions, deletions, and substitutions) needed to make the strings identical, and is a typical example of a sequence dissimilarity measure used in bioinformatics. It is a property of the distance that

$$\text{dist}(A, B) \leq \max\{|A|, |B|\}. \tag{1}$$

We will work with constants $l \in \mathbb{N}$, $\alpha \in \mathbb{R}$ called the *read length* and *coverage*, respectively, which are specific to a particular sequencing experiment. A *read-set* R_A of string A such that

$$|A| \gg l \tag{2}$$

is a multiset of[1]

$$|R_A| = \frac{\alpha}{l}|A| \tag{3}$$

substrings sampled i.i.d. with replacement from the uniform distribution on all the $|A| - l + 1$ substrings of length l of A. Informally, the coverage α indicates the average number of reads covering a given place in A.

Our goal is to propose a distance function $\text{Dist}(R_A, R_B)$ that approximates $\text{dist}(A, B)$ for read-sets R_A and R_B of arbitrary strings A and B. We also want $\text{Dist}(R_A, R_B)$ to be more accurate and less complex to calculate than a natural estimate $\text{dist}(\hat{A}, \hat{B})$ in which the arguments represent putative sequences reconstructed from R_A and R_B using *assembly algorithms* such as [3,11,15].

---

[1] Should the right hand side be non-integer, we neglect its fractional part.

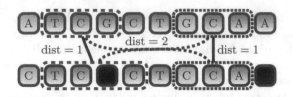

Fig. 1. We calculate read-read distances in order to find matching pairs of reads. For each read from the first sequence we find the least distant read in the second sequence. We see optimal alignment of ATCGCTGCAA and CTCCTCCA. Read TCG is paired with TCC.

2.1 Base Case: Which Reads Belong Together

A natural approach to instantiate $\text{Dist}(R_A, R_B)$ is to exploit the $|R_A||R_B|$ pairwise Levenshtein distances between the reads in R_A and R_B. Most of those values are useless because they match reads from completely different parts of sequences A and B. Therefore we want to account only for those pairs which likely belong together.

If we seek a read from R_B that matches a read $a_i \in R_A$, we make the assumption that the most similar read $b_j \in R_B$ is the one that we look for (see Fig. 1), i.e.,

$$b_j = \arg\min_{b_k \in R_B} \text{dist}(a_i, b_k).$$

To calculate the distance from R_A to R_B, we average over all reads from R_A:

$$\text{Dist}_{\text{ME}}(R_A, R_B) = \frac{1}{|R_A|} \sum_{a_i \in R_A} \min_{b_j \in R_B} \text{dist}(a_i, b_j). \tag{4}$$

This idea was presented in [8] for searching duplicates in database systems. The method is known as the *Monge-Elkan similarity*[2] (hence the ME label) and entails a simple but effective approximation algorithm.

$\text{Dist}_{\text{ME}}(R_A, R_B)$ is non-symmetric in general, which is undesirable given that the approximated distance $\text{dist}(A, B)$ is known to be symmetric. Therefore we define a symmetric version by averaging both directions

$$\text{Dist}_{\text{MES}}(R_A, R_B) = \frac{1}{2}\left(\text{Dist}_{\text{ME}}(R_A, R_B) + \text{Dist}_{\text{ME}}(R_B, R_A)\right). \tag{5}$$

2.2 Distance Scale

Consider duplicating a non-empty string A into AA and assume $R_{AA} = R_A \cup R_A$. Typically for a B similar to A we expect that $\text{dist}(AA, B) > \text{dist}(A, B)$ but the

---

[2] Here we alter the Monge-Elkan similarity into a distance measure. The standard way of using Monge-Elkan is as a similarity measure with min replaced by max and distance calculation by similarity calculation.

(symmetric) Monge-Elkan distance will not change, i.e. $\text{Dist}_{\text{MES}}(R_{AA}, R_B) = \text{Dist}_{\text{MES}}(R_A, R_B)$, indicating a discrepancy that should be rectified.

In fact, Dist_{MES} has the constant upper bound l, which is because it is the average (c.f. (4) and (5)) of numbers no greater than l (see (1)). On the other hand, $\text{dist}(A, B)$ has a non-constant upper bound $\max\{|A|, |B|\}$ as by (1).

To bring $\text{Dist}_{\text{MES}}(A, B)$ on the same scale as $\text{dist}(A, B)$, we should therefore multiply it by the factor $\max\{|A|, |B|\}/l = \max\{|A|/l, |B|/l\}$. By (3) we have $|A| = \frac{l}{\alpha}|R_A|$, yielding the factor $\max\{|R_A|/\alpha, |R_B|/\alpha\}$, in which α is a constant divisor which can be neglected in a distance function. Therefore, we modify the read distance into

$$\underset{\text{MESS}}{\text{Dist}} = \max\{|R_A|, |R_B|\} \underset{\text{MES}}{\text{Dist}}. \tag{6}$$

2.3 Margin Gaps

Consider the situation in Fig. 2 showing two identical sequences each with one shown read. The Levenshtein distance between the two reads is non-zero due to the one-symbol trailing (leading, respectively) gap of the top (bottom) read caused only by the different random positions of the reads rather than due to a mismatch between the sequences. Thus there is an intuitive reason to pardon margin gaps up to certain size t

$$t < \frac{l}{2} \tag{7}$$

when matching reads. Here, t should not be too large as otherwise the distance could be nullified for pairs of long reads with small prefix-suffix overlaps, which would not make sense.

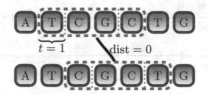

Fig. 2. Because reads locations in sequences are random, we do not want to penalize small leading or trailing gaps.

To estimate a good value for t, consider sequence A and its sampled read-set R_A. We now sample an additional read a of length l from A. Ideally, there should be a zero-penalty match for a in R_A as a was sampled from the same sequence as R_A was. This happens iff there is a read in R_A sampled from the same position in A as a, or from a position shifted by up to t symbols to the left or right as then the induced gaps are penalty-free. Since R_A is an uniform-probability i.i.d. sample from A, the probability that a given read from R_A starts at one of these $1 + 2t$ positions is[3] $\frac{1+2t}{|A|}$. We want to put an upper bound $\varepsilon > 0$ on the probability that this happens for none of the $|R_A| = \frac{\alpha}{l}|A|$ reads in R_A:

$$p = \left(1 - \frac{1 + 2t}{|A|}\right)^{\frac{|A| \cdot \alpha}{l}} \leq \varepsilon \ .$$

Consider the first-order Taylor approximation $(1 + x)^n = 1 + nx + \varepsilon'$ where the difference term $\varepsilon' > 0$ decreases with decreasing $|x|$. Due to Ineq. (7) and (2), $\frac{1+2t}{|A|}$ is small and we can apply the approximation on the above formula for p, yielding

$$p = 1 - \frac{2t + 1}{|A|} \frac{|A| \cdot \alpha}{l} + \varepsilon' = 1 - (2t + 1)\frac{\alpha}{l} + \varepsilon' \leq \varepsilon \ .$$

For simplicity, we choose $\varepsilon = \varepsilon'$. The smallest gap size t for which the inequality is satisfied is obtained by solving $1 - (2t + 1)\frac{\alpha}{l} = 0$, yielding

$$t = \frac{1}{2}\left(\frac{l}{\alpha} - 1\right).$$

This choice of t matches intuition in that with larger read-length l we can allow a larger grace gap t but with larger coverage α, t needs not be so large as there is a higher chance of having a suitably positioned read in the read-set. Another way to look at it is to realize that reads in a read-set are approximately $\frac{l}{\alpha}$ positions from each other. Consider matching read a to reads from R_A. If there is a read $a_1 \in R_A$ requiring gap larger than $\frac{l}{2\alpha}$ to match a, then there will typically be another read $a_2 \in R_A$ requiring gap at most $\frac{l}{2\alpha}$ (see Fig. 3).

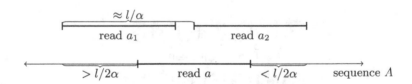

Fig. 3. Illustration to reasoning in Sect. 2.3

We implemented the grace margin gaps into a further version $\text{Dist}_{\text{MESSG}}$ of the constructed distance function, which required only a small change to the standard Wagner-Fischer algorithm [14].[4] When the algorithm is filling the first

---

[3] Strictly speaking, this reasoning is incorrect if read a is drawn from a place close to A's margins, more precisely, if it starts in fewer than t ($t + l$, respectively) symbols from A's left (right) margin, as then not all of the $2t$ shifts are possible. This is however negligible due to Ineq. (2).

[4] The dynamic programming algorithm for calculating the Levenshtein distance [6] is commonly called Wagner-Fischer algorithm [14]. When we refer to sequence alignment problem in bioinformatics, this algorithm is often called Needleman-Wunsch algorithm [9].

or the last row and column of the table, margin gaps up to t symbols are not penalized. Larger margin gaps are penalized in a way that satisfies the constraint that the distance between a word a and an empty word is $|a|$. In particular, the standard linear gap penalty is replaced with a piecewise linear function that gives cost of margin gap at x-th position

$$g(x) = \begin{cases} 0, & \text{if } 0 \leq x \leq t - 1, \\ 2\frac{x-t+1}{l+1-2t}, & \text{if } t - 1 < x \leq l - t, \\ 2, & \text{if } l - t < x < l. \end{cases} \tag{8}$$

2.4 Missing Read

Sometimes there is no good match for read a_i in R_B. During evolution the substring that contained a_i may have been inserted into A or may have vanished from B. Therefore if

$$\text{dist}(a_i, b_j) \geq \theta$$

for some reads a_i and b_j and threshold θ, we consider a_i and b_j to be dissimilar and we force their distance to be l (See Fig. 4).

Threshold θ should be a linear factor of the maximal distance between two sequences of length l, i.e. $\theta = \theta' \cdot l$. Value of θ' should reflect the probability that the read is in one sequence and not in the other. Because the true probability is hidden, it needs to be determined empirically.

The distance function equipped with the missing read detection as described gives rise to the last version denoted as $\text{Dist}_{\text{MESSGM}}$.

3 Theoretical Analysis

3.1 Asymptotic Complexity

Calculating $\text{dist}(A, B)$ for sequences A and B requires $\Theta(|A||B|)$ operations if we use the standard Wagner-Fischer dynamic programming algorithm [14]. This algorithm also requires $\Theta(\min(|A|, |B|))$ memory as we are interested only in distance and not in the alignment. To calculate Dist_{ME} we need to know the distances between all pairs of reads, so we have to evaluate (see (3)) $\frac{\alpha}{l}|A|\frac{\alpha}{l}|B|$ distances where each one requires l^2 operations. Therefore $\alpha^2|A||B|$ operations are required. For the symmetric version Dist_{MES} we make $2\alpha^2|A||B|$ operations, which can be reduced to $\alpha^2|A||B|$ operations and $\Theta(l + \frac{\alpha}{l}(|A| + |B|))$ memory. Further modifications (MESS, MESSG, MESSGM) do not change the asymptotic complexity.

The constants α and l are determined by the sequencing technology and the independent complexity factors are $|A|$ and $|B|$. To calculate the distance in the conventional way as $\text{dist}(\hat{A}, \hat{B})$ requires to reconstruct $\hat{A}$ and $\hat{B}$ from the respective read-sets through an assembly algorithm. This is an NP-hard problem which becomes non-tractable for large $|A|$ and $|B|$, and which is avoided by our approach.

Fig. 4. If the distance between a read and its closest counterpart is greater than threshold θ, we assume that the read matches to a gap in the sequence alignment.

3.2 Metric Properties

Dist_{MES} as well as the later versions are all symmetric and non-negative but none of the proposed versions satisfies the identity condition ($\text{dist}(a, b) = 0$ iff $a = b$) or the triangle inequality, despite being based on the Levenshtein distance dist, which is a metric. For example, let $R_A = \{\text{ATC}, \text{ATC}, \text{GGG}\}$, let $R_B = \{\text{ATA}, \text{GGG}\}$, and let $R_C = \{\text{CTA}, \text{GGG}\}$. Then $\text{Dist}_{\text{MES}}(R_A, R_B) = \frac{7}{12}$, and $\text{Dist}_{\text{MES}}(R_B, R_C) = \frac{1}{2}$ but $\text{Dist}_{\text{MES}}(R_A, R_C) = \frac{14}{12} > \frac{7}{12} + \frac{1}{2}$. While this might lead to counter-intuitive behavior of the proposed distances in certain applications, the violated conditions are not requirements assumed by clustering algorithms.

4 Experimental Evaluation

The **purpose** of the experiments is to compare different methods for estimating the Levenhstein distance $\text{dist}(A, B)$ for various real DNA sequences A, B from their read sets R_A, R_B. The methods include (i) our newly proposed distances (MES, MESS, MESSG, MESSGM) applicable directly on R_A, R_B and implemented in Java with maximum of shared code,[5] (ii) the conventional method based on assembling estimates $\hat{A}, \hat{B}$ of the original sequences A, B using 3 common de-novo gene assemblers (ABySS [11], edena [3] and SSAKE [15]) and then estimating $\text{dist}(A, B)$ as $\text{dist}(\hat{A}, \hat{B})$, (iii) a trivial baseline method estimating $\text{dist}(A, B)$ as $\max\{|R_A|, |R_B|\}$. All the 3 assembly algorithms were configured with the default parameters and the current official C++ version was used. When a result of an assembly procedure consisted of multiple contigs, we selected the longest one.

The evaluation **criteria** consist of (i) the Pearson's correlation coefficient measuring the similarity of the distance matrix produced by the respective methods to the true distance matrix, (ii) The Fowlkes-Mallows index [2] measuring the similarity between the tree produced by a hierarchical clustering algorithm using the true distance matrix, and the tree induced from a distance matrix estimated by the respective method, (iii) runtime needed for assembly (if applicable), distance matrix calculation, and clustering time. For hierarchical clustering, we used the UPGMA algorithm [12] and the neighbor-joining algorithm [10].

---

[5] Implementation is available on https://github.com/petrrysavy/readsIDA2016.

The Fowlkes-Mallows index shows how much the resulting trees differ in structure. Both trees are first cut into k clusters for $k = 2, 3, \ldots, n - 1$. Then the clusterings are compared based on the number of common objects among each pair of clusters. By this procedure we obtain a set of values B_k that show how much the trees differ at various levels.

The testing **data** contain two datasets. The first dataset[6] contains 12 influenza virus genome sequences plus an outgroup sequence. The second dataset[7] contains 17 genomes of different viruses. Furthermore, we used an independent third *training* dataset[8] to tune the value of θ' (see Sect. 2.4). All the sequences were downloaded from the ENA repository http://www.ebi.ac.uk/ena.

In the preliminary **tuning** experiment, the value $\theta' = 0.35$ achieved the best Pearson's correlation coefficient on the training dataset and we carried this value over to the testing experiments. Out of curiosity, we also tried to optimize θ' on the testing datasets obtaining similar values (around 0.3), indicating relative stability of this parameter.

The main experimental **results** are shown in Table 1. The four partitions of the table correspond to the two datasets ($N = 13$, $N = 17$), each used twice with different settings of sequencing coverage α and read length l. The Pearson's correlation coefficient (column 'corr.') demonstrates clear dominance of the MESSG and MESSGM methods (Sects. 2.3 and 2.4), which are the most developed versions of our approach. The MESSG differs from MESSGM only by not discarding poorly matching reads. This finding is generally supported also by the Fowlkes-Mallows index (last four columns) shown for two levels of trees learned by two methods. Figure 5 provides a more detailed insight into the Fowlkes-Mallows values graphically for all the tree levels. One more (rather surprising) observation is that distance estimates achieved by first assembling sequences from read-sets (last 3 lines in every table partition) are systematically worse than the trivial estimate based just on the read-set sizes (first line in every partition).

Columns 1–5 of Table 1 indicate that all the variants of our approach were systematically slower in terms of absolute runtime than the approaches based on sequence assembly, despite the NP-hard complexity of the latter task. The numbers also show that our asymptotic complexity estimate in Sect. 3.1 is generally correct: the ratio between the time spent on calculating the distances on one hand, and the runtime of the reference method on the other hand, is approximately α^2.

---

[6] AF389115, AF389119, AY260942, AY260945, AY260949, AY260955, CY011131, CY011135, CY011143, HE584750, J02147, K00423 and outgroup AM050555. The genomes are available at http://www.ebi.ac.uk/ena/data/view/accession.

[7] AB073912, AB236320, AM050555, D13784, EU376394, FJ560719, GU076451, JN680353, JN998607, M14707, U06714, U46935, U66304, U81989, X05817, Y13051 and outgroup AY884005.

[8] CY011119, CY011127, CY011140, FJ966081, AF144300, AF144300, J02057, AJ437618, FR717138, FJ869909, L00163, KJ938716, KP202150, D00664, HM590588, KM874295, $\alpha = 4$, $l = 40$.

Table 1. Runtime, Pearson's correlation coefficient between distance matrices and Fowlkes-Mallows index for $k = 4$ and $k = 8$. The 'reference' method calculates distances from the original sequences.

Method	assem. ms	distances ms	UPGMA ms	NJ ms	corr.	UPGMA B_4	UPGMA B_8	NJ B_4	NJ B_8
$\alpha = 3, l = 30, N = 13$									
Reference	0	1,587	1	39	1	1	1	1	1
max($|R_A|,|R_B|$)	0	0	1	16	.802	.66	.32	.66	.32
Dist$_{MES}$	0	18,192	0	8	.83	.36	.67	.4	1
Dist$_{MESS}$	0	17,132	1	11	.944	1	1	1	1
Dist$_{MESSG}$	0	35,107	0	7	.99	1	1	1	1
Dist$_{MESSGM}$	0	34,911	1	5	.991	1	1	1	1
ABySS [11]	22,231	7	0	3	.376	.36	.11	.31	.14
edena [3]	3,501	6	1	7	.404	.36	.12	.31	0
SSAKE [15]	6,811	1	1	5	.548	.27	.12	.43	0
$\alpha = 3, l = 30, N = 17$									
Reference	0	23,367	1	38	1	1	1	1	1
max($|R_A|,|R_B|$)	0	1	1	17	.902	.67	.66	.85	.86
Dist$_{MES}$	0	279,965	0	12	.605	.35	.52	.41	.29
Dist$_{MESS}$	0	279,008	1	16	.935	.67	.92	.85	.94
Dist$_{MESSG}$	0	508,947	1	7	.945	.62	.92	1	.96
Dist$_{MESSGM}$	0	546,985	1	15	.95	.62	.92	1	.96
ABySS [11]	30,974	16	0	12	.684	.58	.72	.48	.13
edena [3]	6,287	11	1	91	.666	.58	.63	.46	.13
SSAKE [15]	12,745	2	23	20	.611	.62	.31	.47	.12
$\alpha = 5, l = 100, N = 13$									
Reference	0	1,653	1	19	1	1	1	1	1
max($|R_A|,|R_B|$)	0	1	0	11	.802	.66	.32	.66	.32
Dist$_{MES}$	0	47,238	1	7	.881	.36	.77	.54	1
Dist$_{MESS}$	0	47,703	1	7	.975	1	1	1	1
Dist$_{MESSG}$	0	83,186	0	6	.993	1	1	1	1
Dist$_{MESSGM}$	0	82,973	1	6	.99	1	1	1	.77
ABySS [11]	29,326	814	0	10	.639	.44	.62	.46	.77
edena [3]	6,455	128	1	9	.388	.3	.11	.39	.11
SSAKE [15]	7,258	94	1	7	.706	.54	.21	.72	.11
$\alpha = 5, l = 100, N = 17$									
Reference	0	24,612	1	13	1	1	1	1	1
max($|R_A|,|R_B|$)	0	0	1	16	.903	.67	.66	.85	.86
Dist$_{MES}$	0	687,503	1	13	.648	.35	.52	.41	.37
Dist$_{MESS}$	0	680,522	0	19	.935	.67	.86	1	1
Dist$_{MESSG}$	0	1,254,231	4	6	.94	.62	1	1	1
Dist$_{MESSGM}$	0	1,156,072	1	13	.938	.62	1	.69	.94
ABySS [11]	30,598	5,891	1	12	.576	.51	.66	.47	.15
edena [3]	9,918	363	2	19	.473	.5	.3	.54	.22
SSAKE [15]	28,590	374	1	23	.519	.48	.3	.4	.34

5   Conclusions and Future Work

We have proposed and evaluated several variants of a method for estimating edit distances between sequences only from read-sets sampled from them. In experiments, our approach produced better estimates than a conventional approach based on first estimating the sequences themselves by applying assembly algorithms on the read-sets.

A further observation was that the conventional approach was surprisingly fast despite involving the NP-hard assembly problem, and resulted in surprisingly low-quality estimates. A possible explanation for this is that the assemblers

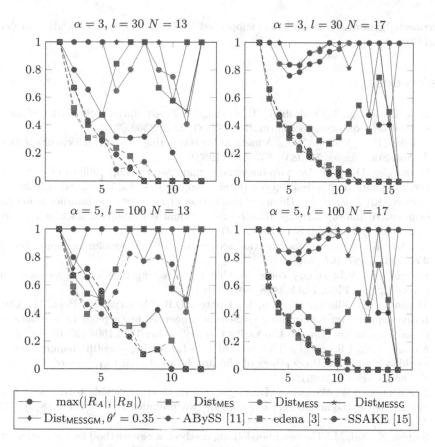

Fig. 5. Plots of Fowlkes-Mallows index B_k versus k. The index compares trees generated by the neighbor-joining algorithm. The tree is compared with the tree generated from the original sequences. If all values are equal to 1, the structures of the trees are the same.

require a higher coverage α to produce good sequences compared to the values we chose ($\alpha = 3, \alpha = 5$).

In our opinion, the most urgent goal for follow-up work is to conduct a more thorough experimental evaluation, and specifically, find the sequences lengths for which the exponential runtime of the assembly-based approach reaches the quadratic runtime of our approach, and then re-evaluate the approximation qualities. Furthermore, our approach offers many directions for technical improvements. For example, one may consider a *partial assembly* approach, in which sets of a few (up to a constant) reads would be pre-assembled and the Monge-Elkan distance would be applied on such partial assemblies. This view would open a 'continuous' spectrum between our approach on one hand, and the conventional assembly-based approach, on which the optimal trade-off could be identified.

Acknowledgment. This work was supported by Czech Science Foundation project 14-21421S.

References

1. Bao, E., Jiang, T., Kaloshian, I., Girke, T.: Seed: efficient clustering of next-generation sequences. Bioinformatics **27**(18), 2502–2509 (2011)
2. Fowlkes, E.B., Mallows, C.L.: A method for comparing two hierarchical clusterings. J. Am. Stat. Assoc. **78**(383), 553–569 (1983)
3. Hernandez, D., et al.: De novo bacterial genome sequencing: millions of very short reads assembled on a desktop computer. Genome Res. **18**(5), 802–809 (2008)
4. Jalovec, K., Železný, F.: Binary classification of metagenomic samples using discriminative dna superstrings. In: 8th International Workshop on Machine Learning in Systems Biology, MLSB 2014 (2014)
5. Lander, E.: Initial impact of the sequencing of the human genome. Nature **470**(7333), 187–197 (2011)
6. Levenshtein, V.I.: Binary codes capable of correcting deletions, insertions, and reversals. Sov. Phys. Dokl. **10**(8), 707–710 (1966)
7. Malhotra, R., Elleder, D., Bao, L., Hunter, D.R., Acharya, R., Poss, M.: Clustering pipeline for determining consensus sequences in targeted next-generation sequencing. arXiv (Conrell University Library) arXiv:1410.1608 (2016)
8. Monge, A.E., Elkan, C.P.: The webfind tool for finding scientific papers over the worldwide web. In: Proceedings of the 3rd International Congress on Computer Science Research (1996)
9. Needleman, S.B., Wunsch, C.D.: A general method applicable to the search for similarities in the amino acid sequence of two proteins. J. Mol. Biol. **48**(3), 443–453 (1970)
10. Saitou, N., Nei, M.: The neighbor-joining method: a new method for reconstructing phylogenetic trees. Mol. Biol. Evol. **4**(4), 406–425 (1987)
11. Simpson, J.T., et al.: ABySS: a parallel assembler for short read sequence data. Genome Res. **9**(6), 1117–1123 (2009)
12. Sokal, R.R., Michener, C.D.: A statistical method for evaluating systematic relationships. Univ. Kansas Sci. Bull. **38**, 1409–1438 (1958)
13. Železný, F., Jalovec, K., Tolar, J.: Learning meets sequencing: a generality framework for read-sets. In: 24th International Conference on Inductive Logic Programming, Late-Breaking Papers, ILP 2014 (2014)
14. Wagner, R.A., Fischer, M.J.: The string-to-string correction problem. J. ACM **21**(1), 168–173 (1974). http://doi.acm.org/10.1145/321796.321811
15. Warren, R.L., et al.: Assembling millions of short DNA sequences using SSAKE. Bioinformatics **23**(4), 500–501 (2007)
16. Weitschek, E., Santoni, D., Fiscon, G., Cola, M.C.D., Bertolazzi, P., Felici, G.: Next generation sequencing reads comparison with an alignment-free distance. BMC Res. Notes **7**(1), 869 (2014)

Widened Learning of Bayesian Network Classifiers

Oliver R. Sampson$^{(\boxtimes)}$ and Michael R. Berthold

Chair for Bioinformatics and Information Mining,
Department of Computer and Information Science,
University of Konstanz, Konstanz, Germany
oliver.sampson@uni-konstanz.de

Abstract. We demonstrate the application of *Widening* to learning performant Bayesian Networks for use as classifiers. Widening is a framework for utilizing parallel resources and diversity to find models in a hypothesis space that are potentially better than those of a standard greedy algorithm. This work demonstrates that widened learning of Bayesian Networks, using the Frobenius Norm of the networks' graph Laplacian matrices as a distance measure, can create Bayesian networks that are better classifiers than those generated by popular Bayesian Network algorithms.

1 Introduction

Widening [2,18] formalizes a method for executing a greedy learning algorithm in parallel while using diversity to guide the parallel refinement paths through a hypothesis[1] space. This enables the system as a whole to avoid local optima and potentially find better models than the greedy learning algorithm would otherwise find. Previous work [13,29] has demonstrated its viability on real world algorithms. This work builds on that with an application to the superexponentially-sized [28] hypothesis space of learning Bayesian Networks. Bayesian Networks [26] are probabilistic graphical networks, which describe relationships of conditional dependence between the features of a dataset. Perhaps the best known of these graphical networks is the network defined by the NAÏVE BAYES algorithm [11,23]. This paper describes the application of Widening to the learning of Bayesian Networks for use as classifiers.

The ultimate goal of Widening is not just to provide better solutions using parallel resources, but to provide better solutions in the same time or less than the canonical greedy algorithm. To enable this, *communication-free* Widening would allow the model refinement paths, separated by some measure of diversity, to be followed through the solution space until some stopping criterion is met. The difficulty in that effort has been finding a suitable measure of distance, i.e.,

---

[1] We freely mix the use of "solution space" and "hypothesis space" throughout this paper, referring essentially to the same space, but drawing attention to whether it is the evaluation of the hypothesis or the hypothesis itself that is important.

H. Boström et al. (Eds.): IDA 2016, LNCS 9897, pp. 215–225, 2016.
DOI: 10.1007/978-3-319-46349-0_19

diversity. Here, we show that the Frobenius Norm of Bayesian Networks' graph Laplacians is a useful measure of diversity for comparing Bayesian networks in the Widening framework, albeit not in a communication-free framework.

2 Background

2.1 Learning and Scoring Bayesian Networks

A Bayesian network, B, derived from a dataset, $\mathcal{D}$, is a triple, $\langle \mathcal{X}, G, \Theta \rangle$, where $\mathcal{X}$ is the set of features or random variables in the dataset, G is a directed-acyclic-graph (DAG), and Θ is the set of conditional probability tables (CPT) for the features in $\mathcal{X}$. The graph $G = (\mathcal{X}, \mathcal{E})$, is an ordered pair, where each node, $X \in \mathcal{X}$, is a feature from the dataset and where each edge, $E = \{X_i, X_j\} \in \mathcal{E}$, is directed according to the dependency of one feature on another.

There are four general categories of algorithms for learning Bayesian networks: *search-and-score, constraint-based, hybrid* [19] and *evolutionary algorithms* [20]. Search-and-score methods such as K2 [9] and GREEDY EQUIVALENCE SEARCH (GES) [8] rely on heuristics to sequentially add, remove, or change the direction of the edges in the graph, G, to which a scoring method is applied. Edges that improve the score are kept in the graph for the next iteration of add, delete, or change. Constraint-based methods such as PC ALGORITHM [32] or CBL [7] rely on some assumptions about the dependency relationships of the features, from which a partially-directed-acyclic-graph is generated. This "skeleton" of a graph describes the neighbors of each of the feature nodes within the graph, but not necessarily the direction of the edges between the nodes. After determining the skeleton, search-and-score methods are used to find better networks, i.e., networks with a higher score, by flipping the direction of the edges and re-evaluating the score. Hybrid methods such as MAX-MIN HILL-CLIMBING [35] incorporate techniques from both the search-and-score and the constraint-based methods. Algorithms based on evolutionary techniques randomly change and combine networks and evaluate them with a fitness function.

Several scoring functions have been proposed for the use of learning Bayesian networks. For an extensive overview and comparison, the reader is referred to [6]. Scoring functions can be grouped into two categories, *Bayesian* and *information-theoretic*. Bayesian scoring methods calculate the posterior probability distribution based on the prior probability distribution conditioned on $\mathcal{D}$. Some examples of Bayesian scoring functions are K2 [9], Bayesian Dirichlet (BD) [17], BD with equivalence assumption (BDe) [9], and BD with equivalence and uniform assumptions (BDeu) [5].

Information-theoretic score functions are based on Shannon entropy and the amount of compression possible for a Bayesian network. The Log-Likelihood (LL) score is based on the logarithm of the likelihood of $\mathcal{D}$ given B, i.e., $\log(P(\mathcal{D}|B))$. The LL score is better, in general, for complete networks, and for this reason alternate scoring functions have been proposed that penalize the LL according to some factor. The Minimum Description Length [34] (MDL), the Akaike Information Criterion [1] (AIC), and Bayesian Information Criterion [30] (BIC) (roughly

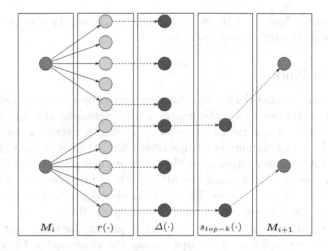

Fig. 1. Widening. Each $m_{k,i} \in M_i$ (green) is refined to five models (yellow). In each of these sets, the three most diverse from another are determined (red). The two best performing models (blue) are selected and used for the next iteration, $m_{k,i+1} \in M_{i+1}$ (green). (Color figure online)

equivalent to MDL) all adjust the LL by different proportions of the network complexity.

Less commonly, and more often associated with evolutionary algorithms such as [27,31], the performance of the networks as a classifier is used as the scoring, i.e., fitness, function. In this vein, the work described in this paper also uses accuracy as the scoring function.

When used as a classifier, the relevant portion of the network contains the parents of the target node, the children of the target node, and the children's other parents. This is termed the *Markov blanket* [26, p. 97].

2.2 Widening

The Widening framework [2,18] (See Fig. 1.) describes a general process for improving greedy learning algorithms where models, $m \in M$, are iteratively refined and scored in parallel. Each refinement path follows a different route through the hypothesis space. The models at each refinement step are separated using a diversity measure, Δ, which enforces a distance between the models' respective refinement paths.

More formally, a refinement operator, $r(\cdot)$, applied to a model, m, generates a set of models, M', from the set of all possible models, M, in the hypothesis space. A selection operator, $s(\cdot)$, when applied to a set of models, selects a subset according to a performance metric. In Widening's most rudimentary form, the best k performing models from a refinement step are selected, $s_{top-k}(M')$, which in turn are further refined and selected until a stopping criterion is met. The $s_{top-k}(\cdot)$ operator has a similarity to a BEAM SEARCH, [22] but instead of a selection operator based solely on performance, the selection is also based on diversity, due to the refinement operator.

2.3 Related Work

Learning Bayesian Networks for classification, either by modifying networks created by NAÏVE BAYES or by the generation of networks through completely different methods, is a very active research area. An excellent survey can be found in [4]. In [14], Friedman et al. describe TREE AUGMENTED NAÏVE BAYES NETWORK (TAN) where edges are added between child nodes of a Naïve Bayes network in a greedy search using the MDL scoring function, and whose structure is limited to that of a tree. The authors also describe learning an "unrestricted" BAYESIAN NETWORK AUGMENTED NAÏVE BAYES (BAN), but these networks do not include networks with nodes as parents for the target nodes, but rather just more complex relationships among the child nodes. Cheng et al. in [7] describe an algorithm (CBL) for finding General Bayesian Networks (GBN) based on conditional independence tests using Mutual Information (MI).

In [25] Nielsen et al. present K-GREEDY EQUIVALENCE SEARCH (KES) which is a modification to the GES, where a random subset of models from the entire set is chosen and evaluated. They describe this as a method specifically to avoid the local optima encountered by GES in [8].

Su and Zhang describe in [33] what they call Full Bayesian Networks (FBN), which are TANs where all child nodes of the target are connected to a maximal subset of the other child nodes based on an ordering using MI. This structure is in turn used to learn a Decision Tree-like structure for learning CPT-Trees.

The work presented here is similar to the TAN in [14], in that we perform a greedy search for better networks starting with a network generated by NAÏVE BAYES. It is similar to the work in [25], in that a subset of models is chosen and evaluated specifically to avoid local optima. It differs from these two, in that (1) any configuration of Bayes Network is allowed, (2) diversity between networks rather than randomness is used to select models, and (3) classification accuracy is employed for the scoring function.

3   Widened Bayesian Networks

3.1   Application of the Widening Framework

The simplest search-and-score method (HILL-CLIMBING or GREEDY SEARCH) refines a Bayesian Network model by changing a randomly or heuristically chosen edge, E, and scores the network according to one of the scoring functions discussed in Sect. 2.1. The algorithm greedily keeps the changed edge if it improves the score. Using the Widening notation, the greedy search-and-score method is $B_{i+1} = s_{top-k=1}(r(B_i))$, where i refers to the current search-and-score iteration. The process stops when no further improvement is seen.

The application of Widening to this process is to refine a set of different Bayesian networks at each stage, $B_{i+1} = r(B_i)$. Each model is refined to a number, l, of refinements. From this set, k models are selected by the selection operator, $s_{top-k}(\cdot)$. $k \times l$ models are generated during each refinement iteration, with the exception of the initial one. Additionally, the application of a diversity measure, Δ, is used by the refinement operator, and therefore notated as $r_\Delta(\cdot)$. The refinement operator ensures that the models are different enough to explore disparate regions of the hypothesis space.

Scoring Bayesian Networks by using classification accuracy is common only with the evolutionary algorithms, even though, for example, Friedman et al. in [14] explicitly say that one of the reasons that their TAN ALGORITHM did not always provide superior solutions was that the structural score may not have been a good analog for the use of the network in its role as a classifier.

In summary, each step in the top-k Widening process is described as

$$B_{i+1} = s_{top-k}(r_\Delta(B_i)) \tag{1}$$

3.2   Refinement Operator

The refinement operator creates a list of all possible pairs of nodes, i.e., all possible edges. Each edge compared with the current model and up to two additional models are created based on the edge. (See Fig. 2.)

1. If it is possible to add the edge to the initial model (Fig. 2a), i.e., its presence would not contravene the definition of DAG by creating a loop, it is added. (See Fig. 2f.)
2. If it is present in the model, it is removed. (See Figs. 2b and d.)
3. If it is present in the model, and the reversal of its direction would not create a loop, it is reversed. (See Figs. 2c and e.)

A distance matrix of all distances between network model pairs is then calculated.

3.3   Diversity

There are a variety of measures for comparing two labeled DAGs. Early experiments indicated that the Hamming distance [16] does not measure diversity in a way that scales well to larger networks. For this work, we have chosen the Frobenius Norm of the difference between the graphs' Laplacian matrices. The Frobenius Norm is sometimes referred to as the Euclidean norm and provides a "measure of distance on the space of matrices". [15] The Frobenius Norm for a matrix, $A \in \mathbb{R}^{m \times n}$ is defined as

$$||A||_F = \sqrt{\sum_{i=1}^{m} \sum_{j=1}^{n} |a_{ij}|^2} [15] \tag{2}$$

where, a_{ij} are elements of matrix A.

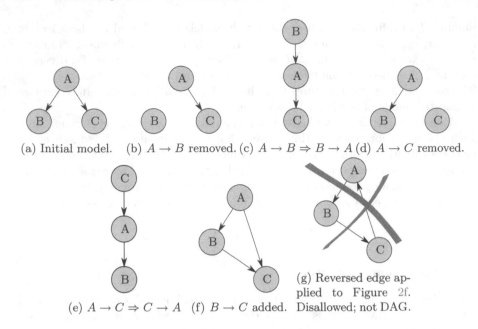

(a) Initial model.   (b) $A \to B$ removed. (c) $A \to B \Rightarrow B \to A$ (d) $A \to C$ removed.

(e) $A \to C \Rightarrow C \to A$ (f) $B \to C$ added.

(g) Reversed edge applied to Figure 2f. Disallowed; not DAG.

Fig. 2. Example possible refinements for the three edges $\{\langle A, B \rangle, \langle A, C \rangle, \langle B, C \rangle\}$

The Laplacian matrix of a graph is given by the formula $L = D - A$, where D is the out-degree matrix, and A is the adjacency matrix. Here we use the Frobenius Norm of the difference of each pair of Bayesian networks' Laplacian matrices, i.e.,

$$\Delta_{\text{Frobenius}} = ||L_{B_p} - L_{B_q}||_F : B_p, B_q \in \boldsymbol{B}_i \tag{3}$$

and $\boldsymbol{B}_i$ is the set of refined Bayesian networks from Eq. 1.

The P-DISPERSION PROBLEM describes selecting a subset of points from a larger set, where the subset's minimum pairwise distances are maximized. There are several diversity measures used commonly with the P-DISPERSION PROBLEM, including *sum* and *min-sum*. *p-dispersion-sum* simply maximizes the sum of the distances between any two points in the subset, whereas *p-dispersion-min-sum* maximizes the sum of the minimum distances between two points. *p-dispersion-sum* has the property of pushing the resultant subset to the margins of the original set, whereas the subset derived using *p-dispersion-min-sum* is more representative of the dataset as whole [24]. Because of this property, and based on the results in [29], we favor *p-dispersion-min-sum* as the diverse subset selection method.

Definition 1 *p-dispersion-min-sum.*[2] *Given a set* $\boldsymbol{B} = \{B_1, \cdots, B_n\}$ *of* n *distinct Bayesian networks and* p, *where* $p \in \mathbb{N}$ *and* $p \leq n$, *and a distance*

---

[2] In this application, it would be correctly termed "*l*-dispersion-min-sum," but the notation is written here as "*p*" to be consistent with the literature.

measure $d(B_i, B_j)$: $B_i, B_j \in \boldsymbol{B}$ between Bayesian networks B_i and B_j, the p-*diversity-min-sum* problem is to select the set $\hat{\boldsymbol{B}} \subseteq \boldsymbol{B}$, such that

$$\hat{\boldsymbol{B}} = \underset{\substack{B' \subseteq B \\ |B'|=p}}{\arg\max} f(\boldsymbol{B'}), \text{ where}$$

$$f(\boldsymbol{B'}) = \sum_{i=1}^{p} \min_{1 \leq i,j \leq n, i \neq j} d(B_i, B_j) : B_i, B_j \in \boldsymbol{B} [24] \tag{4}$$

The P-DISPERSION PROBLEM is known to be $\mathcal{NP}$-complete, and when adjusting the diversity criterion to be *min-sum*, the problem is $\mathcal{NP}$-hard [12].

3.4 Selection Operator

The selection operator presented in this work is simply the performance metric of the Bayesian network as a classifier, similar to that of [31]. When compared to the scoring methods described in Sect. 2.1, this has the advantage of being directly related to the network's use as a classifier, and networks that perform poorly as classifiers are eliminated from the refinement paths. The calculation for determining the target winner is similar to that of NAÏVE BAYES, except the probabilities of the parents of the target node and of the other parents of the target's child nodes are considered.

$$\hat{C} = \underset{j=1,\ldots,|C|}{\arg\max} P(C_j, \boldsymbol{X}_m) = \underset{j=1,\ldots,|C|}{\arg\max} P(C_j|\boldsymbol{pa}(C)) \prod_{i=1}^{m} P(x_i|\boldsymbol{pa}(x_i)) [4] \tag{5}$$

where $\boldsymbol{X}_m \subseteq \mathcal{X}$ is the subset of features contained in the Markov-blanket of the target node, C, and $\boldsymbol{pa}(\cdot)$ is the set of parents of a child of C in the Bayesian network.

4 Experimental Results

The experiments were performed in KNIME Analytics Platform [3]. The datasets from the UCI Machine Learning repository[3] [21] were discretized using the LUCS-KDDN software.[4] Unlike algorithms such as K2 or CBL, no assumptions were made concerning the ordering of the features within the dataset. Datasets with missing values or continuous values were not considered, because we are interested in testing the Widened learning process and not the robustness of the algorithm to various data types. The refinement operator placed no restrictions on the number of parents a node may have. The stopping criterion was set to stop the iterations when improvement in the best model compared to its performance in the previous iteration was less than 0.01 %. The records in the datasets were shuffled between each widening trial of a different breadth and width.

---

[3] http://archive.ics.uci.edu/ml/.
[4] http://www.csc.liv.ac.uk/~frans/KDD/Software/LUCS_KDD_DN/.

Table 1. Accuracy ($\mu \pm 2\sigma$) comparison of all tested algorithms with 5-fold cross-validation.

Dataset	$|\mathcal{D}|$	$|\mathcal{X}|$	$|C|$	WIDENED BAYES	MMHC	TABU	HILL-CLIMBING
ecoli	336	7	8	**0.747 ± 0.032**	0.430 ± 0.123	0.593 ± 0.119	0.647 ± 0.057
flare	1389	10	9	**0.843 ± 0.015**	**0.843 ± 0.013**	**0.843 ± 0.013**	**0.843 ± 0.013**
glass	214	9	7	**0.649 ± 0.137**	0.457 ± 0.151	0.564 ± 0.133	0.536 ± 0.111
nursery	12960	8	8	**0.935 ± 0.047**	0.570 ± 0.150	0.621 ± 0.214	0.632 ± 0.246
pageBlocks	5473	10	5	0.898 ± 0.015	**0.913 ± 0.011**	**0.913 ± 0.004**	0.910 ± 0.023
pima	768	8	2	0.710 ± 0.068	0.721 ± 0.136	**0.757 ± 0.098**	0.736 ± 0.143
waveform	5000	21	3	**0.790 ± 0.025**	0.342 ± 0.014	0.619 ± 0.020	0.620 ± 0.021
wine	178	13	3	**0.939 ± 0.091**	0.746 ± 0.150	0.798 ± 0.116	0.747 ± 0.184

The initial state could be any network configuration that satisfies the definition of a DAG, including a network without any edges. Because our effort is to prove the ability of Widening to find superior solutions to traditional greedy methods, we chose a Naïve Bayes configuration, where all of the non-target features are dependent on the target variable, as the initial state. This was a pragmatic decision in the sense that finding a network out of all possible networks that is tuned to the target node is impractical. Additionally, NAÏVE BAYES performs remarkably well given its simplicity for a large number of datasets and is a measuring stick for many new algorithms.

We tested eight datasets, ecoli, flare, glass, nursery, pageBlocks, pima, waveform and wine against three standard Bayesian Network learning algorithms, MAX-MIN HILL-CLIMBING (MMHC), TABU, and HILL-CLIMBING, from the **R** bnlearn[5] package, version 3.8.1. MMHC and HILL-CLIMBING used parameters test = mi, restart = 100, and perturb = 100. These values were chosen experimentally as values that provide good results for all datasets.

WIDENED BAYESIAN NETWORKS (WBN) significantly outperformed the other three reference implementations in five of the eight datasets, tied in one, and performed slightly worse in two (Table 1).

The results in Fig. 3 show a two responses to Widening. In general, with Widening we expect a gradual improvement of average performance with the width, i.e., the number of parallel paths in the solution space. Additionally, we expect a decrease in the variance of the results as the many paths push themselves towards better solutions. ecoli, glass, nursery, pima, waveform, and wine show this behavior nicely. pageBlocks and flare demonstrate how some solution space topologies cannot be explored with the refine-and-select process presented here, even though the results for the comparison algorithms for flare indicate that the resultant Bayesian network is a best fit. The non-responsive nature of pageBlocks however, invites further research into other refining-and-select strategies and/or diversity measures.

---

[5] http://www.bnlearn.com/.

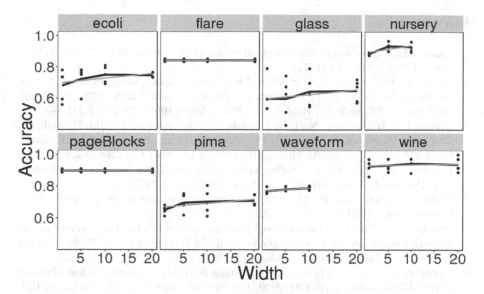

Fig. 3. Widened Bayesian Networks accuracy progression verses *width* with 5-fold cross-validation.

5   Conclusion and Future Work

This paper demonstrates the successful initial application of Widening to learning Bayesian Networks for use classifiers and demonstrates classification scoring techniques with the search-and-score greedy heuristic. The technique was able to find superior solutions when compared to standard Bayesian Network learning algorithms from the **R bnlearn** package. Although the results are similar or superior to established Bayesian Network learning algorithms on some datasets, the execution time does not meet the specified goal of finding better solutions in the same time or less as the greedy algorithm. The primary impediment to this goal, as it is demonstrated here, is the use of *p-dispersion-min-sum* for finding a maximally diverse subset of networks for refinement. Methods that allow for diverse subsets to be calculated without communication between the parallel workers would be better. (See [18] for details.) Additionally, the refinement operator considers the entire space of possible networks, where only the refinements to the Markov blanket are actually necessary. Significantly, the use of the Frobenius Norm of the difference of the Bayesian networks's graph Laplacians is very encouraging and suggests further research into distance measures based on graph features such as those derived from Spectral Graph Theory. Experiments with alternate starting states based on conditional information, in a manner similar to the PC ALGORITHM and CBL, or constraint-based algorithms like Incremental Association or HITON, or even to those claiming to find the exact network structure [10] could also be promising.

References

1. Akaike, H.: A new look at the statistical model identification. IEEE Trans. Autom. Control **19**(6), 716–723 (1974)
2. Akbar, Z., Ivanova, V.N., Berthold, M.R.: Parallel data mining revisited. better, not faster. In: Hollmén, J., Klawonn, F., Tucker, A. (eds.) IDA 2012. LNCS, vol. 7619, pp. 23–34. Springer, Heidelberg (2012). doi:10.1007/978-3-642-34156-4_4
3. Berthold, M.R., Cebron, N., Dill, F., Gabriel, T.R., Kötter, T., Ohl, P., Sieb, C., Thiel, K., Wiswedel, B.: KNIME: the konstanz information miner. In: Preisach, C., Burkhardt, H., Schmidt-Thieme, L., Decker, R. (eds.) Data Analysis, Machine Learning and Applications. Studies in Classification, Data Analysis, and Knowledge Organization, pp. 319–326. Springer, Heidelberg (2007)
4. Bielza, C., Larranaga, P.: Discrete Bayesian network classifiers: a survey. ACM Comput. Surv. (CSUR) **47**(1), 5 (2014)
5. Buntine, W.: Theory refinement on Bayesian networks. In: Proceedings of the Seventh Conference on Uncertainty in Artificial Intelligence, pp. 52–60. Morgan Kaufmann Publishers Inc., Los Angeles (1991)
6. Carvalho, A.M.: Scoring functions for learning Bayesian networks. Technical report INESC-ID Technical report 54/2009, Instituto superior Téchnico, Technical University of Lisboa, April 2009
7. Cheng, J., Bell, D.A., Liu, W.: An algorithm for Bayesian belief network construction from data. In: Proceedings of AI & STAT 1997, pp. 83–90 (1997)
8. Chickering, D.M.: Optimal structure identification with greedy search. J. Mach. Learn. Res. **3**, 507–554 (2002)
9. Cooper, G.F., Herskovits, E.: A Bayesian method for the induction of probabilistic networks from data. Mach. Learn. **9**(4), 309–347 (1992)
10. De Campos, C.P., Zeng, Z., Ji, Q.: Structure learning of Bayesian networks using constraints. In: Proceedings of the 26th Annual International Conference on Machine Learning, pp. 113–120. ACM (2009)
11. Duda, R.O., Hart, P.E.: Pattern Classification and Scene Analysis. Wiley, New York (1973)
12. Erkut, E.: The discrete p-dispersion problem. Eur. J. Oper. Res. **46**(1), 48–60 (1990)
13. Fillbrunn, A., Berthold, M.R.: Diversity-driven widening of hierarchical agglomerative clustering. In: Fromont, E., Bie, T., Leeuwen, M. (eds.) IDA 2015. LNCS, vol. 9385, pp. 84–94. Springer, Heidelberg (2015). doi:10.1007/978-3-319-24465-5_8
14. Friedman, N., Geiger, D., Goldszmidt, M.: Bayesian network classifiers. Mach. Learn. **29**(2–3), 131–163 (1997)
15. Golub, G.H., van Loan, C.F.: Matrix Computations, 4th edn. The Johns Hobpkins University Press, Baltimore (2013)
16. Hamming, R.W.: Error detecting and error correcting codes. Bell Syst. Tech. J. **29**(2), 147–160 (1950)
17. Heckerman, D., Geiger, D., Chickering, D.M.: Learning Bayesian networks: the combination of knowledge and statistical data. Mach. Learn. **20**(3), 197–243 (1995)
18. Ivanova, V.N., Berthold, M.R.: Diversity-driven widening. In: Tucker, A., Höppner, F., Siebes, A., Swift, S. (eds.) IDA 2013. LNCS, vol. 8207, pp. 223–236. Springer, Heidelberg (2013). doi:10.1007/978-3-642-41398-8_20
19. Koski, T.J., Noble, J.M.: A review of Bayesian networks and structure learning. Math. Applicanda **40**(1), 53–103 (2012)

20. Larrañaga, P., Karshenas, H., Bielza, C., Santana, R.: A review on evolutionary algorithms in Bayesian network learning and inference tasks. Inf. Sci. **233**, 109–125 (2013)
21. Lichman, M.: UCI Machine Learning Repository (2013)
22. Lowerre, B.T.: The HARPY speech recognition system. Ph.D. thesis, Carnegie Mellon University, Pittsburgh, PA, USA (1976)
23. Maron, M.E., Kuhns, J.L.: On relevance, probabilistic indexing and information retrieval. J. ACM (JACM) **7**(3), 216–244 (1960)
24. Meinl, T.: Maximum-score diversity selection. Ph.D. thesis, University of Konstanz, July 2010
25. Nielsen, J.D., Kočka, T., Peña, J.M.: On local optima in learning Bayesian networks. In: Proceedings of the Nineteenth Conference on Uncertainty in Artificial Intelligence, pp. 435–442. Morgan Kaufmann Publishers Inc., San Francisco (2003)
26. Pearl, J.: Probabilistic Reasoning in Intelligent Systems: Networks of Plausible Inference. Morgan Kaufmann Publishers Inc., San Francisco (1988)
27. Pernkopf, F.: Bayesian network classifiers versus k-NN classifier using sequential feature selection. In: AAAI, pp. 360–365 (2004)
28. Robinson, R.W.: Counting unlabeled acyclic digraphs. In: Little, C.H.C. (ed.) Combinatorial Mathematics V. LNM, vol. 622, pp. 28–43. Springer, Heidelberg (1977)
29. Sampson, O., Berthold, M.R.: Widened KRIMP: better performance through diverse parallelism. In: Blockeel, H., Leeuwen, M., Vinciotti, V. (eds.) IDA 2014. LNCS, vol. 8819, pp. 276–285. Springer, Heidelberg (2014). doi:10.1007/978-3-319-12571-8_24
30. Schwarz, G.: Estimating the dimension of a model. Ann. Stat. **6**(2), 461–464 (1978)
31. Sierra, B., Larrañaga, P.: Predicting the survival in malignant skin melanoma using Bayesian networks. an empirical comparison between different approaches. Artif. Intell. Med. **14**(1–2), 215–230 (1998)
32. Sprites, P., Glymour, C., Scheines, R.: Causation, Prediction, and Search. MIT Press, Cambridge (1993)
33. Jiang, S., Zhang, H.: Full Bayesian network classifiers. In: Proceedings of the 23rd International Conference on Machine Learning, pp. 897–904. ACM (2006)
34. Suzuki, J.: A construction of Bayesian networks from databases based on an MDL principle. In: Proceedings of the Ninth International Conference on Uncertainty in Artificial Intelligence, pp. 266–273. Morgan Kaufmann Publishers Inc., San Francisco (1993)
35. Tsamardinos, I., Brown, L.E., Aliferis, C.F.: The max-min hill-climbing Bayesian network structure learning algorithm. Mach. Learn. **65**(1), 31–78 (2006)

Vote Buying Detection via Independent Component Analysis

Antonio Neme[1](✉) and Omar Neme[2]

[1] Institute of Biomedicine, University of Eastern Finland,
Room 3180, Kuopio, Finland
antonio.nemecastillo@uef.fi
[2] Instituto Politécnico Nacional, Escuela Superior de Economia, Mexico City, Mexico

Abstract. Electoral fraud can be committed along several stages. Different tools have been applied to detect the existence of such undesired actions. One particular undesired activity is that of vote-buying. It can be thought of as an economical influence of a candidate over voters that in other circumstances could have decided to vote for a different candidate, or not to vote at all. Instead, under this influence, some citizens cast their votes for the suspicious candidate. We propose in this contribution that intelligent data analysis tools can be of help in the identification of this undesired behavior. We think of the results obtained in the affected ballots as a mixture of two signals. The first signal is the number of votes for the suspicious candidate, which includes his/her actual supporters and the voters affected by an economic influence. The second mixed signal is the number of citizens that did not vote, which is affected also by the bribes or economic incentives. These assumptions allows us to apply an instance of blind source separation, independent component analysis, in order to reconstruct the original signals, namely, the actual number of voters the candidate may have had and the actual number of no voters. As a case of study we applied the proposed methodology to the case of presidential elections in Mexico in 2012, obtained by analyzing public data. Our results are consistent with the findings of inconsistencies through other electoral forensic means.

1 Introduction

Elections are a fundamental activity in healthy democratic societies. Different aspects of electoral processes have been studied, and many of those analyses come from a mathematical and statistical mechanics perspective [1]. In a seminal work by [2] statistical mechanics of voting were firmly presented and from that work, several other works have identified general patterns in election outcomes. For example, power laws have been identified in the outcome of several electoral processes [3], whereas patterns in the probability of success of candidates have also been described [4].

Free and fair elections are a major pursuit in all democratic regimes. An increasing number of professionals and public as well as private non-governmental organizations are focused in verifying that such elections are run in

H. Boström et al. (Eds.): IDA 2016, LNCS 9897, pp. 226–236, 2016.
DOI: 10.1007/978-3-319-46349-0_20

solid basis, and that no undesired influences percolate into the process. Unfortunately, the evidence of anomalies, fraud, and other undesired activities are common in several countries [5]. Different tools have been applied in order to detect anomalies in electoral results. Traditionally, social scientists mainly use questionnaires and interviews [6], whereas journalists struggle to document evidence of those misconducts via intrincate paths linking politicians and misconducts [8,9,13,14,24]. In a different route, physicists and mathematicians tend to apply statistical mechanics to study abnormalities in the distribution of expected results [12], or to use Benford's law [10,11].

Computer scientists and engineers have scarcely been involved in actively investigating abnormal behaviors in elections. In a recent contribution [20] offer a panoramic overview of how machine learning algorithms, such as k-means and Bayesian methods, can be of help in detecting ballot stuffing and vote stealing. However, the impact of computational methods still have much to offer in the field of electoral forensics.

A particularly difficult behavior to corroborate is that of vote-buying (VB) [21]. This undesired situation occurs when one or more of the contending candidates offer economic incentives in exchange of evidence that citizens exerted the vote in favor of a certain candidate. We describe in this contribution a methodology that allows the identification of signs that this undesired behavior is present in some of the ballots. The assumptions behind the model we apply are not more stringent than those in the traditional electoral forensics tools.

In this contribution we think of ballots as composed by two signals: the first one is the number of votes the suspicious candidate received, which we will call E, and the number of citizens that did not cast their vote (no voters), which we will call M. Each ballot has assigned a number of voters L_i So, a particular ballot i can be described by the two signals E_i/L_i and M_i/L_i. The hypothesis we follow in this contribution is that the actual number of votes e that the suspicious candidate may had received, as well as the number of no voters m, can be extracted from E and M, via a mixing matrix. Note that if we replace E by X_1 and M by x_2, and e by s_1 and m by s_2 then this is formulation is equivalent to the problem that blind source separation (BSS), and in particular one of its instances, independent component analysis (ICA).

The rest of this contribution goes as follows. In Sect. 2 we briefly describe the ICA algorithm and its main properties. Then, in Sect. 3 we explain how some electoral anomalies can be studied by ICA, imposing the assumptions required by ICA. Once we have described ICA and its potential use in detecting vote-buying, we present in Sect. 4 the case study of detecting vote buying in the Presidential Mexican elections of 2012 sustaining our investigation in public data from the electoral authorities. Finally, in Sect. 5 we present some conclusions including the confirmation through ICA that there is evidence of vote buying in those elections, as previously supported from several and solid signs derived from both statistical mechanics and journalists reports.

2   Independent Component Analysis

Signals of scientific interest are normally registered as mixed in the detection devices either because the detector itself is not error-free, or because of the nature of the sampling environment. In the traditional problem of the noisy cocktail party, an individual is listening to her interlocutor and still, regardless of the noisy and mixed conversations in the surroundings, she is still able to understand (reconstruct) the words being said to her [18]. The task of recovering the signal of interest from a mixture of other signals is commonly known as blind source separation [19,22]. Independent component analysis (ICA) is a general case of projection pursuit (PP). PP is a set of techniques that allows the reconstruction of a specific mixed signal from a group of signals, in a serial fashion. ICA reconstruct all mixed signals in a parallel fashion, which is translated in a more efficient algorithm and robust results [18].

Let the mixed or observed variables or signals be $x_i(t)$, with $i = 1, ..., n$, and $t = 1, ..., T$, where n is the number of signals and T is the index, that might represent time or location of the measurement [15]. Let the original or source signals with index t be $s_i(t)$. The general assumption is that x_i can be modeled as linear combinations of the original (hidden or latent) variables by the following equation:

$$x_i(t) = \sum a_{ij} s_i(t) \tag{1}$$

where the parameter a_{ij} is a mixing coefficient. In a matrix notation, we can represent the latent and observed signals as:

$$\mathbf{X} = \mathbf{AS} \tag{2}$$

The problem of reconstructing the original signals is then to estimate the mixing coefficients. In geometric terms, the problem is reduced to finding the spatial transformation which maps a set of observed signals to a set of source signals [18]. There are several possible algorithms to find such transformations, and one of the most successful ones is that of FastICA. Both ICA, and FastICA rely on the following conditions in order for the components to be approximated [16]:

- 1. Components s_i have to be statistically independent.
- 2. Components s_i are described by non-Gaussian distributions.
- 3. The mixing matrix A is square and invertible.

The requirement that the components s_i need to be statistically independent is fulfilled in many of situations as the signals are assumed to be originated in independent systems, for example, the simultaneous conversations in the cocktail party are not linked to each other. However, it has been shown that either when this stringent condition is not totally satisfied, it is still possible to reconstruct the latent variables.

The general route for FastICA to approximate the components is via looking for orthogonal rotation of prewhitened data, through a fixed-point iteration scheme, in order to maximize a given measure of non-normality of those rotated components. Commonly, such a measure is the kurtosis. Non-normality substitutes the stringent condition of statistical independence, which is computationally very demanding [17]. Components s_i have to be non-Gaussian distributions in order to identify a "direction" of rotation. If the data is *iid*, then the number of possible rotations is infinite.

ICA identifies independent components such that the following two constraints are present:

- 1. Components are not identified in any particular order.
- 2. Components are identified up to a scaling factor.

In short, the assumptions for ICA to hold are that the probability density function (pdf) of the joint components has to be factorable and that the pdfs of the individual components have to be non-normal.

In the next section we detail how ICA can be applied to the problem of detecting vote buying in elections.

3   ICA for Vote Buying Detection

Changing the ordering of values in a given signal has no effect on the pdf of that signal, and thus, allows the use of ICA in datasets that are not time series, such as images or, actually, as votes in ballots.

In order to detect vote buying (VB) using ICA, we made a few assumptions, some of them as strong as those assumed in the statistical mechanics tools used for the same purposes. First, it is important to note that a change in the ordering of the values of a variable has no effect in the pdf.

First, we assumed that the group of citizens that do not normally vote, were compelled to change their behavior and casted their vote by an unique candidate, the one suspicious of VB, and hereafter labeled as SC. That is, our assumption is that the VB existed, although we are aware of the dificulties in documenting such behavior.

Our second assumption is that voters for candidates other than the SC were not convinced of changing their minds to vote for him/her. Our third assumption was that hardcore supporters of the SC and no voters are independent and well differentiated groups of citizens. People that do not normally vote may do so because of lack of interest, or as a result of political concerns. People that normally vote for one candidate tend to do so because they are convinced of her/his ideas, or because they were convinced by bribes. We think of those two groups as independent entities. This assumption is closely related to the requirement of statistical independence of the components.

A fourth assumption is that VB was not homogeneously distributed, that is, not all ballots present evidence of it.

Let X_i be the variable that represents the number of votes received by the SC along the ballots i. Let Y_i be the number of people that did not show up to cast their vote, that is, the no-voters. As the number of voters assigned to each ballot is fixed (at least in the case presented in the next section), we have to take extra measures to guarantee as much as possible the independence of X and Y. Let L_i be the number of voters assigned to ballot i. If only one candidate contending in the elections, there would not be independence between X_i and Y_i as a consequence of the fact that the maximum number of votes is delimited by $L_i = X_i + Y_i$. However, if the number of candidates is greater than 1, then there is room for independence between X and Y, since in general $L_i > X_i + Y_i$.

For simplicity, we will work with the proportion of votes instead of the number of votes. Let $E_i = X_i/L_i$ be the fraction of votes received by the SC in ballot i and $M_i = Y_i/L_i$ the percentage of no voters in i. The final goal would be to identify the actual number of votes the SC would had achieved if no VB occurred. Let e_i be that number, and let m_i be the actual number of no voters that would had existed under the same conditions. See Fig. 1-b.

If we follow the attributes of the results offered by ICA to its last logical consequences (see previous section), we have to admit two undesired consequences: Since the resulting components are scaled and presented in no particular order, we would not be able to distinguish e and m. This avoids us to actually identify the real number of no voters that were convinced to vote for the SC, but still, it offers evidence that VB is present in the elections. Also, postprocessing stages in order to isolate the component that represents those numbers would be possible (as an extension of this work). A second logical result is that by VB, some hardcore voters of the SC would had been convinced of not voting. This assumption may not be totally consistent with the reality.

Regardless of the apparent strong assumptions and a pair of undesired consequences of ICA to detect VB, we present in the next section the results of inspecting the Presidential Mexican elections of 2012. There, we offer evidence of VB since we were able to identify conditions in which the signals e and m were reconstructed in a way that the conditions sustaining ICA were maintained.

4   Results

In the Presidential Mexican elections of 2012, more than 77 million citizens had the right to vote, from which over 48 millions did so. There were 149500 ballots, and the range of voters assigned to each was $[71, 801]$ [23]. There were four candidates, although one of them obtained less that 2% of the votes and is no further regarded.

Figure 2 shows some statistics about the three candidates, including the SC (A). A first difference is already clear here, since it is peculiar that candidate A has very few ballots in which he only received a very small percentage of votes. This, however, is just a peculiarity and it might be possible that this effect is caused by legitimate actions. Following the method described in [12], we constructed an electoral signature for the suspicious candidate, shown in the bottom part of Fig. 2. This electoral signature consists of a series of histograms for all ballots indicating the percentage of turnout and the percentage of votes achieved by the suspicious candidate. If bimodal distributions or the percentage

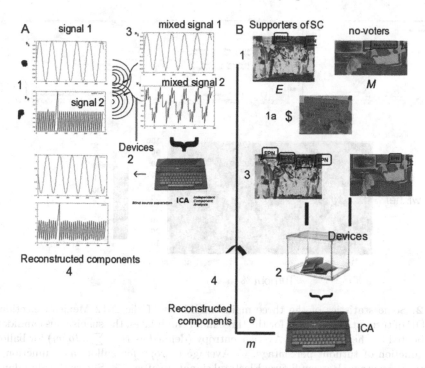

Fig. 1. An overview of ICA (A). Independent signals are generated (1) and mixed by the environment (2), where they are registered with a device (2). The mixed signals (3) are processed by ICA and the independent components (signals) are computed. The same procedure is applied for vote buying detection (B). The original signals represent supporters of the suspicious candidate and no voters (1). When no voters are exposed to bribes, some of them cast their vote towards the candidate under suspicious. The ballot (2) is the counterpart of the measuring device, from which two mixed signals E and M (3) are recovered and fed to ICA. ICA reconstructs the original signals e and m.

of votes for the winning candidate is very high and very similar to the percentage of turnout, then there is evidence of ballot stuffing. In the case of Mexican elections, there is no evidence of such behavior. However, it is again peculiar that the percentage obtained by the suspicious candidate is never below 10–15 %. Accordingly to [12], this may constitute a cause for concern.

Our objective was to find evidence of VB in this elections using ICA and its formalisms. Evidence is found once two signal are reconstructed such that their joint pdf is factorable and their marginal pdfs are non-normal. In the case here investigated, it means we have to identify, from the universe of 149000 ballots, those that satisfy the required conditions by ICA. For example, if all ballots were assumed to be influenced by VB, the components obtained by ICA show Gaussian distributions (see Fig. 3). A remaining question is to identify those ballots whose reconstructed signals.

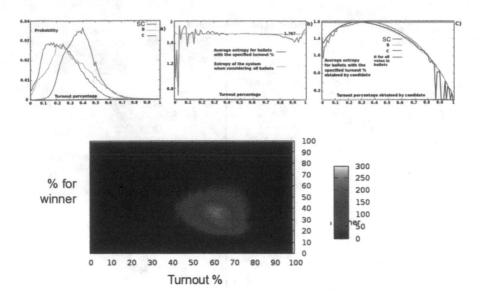

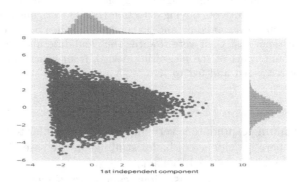

Fig. 2. Some statistics of the three main candidates of the 2012 Mexican elections. A: PDF of turnout percentages for the three main candidates, the suspicious candidate is indicated by the label SC. B: Average entropy (defined as $H = \sum p_i log p_i$) for ballots as a function of turnout percentage. C: Average entropy for ballots as a function of turnout percentage. Bottom figure: Electoral signature after [12]. Shows the abundance of ballots with the specified turnout percentage and the percentage of votes for the suspicious candidate.

Fig. 3. Components obtained by ICA from all 149000 ballots. It is observed that the components are Gaussian (p value = 0.32 and 0.24 from a Shapiro-Wilk test for the two components)

E and M agree with the constraints of ICA. Two possibilities arose from this task.

The first and näive way to identify ballots that stick to the ICA constraints of non-Gaussianity and independence is to hand pick (perhaps through an incremental brute force algorithm) those that fullfil the requirement. This would be

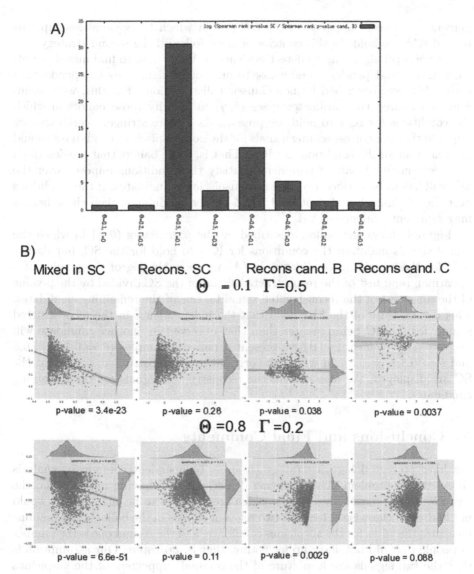

Fig. 4. The ballots described by parameters θ, γ for which the conditions behind ICA hold, for the SC, and at the same time, are not valid for the remaining candidates. In A we show the log of the p-value of a Spearman rank test for the reconstructed signals of the SC divided by the p-value of the same test over the corresponding signals for one of the remaining candidates. In B it is shown the mixed and reconstructed signals for the SC, as well as the reconstructed signals for the remaining two candidates

time consuming, not to mention biased, since we would be guiding the search towards the results we desire. A second and more convenient alternative is to consider only ballots with at least a given turnout percentage (θ) and a given percentage of votes assigned to the SC (γ), in which the general assumptions behind ICA are hold. In this contribution, we followed the second strategy.

The general algorithm to detect evidence of VB is then to find mixed signals with statistical dependence and whose reconstructed signals are independente of each other, and described by non-Gaussian distributions. For this, we systematically explored the parameter space (θ, γ) to identify those ballots in which the conditions for ICA to hold are present. As a more stringent condition, we imposed that the corresponding signals for the non-suspiscious candidates should not maintain all the conditions for ICA. That is, if the ballots that are described by a certain (θ, γ) pair of parameters satisfy the conditions imposed over the SC, still has to be verified over the non-suspiscious candidates. If the conditions over the SC hold, but do not hold for the other candidates, then those ballots may represent evidence of VB.

Figure 4 shows the ballots described by the parameters (θ, γ) in which the mixed signals maintain the conditions for ICA to hold for the SC, but do not hold for the remaining candidates. In (A) we show the log of the p-value of a Spearman rank test of the reconstructed signal of the SC divided by the p-value of the same test of the reconstructed signal for one of the remaining candidates. This is a measure of the likelihood that the independence in the reconstructed signals for the SC being greater than that observed in another candidate will be caused by chance. In (B) we present a pair of examples ($\theta = 0.1, \gamma = 0.5$ and $\theta = 0.8, \gamma = 0.2$), in which the original and reconstructed signals for the SC are displayed. We also present the reconstructed signals for the remaining candidates.

5   Conclusions and Final Comments

We have presented here a strategy to apply independent component analysis (ICA) to the problem of vote buying detection. Vote buying occurs when the suspicious candidate bribes voters, which are assumed to be citizens that would not had vote otherwise. There are two relevant signals in the ballots: The first one measures the votes for the suspicious candidate, and the second one is the number of people that did not show up for voting, that is, no voters. Our assumption is that the two signals are a mixture of the original supporters of the suspicious candidate and the normally no-voters.

Several assumptions had to be made in order to apply ICA in the vote buying problem, such as that of independence between hardcore supporters of the suspicious candidate and that of no voters. However, the assumptions made in this work are not more stringent than those in other models. The use of ICA detected groups of ballots in which the reconstructed signals satisfy the requirements behind ICA, namely, non-Gaussian components, and independent pdfs.

Although it is extremely difficult to corroborate our findings by means other than from statistical mechanics tools or journalist works, we think ICA offers an additional tool to be considered in electoral forensics.

References

1. Ball, P.: Critical Mass: How One Thing Leads to Another. Farrar, Straus and Giroux, New York (2006)
2. Galam, S.: Application of statistical physics to politics. Phys. A Stat. Mech. Appl. **274**, 132–139 (1999)
3. Costa Filho, R.N., Almeida, M.P., Moreira, J.E., Andrade Jr., J.S.: Brazilian elections: voting for a424 scaling democracy. Phys. A **322**, 698–700 (2003)
4. Fortunato, S.: Physics peeks into the ballot box. Phys. Today **65**(10), 74–76 (2012). doi:10.1063/PT.3.1761
5. Alvarez, R.M., Hall, T., Hyde, S.: Election Fraud. Detecting and Deterring Electoral Manipulation. Brookings Institution Press, Washington, DC (2012)
6. Vicente, P., Wantchekon, L.: Clientelism and vote buying: lessons fromfield experiments in African elections. Oxford Rev. Economicpolicy **25**(2), 292–305 (2009). doi:10.1093/oxrep/grp018
7. http://www.guardian.co.uk/world/2012/jul/04/mexico-elections-shadow-pena-nieto. Accessed 20 Dec 2012
8. Taibo, P.I. II, Poniatowska, E., Diaz-Polanco, H., Mejia-Madrid, F., Vasconcelos, H., Martinez, S., Miguel, P., Ramirez-Cuevas, J., Suarez del Real, J.A.: Fraude 2012. In: Para Leer en Libertad AC (ed.) Movimiento de Regeneracion Nacional (2012)
9. Ribando Seelke, C.: Mexico 2012 Elections. Congressional Research Service 2012, 7-7500 (2012)
10. Diekmann, A., Jann, B.: Benford's Law and Fraud Detection: Facts and Legends (2013)
11. Deckert, J., Myagkov, M., Ordeshook, P.: Benfords law and the detection of election fraud. Polit. 427 Anal. **19**, 245–268 (2011)
12. Klimek, P., Yegorovb, Y., Hanela, R., Thurner, S.: Statistical detection of systematic election 430 irregularities. In: PNAS 2012, pp. 1–5 (2012)
13. Vergara, R.: Elección Comprada: el Escándalo Peña Nieto-Soriana. Revista Proceso. 7 de Julio de 2012. http://www.proceso.com.mx/?p=313518. Retrieved on 9 July 2012
14. Hernández, J.: Astillero, diario La Jornada. Consulted on 8 July 2012. http://www.jornada.unam.mx/2012/06/11/opinion/004o1pol
15. Hyvärinen, A., Karhunen, J., Oja, E.: Independent Component Analysis. Wiley, New York (2001)
16. Hyvärinen A. Independent component analysis: recent advances. Phil. Trans. Royal Soc. (2011)
17. Hyvarinen, A.: Fast and robust fixed-point algorithms for independent component analysis. IEEE Trans. Neural Netw. **10**(3), 626–634 (1999). doi:10.1109/72.761722
18. Stone, J.: Independent Analysis Component: A Tutorial Introduction. MIT Press, Cambridge (2004)
19. Choi, S., Cichocki, A., Park, H., Lee, S.Y.: Blind source separation. Neural Inf. Process. Lett. Rev. **6**(1), 1–57 (2005)
20. Levin, I., Pomares, J., Alvarez, R.M.: Using machine learning algorithms to detect election fraud. Alvarez, R.M. (Ed.) Computational Social Science: Discovery and Prediction. Cambridge University Press, Cambridge
21. Nichter, S.: Vote buying or turnout buying? Machine politics and the secret ballot. Am. Polit. Sci. Rev. **102**(1), 19–31 (2008)

22. Comon, P., Jutten, C.: Handbook of Blind Source Separation. Independent Component Analysis and Applications. Academic Press, Oxford (2010)
23. http://74.200.195.178/prep/NACIONAL/PresidenteNacionalEdoVPC.html
24. http://www.theguardian.com/world/2012/jul/04/mexico-elections-shadow-pena-nieto

Unsupervised Relation Extraction in Specialized Corpora Using Sequence Mining

Kata Gábor[1]([🖂]), Haïfa Zargayouna[1], Isabelle Tellier[2], Davide Buscaldi[1], and Thierry Charnois[1]

[1] LIPN, CNRS (UMR 7030), Université Paris 13, Villetaneuse, France
{kata.gabor,haifa.zargayouna,davide.buscaldi,
thierry.charnois}@lipn.univ-paris13.fr
[2] LaTTiCe, CNRS (UMR 8094), ENS Paris, Université Sorbonne Nouvelle - Paris 3,
PSL Research University, Université Sorbonne Paris Cité, Paris, France
isabelle.tellier@univ-paris3.fr

Abstract. This paper deals with the extraction of semantic relations from scientific texts. Pattern-based representations are compared to word embeddings in unsupervised clustering experiments, according to their potential to discover new types of semantic relations and recognize their instances. The results indicate that sequential pattern mining can significantly improve pattern-based representations, even in a completely unsupervised setting.

1 Introduction

Relation extraction and classification deal with identifying the semantic relation linking two entities or concepts based on different kinds of information, such as their respective contexts, their co-occurrences in a corpus and their position in an ontology or other kind of semantic hierarchy. It includes the task of finding the instances of the semantic relations, i.e. the entity tuples, and categorizing their relation according to an ontology or a typology. Most systems focus on binary semantic relations. In supervised learning approaches, relation extraction is usually performed in two steps: first, the entity couples corresponding to concepts are extracted or generated, and a binary classification is performed to distinguish those couples which are instances of any semantic relation. Second, the relation itself is categorized according to its similarity to other, known relation types. Unsupervised relation extraction has received far less attention so far. In an unsupervised framework, relation types are inferred directly from the data and, instead of a pre-defined list of relations, new types can be discovered in parallel with the categorization of relation instances.

Unsupervised extraction is often applied to specialized domains, since the manual construction of knowledge bases or training examples for such domains is costly in terms of effort and expertise. The research we present is concerned with unsupervised relation extraction in the scientific domain. The types of relations we expect to extract allow to provide a deeper semantic model and understanding

© Springer International Publishing AG 2016
H. Boström et al. (Eds.): IDA 2016, LNCS 9897, pp. 237–248, 2016.
DOI: 10.1007/978-3-319-46349-0_21

of scientific papers and, in the long term, contribute to automatically build the state of the art of a given domain from a corpus of articles relevant to it. The deep semantic analysis of scientific literature allows the identification of inter-document links to facilitate inter-textual navigation and the access to semantic context [28]. It also allows to study the evolution of a scientific field over time [6,26], as well as the access to scientific information through information retrieval [30].

Within this research context, a typology of domain-specific semantic relations was first created and used for corpus annotation in order to confirm the feasibility of the extraction task and to perform extrinsic evaluation of the results [12]. From this point on, our approach to the relation extraction task is completely unsupervised: we do not rely on any of the manually annotated or categorized data. The most important issue consists in defining an approach which is independent from both the domain and the corpus: we do not want to impose any constraint on the types of the relations to be extracted. Different types of information, such as pattern-based representations and word embeddings, are used as input to the classification of the entity couples according to the semantic relation. After performing a range of clustering experiments, we conclude that pattern-based clustering can be significantly improved using sequence mining techniques, yielding the best results in every clustering algorithm we tested.

In what follows, we present the state of the art and explain the specificities of our approach as compared to previous work (Sect. 2). We then describe the data we used (Sect. 3), the input features (Sect. 4) and the algorithms (Sect. 5). Subsequently, we present the evaluation (Sect. 6) and discuss the results (Sect. 6.3). Section 7 concludes the paper and indicates the lines of future work.

2   Related Work

Semantic relation extraction and classification is an important task in the domain of information extraction, knowledge extraction and knowledge base population. A plethora of approaches have been applied to relation extraction, among which we can distinguish tendencies according to:

- whether the method aims to classify entity couples in a given set of relations or to discover new types of relations,
- the approach to be used: symbolic approaches through e.g. hand-crafted extraction patterns, or machine learning approaches (classification/clustering or sequence labeling methods),
- the input information and the representation used: pattern-based, lexical, syntactic features, distributional vectors, etc.

Supervised systems rely on a list of pre-defined relations and categorized examples, as described in the shared tasks of MUC and ACE campaigns [15]. Using small, manually annotated training corpora, these systems extract different kinds of features eventually combined with external knowledge sources, and build classifiers to categorize new relationship mentions [38]. Symbolic systems, similarly to supervised learning algorithms, are specific to the list of relations they are designed to recognize - based on hand-crafted linguistic rules created by linguists or domain experts [13]. On the other hand, with the proliferation of available corpora, a new task emerged: *Open Information Extraction (OpenIE)* [1,8]. It aims at developing unsupervised or distantly supervised systems with a double objective: overcome the need of scarcely available annotated data, and ensure domain-independence by being able to categorize instances of new relation types. In this kind of work, applications are not limited to a given set of relations and become able to cope with the variety of domain-specific relations [1,10]. Such experiments can also be beneficial for the automated population of ontologies [27] or thesauri [10]. Our work belongs to that second line of research.

According to the type of input features which serve as a base for classification, we can distinguish pattern-based approaches from classification approaches relying on diverse quantifiable attributes. The hypothesis behind pattern-based approaches is that the semantic relation between two entities is explicit in at least a part of their co-occurrences in the text, and therefore relation instances can be identified based on text sequences between/around the entities. Such characteristic patterns are usually manually defined, incorporating linguistic and/or domain knowledge in rule-based approaches [13,34]. Patterns are not limited to sequences of words, they can contain a combination of lexical and syntactic information [9]. Patterns can also be used indirectly as inputs to supervised classifiers [1] or for calculating similarities between entities' distribution over patterns [20,32]. Most of these approaches rely on hand-crafted lists of patterns. In [4], sequential pattern mining is used to discover new linguistic patterns within the framework of a symbolic approach.

Another way of including quantifiable context features for relation extraction is to use distributional word vectors, either as "count models" [3] or as word embeddings [21]. Entity couples can be represented by a vector built from the vectors associated with each of its members; popular methods include concatenating the two vectors [2] or taking their difference [33]. These representations will then serve as input for a supervised classifier. However, it has recently been argued in [19] that both concatenation and difference are "clearly ignoring the relation between x and y" (i.e. what links the entities): they only provide information on the type of the individual entities. In this article, the conclusion was that "contextual features might lack the necessary information to deduce how one word relates to another".

Finally, certain biclustering or iterative clustering methods are sometimes applied to divide not only the objects (word or entity couples), but also the dimensions (patterns or features) in parallel. Generative models are more prevalent in this framework. In [35] Latent Dirichlet Allocation (LDA) is adapted to the task of unsupervised relation classification. In [16] Markov logic is used, while in [24] an iterative soft clustering algorithm is applied, based on a combination of distributional similarities and a heuristic method for mining hypernyms in the corpus.

The approach we put forward belongs to the unsupervised/OpenIE framework. We do not rely on any manually classified data or typology of relations. Our experiments rely on unsupervised clustering using two types of representations: text patterns and word embeddings. Moreover, we make use of sequential pattern mining in order to enrich our couples of entities/text patterns matrix and address data sparsity. Our experiments were conducted on the ACL Anthology Corpus of computational linguistics papers, but they can be applied to any field in the scientific domain. In the context of our work, the final purpose is to extract the state of the art of a scientific domain, therefore the constitution of the corpus and the evaluations are focused on the relation types relevant for this kind of information; however, this context does not directly influence our choice of representation and clustering algorithm. Our approach differs from standard relation classification tasks, as defined e.g. in SemEval campaigns [14] in two respects. First, we do not target relations belonging to a pre-defined set. Second, the semantic relations considered in SemEval were lexical by nature, e.g.:

Component-Whole *Example: My apartment has a large kitchen.*
Member-Collection *Example: There are many trees in the forest.*

On the contrary, the relations we hope to extract are largely contextual. The same couple of entities can instantiate several distinct relations in the same corpus in different contexts:

Uses_information: (...) **models** extract rules using **parse trees** (...)
Used_for: (...) **models** derive **discourse parse trees** (...)

3 Data and Resources

For the purpose of these experiments, we used a corpus where concepts in the scientific domain are annotated. The corpus is extracted from the ACL Anthology Corpus [29]. We decided to focus on the abstract and introduction parts of scientific papers since they express essential information in a compact and often repetitive manner, which makes them an optimal source for mining sequential patterns. The resulting corpus of abstracts and introductions contains 4,200,000 words from 11,000 papers.

Entity annotation was done in two steps. First, candidates were generated with the terminology extraction tool TermSuite [7]. The list of extracted terms was then mapped to different ontological resources: the knowledge base of Saffron Knowledge Extraction Framework [5], and the BabelNet ontology [25]. If a term was validated as a domain concept (i.e., found in at least one of the resources), it was annotated in the text. The reader is referred to [11] for further information on the annotated corpus.

4 Input Representations

The goal of this part is to represent each co-occurring entity couple in a vector space which allows to calculate a similarity between them. Three distinct types of vector spaces were used as representation bases for our clustering experiments. The first two are pattern-based: they rely on the assumption that couples of entities linked by the same semantic relation will be characterized by similar patterns in at least a part of their co-occurrence contexts (i.e. in the text between the two elements of the couple). One of the representations uses complete text sequences as they are found in the corpus, while the other one relies on *patterns* that were extracted from these sequences using sequential pattern mining.

The expected advantages of identifying patterns inside the sequences are similar to those using distributed representations. First, using the complete sequences as features leads to data sparsity. Although patterns are basically sub-sequences of the sequences in the string representation and thus we can expect the size of the feature space to grow, the same sequence can belong to more than one pattern, and thus the number of *frequent* features is also expected to grow. Second, while adding some words to a sequence may not modify its meaning and the relationship between the two entities, it will still result in separate features in the full sequence representation. A pattern-based representation can capture and quantify the elements of similarity between close, but not identical sequences. Finally, sub-sequences can encode different types of information, e.g. grammatical words can be relevant for the relation between the entities, while content words will provide information about the topic of the context, and both kinds of information are expected to bring us closer to characterizing the semantic relation.

The third representation uses word embeddings of the entities considered separately and hypotheses that their semantic relation is mainly context-independent. By calculating the pairwise similarity between the entities, we expect to quantify the similarity between relation instances. This representation is similar to the one used in [2], though the scope of the experiments and the classification method are different.

4.1 Pattern-Based Representations

In the pattern-based representation, attributes correspond to text sequences that are found between co-occurring entity couples. We extracted from our corpus every entity couple occurring in the same sentence, together with the text between them. Text sequences can contain other entities, but their length is limited to ≤ 8 words. This results in 998,000 instances extracted.

String Representation. Using these co-occurrence data, we first built a sparse matrix M with lines corresponding to entity couples $e = (e_1, e_2)$ and columns corresponding to text sequences $p \in P$. The cells $M_{e,p}$ contain an association value between e et p. One of the representations uses raw co-occurrence count, while the other one uses $PPMI_\alpha$ weighting. This weight is a variant of Pointwise Mutual Information (PMI) in which values below 0 are replaced by 0. Moreover, the context distribution smoothing method proposed by [18] is applied to the positive PMI weighting. This smoothing, inspired by the success of neural word embeddings in semantic similarity tasks [3, 21], allows to reduce the bias of PMI towards rare contexts. Context words' co-occurrence counts are raised to the power of α (in Eq. (2)). Its optimal value is reported to be $\alpha = 0{,}75$ according to the experiments of [22]. This finding was directly adapted to PMI [18]:

$$PPMI_\alpha(e, p) = max(log_2 \frac{P(e, p)}{P(e) \times P_\alpha(p)}, 0) \qquad (1)$$

$$P_\alpha(p) = \frac{freq(c)^\alpha}{\sum_c freq(c)^\alpha} \qquad (2)$$

We will refer to the vectors built as such as the **string** representation.

Sequential Pattern Representation. For the second experiment, we applied sequential pattern mining techniques [31] to discover relevant patterns which are specific to semantic relations. The extraction is completely unsupervised: frequent sequential patterns which fulfill a certain number of constraints are automatically extracted from the input. A sequence, in this context, is a list of literals (items) and an item is a word in the corpus. The input corpus was made of all the sequences extracted from co-occurring entities (i.e. the feature space for the **string** representation). The pattern mining process is applied to word forms without using any additional linguistic information.

The sequence mining tool [4] we used allows distinct options to add constraints on the extracted sequences. We selected contiguous sequences of length between 2 and 8 words and a minimum support of 10. The support of a sequential pattern in a sequence database is the number of sequences in the database containing the pattern. Only *closed* sequential patterns were considered, i.e. patterns which are not sub-sequences of another sequence with the same support.

To construct the matrices, we filled the cells with raw co-occurrences (how many times a pattern occurs somewhere between the two entities) and, for a second matrix, with the $PPMI_\alpha$-weighted values. We will refer to this representation as **pattern** representation.

4.2 Distributional Representation

This type of feature space also uses contextual information, but it is computed independently for the two entities. We used word2vec [21] to create the distributional vectors, as it proved to be particularly well adapted for semantic similarity tasks and is presumed to encode analogies between semantic relations [23]. word2vec was trained on the whole ACL Anthology Corpus using the skip-gram model [21] and the resulting word embeddings (size = 200) were used to represent each entity.

The vector of an entity couple is simply made of the concatenation of the vectors of each entity[1] [2]. We expect this representation to capture very specific

---

[1] Entities corresponding to multiword expressions will have their unique vector, since word2vec includes an internal module for recognizing multiword expressions.

relation types, where the potential arguments belong to a restricted semantic class.

5 Clustering

Two methods of hierarchical clustering were tested using cosine similarity and Cluto's [37] clustering functions. The first one is a top-down clustering based on repeated bisections: at every iteration, a cluster is split in two until the desired number of clusters is reached. This number has to be pre-defined: experiments were performed using different values. The cluster to be divided is chosen so that it maximizes the sum of inter-cluster similarities for each resulting cluster. We will refer to this method as **divisive**.

The second method is a hierarchical agglomerative clustering with a bisective initialization [36]: a clustering in $\sqrt{n}$ clusters is first calculated (where n is the number of clusters to be produced) through repeated bisections. The vector space is then augmented with $\sqrt{n}$ new dimensions that represent the clusters calculated at the first step, and the values of these dimensions are given by the distance of each object from the centroids of the clusters produced at the initiation stage. The agglomerative clustering is then performed on this augmented vector space. This method was created to combine advantages of divisive (global) and agglomerative (local) algorithms by reducing the impact of errors from initial merging decisions, which tend to be multiplied as the agglomeration progresses [36].

We will refer to this algorithm as **agglo**.

6 Evaluation

6.1 Standard Classification

For the sake of the experiments, we selected a sample of 500 abstracts (about 100 words/abstract) and manually annotated relevant relations occurring in this sample. The typology of relations was data-driven: it was established in parallel with the categorization of the examples. An illustration of the relations we identified is shown in Table 1, for a complete description of the manual annotation work and the typology, see [12]. The relations are not specific to the natural language processing domain; they can be used for any scientific corpora.

As a second step, a sample of 615 entity couples which co-occur in the corpus was manually categorized according to this typology. This sample was used as a gold standard for clustering evaluation.

6.2 Baseline and Evaluation Measures

The clustering results were compared to the standard one as a series of decisions: whether to classify two couples in the same group or in different groups. This evaluation is less influenced by structural differences between two clustering solutions and allows to quantify results in terms of precision and recall.

Table 1. Extract of the typology of semantic relations

char	ARG1: *observed characteristics* of an observed ARG2: *entity*
composed_of	ARG1: *database/resource* ARG2: *data*
methodapplied	ARG1: *method* applied to ARG2: *data*
model	ARG1: *abstract representation* of an ARG2: *observed entity*
phenomenon	ARG1: *entity, a phenomenon* found in ARG2: *context*
propose	ARG1: *paper/author* presents ARG2: *an idea*

We also calculated APP *Adjusted Pairwise Precision* [17]: this measure quantifies average cluster purity, weighted by the size of the clusters. This provides additional information on the proportion of the relevant clusters.

$$APP = \frac{1}{|K|} \sum_{i=1}^{|K|} \frac{nb \; correct \; pairs \; in \; k_i}{nb \; pairs \; in \; k_i} \times \frac{|k_i| - 1}{|k_i| + 1} \tag{3}$$

For each experiment with respect to cluster size, we created a corresponding random clustering to estimate the difficulty of the task and the contribution of our approaches.

6.3 Results and Discussion

The evaluation was conducted so as to allow comparisons between the two clustering algorithms, the three input representations and the two weighting systems. Cluster sizes have an important effect on the results because they are correlated with the number of classes in the standard (21 in our case). On the other hand, a different cluster structure e.g. with finer grained distinctions, may also be semantically justified. The real validity of the clusters must therefore be established by human inspection.

Table 2 shows the results of the **divisive** method. The **string** and **pattern** representations with raw frequency counts are compared with the baseline and **word2vec** vector representations (where weights are implicitly included in the language model learning). Although the word2vec representation yields a very good performance with respect to both precision measures, it comes at the cost of a very low recall. Since this representation is solely based on the similarity between individual entities, this means that mainly couples having nearly identical entities end up in the same cluster, e.g.: *parsing - sentences, parses - sentences, parse - sentence*. In agreement with [19], this result reveals that this representation is not good at capturing relational similarities.

Another interest of Table 2 is the improvement in precision (for both measures) brought by the **pattern** representation, compared to the **string** representation. This improvement is accompanied by slight decreases in recall. It is also interesting to note that, as shown by Table 3, $PPMI_\alpha$ weighting transforms

Table 2. Clustering results with the divisive algorithm

Input	#clusters	Algorithm	Weight	APP	Prec	Recall	F-measure
baseline	100	random	N/A	0.0813	0.0955	0.0097	0.0176
baseline	50	random	N/A	0.0883	0.1036	0.0198	0.0332
baseline	25	random	N/A	**0.0979**	**0.1040**	**0.0410**	**0.0588**
string	100	divisive	freq	0.2498	0.3037	0.1030	0.1538
pattern	100	divisive	freq	0.2823	**0.3718**	0.0993	0.1568
string	50	divisive	freq	0.2985	0.2805	0.1302	0.1778
pattern	50	divisive	freq	0.3265	0.3159	0.1235	0.1776
string	25	divisive	freq	0.3941	0.2219	**0.1904**	0.2050
pattern	25	divisive	freq	**0.3947**	0.2776	0.1773	**0.2164**
word2vec	100	divisive	incl	0.3396	**0.5734**	0.0527	0.0965
word2vec	50	divisive	incl	0.3541	0.4761	0.0890	0.1499
word2vec	25	divisive	incl	**0.3545**	0.4182	**0.1539**	**0.2250**

Table 3. The effect of the PPMI$_\alpha$ weighting with the divisive algorithm

Input	#clusters	Algorithm	Weight	APP	Prec	Recall	F-measure
string	100	divisive	PPMI$_\alpha$	0.3112	**0.4905**	0.0462	0.0844
string	50	divisive	PPMI$_\alpha$	**0.3625**	0.3789	0.0799	0.1320
string	25	divisive	PPMI$_\alpha$	0.3555	0.3133	**0.1400**	**0.1936**

sequence-based scores the same way as to what we observe with **word2vec** representations: very high precision with very low recall -despite the fact that the semantics captured by the input representations are different in both cases.

Table 4 presents the results of the **agglo** clustering method. This algorithm works better for every type of representation we considered. The scores reported here are obtained on PPMI$_\alpha$-weighted string and pattern representations. The **pattern** representation comes out as the absolute winner, with important improvements over **string** both in terms of precision (6–10%) and recall (3.5–6.5 %). Scores marked by * indicate statistically significant improvements according to a 10-fold cross-validation on the **string** and **pattern** clustering solutions. The **pattern** representation also beats the precision of **word2vec** in two out of the three settings. Although the recall obtained with the word2vec vectors is also improved by the agglomerative method, the **pattern** representation still holds an important advantage.

Table 4. Clustering results with the agglomerative algorithm

Input	#clusters	Algorithm	Weight	APP	Prec	Recall	F-measure
string	100	agglo	$PPMI_\alpha$	**0.3020**	0.4184	0.1582	0.2296
pattern	100	agglo	$PPMI_\alpha$	0.2810	**0.4758**	0.1936	0.2752
string	50	agglo	$PPMI_\alpha$	0.2535	0.3246	0.2142	0.2581
pattern	50	agglo	$PPMI_\alpha$	0.2697	0.4200*	0.2657*	0.3268
string	25	agglo	$PPMI_\alpha$	0.2585	0.2898	0.2277	0.2550
pattern	25	agglo	$PPMI_\alpha$	0.2460	0.3777*	**0.2914**	**0.3290**
word2vec	100	agglo	incl	**0.3630**	**0.5285**	0.1316	0.2107
word2vec	50	agglo	incl	0.2966	0.3694	0.1938	0.2542
word2vec	25	agglo	incl	0.2972	0.3330	**0.2399**	**0.2789**

7   Conclusion and Future Work

We presented an approach to extract new types of semantic relations and instances of relations from specialized corpora using unsupervised clustering. Two types of representations were compared: pattern-based vectors and word embeddings. In agreement with previous results, we found that concatenated word embeddings tend to have a limited contribution to discovering new relation types.

An important finding is that sequential pattern mining contributes to create a much more adapted feature space, as shown by the significant improvement both in terms of precision and recall. This confirms our expectation that sequential patterns are better than full sequences in capturing relational similarities. Another advantage is that the pattern mining process is completely unsupervised.

We plan to conduct a manual evaluation of the resulting clusters. This would allow to have a better insight on the nature of the resulting clusters. Biclustering methods can also be tested on the data: they have the potential to automatically identify the most relevant patterns for each relation type.

Acknowledgments. This work is part of the program "Investissements d'Avenir" overseen by the French National Research Agency, ANR-10-LABX-0083 (Labex EFL). The authors would like to thank the anonymous reviewers for their valuable comments.

References

1. Banko, M., Cafarella, J., Soderland, S., Broadhead, M., Etzioni, O.: Open information extraction from the web. In: IJCAI, pp. 2670–2676 (2007)
2. Baroni, M., Bernardi, R., Do, N.-Q., Shan, C.-C.: Entailment above the word level in distributional semantics. In: ACL 2012 (2012)

3. Baroni, M., Dinu, G., Kruszewski, G.: Dont count, predict! A systematic comparison of context-counting vs. context-predicting semantic vectors. In: ACL 2014 (2014)
4. Béchet, N., Cellier, P., Charnois, T., Crémilleux, B.: Discovering linguistic patterns using sequence mining. In: Gelbukh, A. (ed.) CICLing 2012. LNCS, pp. 154–165. Springer, Heidelberg (2012). doi:10.1007/978-3-642-28604-9_13
5. Bordea, G., Buitelaar, P., Polajnar, T.: Domain-independent term extraction through domain modelling. In: TIA 2013 (2013)
6. Chavalarias, D., Cointet, J.-P.: Phylomemetic patterns in science evolution - the rise and fall of scientific fields. PLOS ONE 8(2), e54847 (2013)
7. Daille, B.: Building bilingual terminologies from comparable corpora: the TTC termsuite. In: 5th Workshop on Building and Using Comparable Corpora, Co-located with LREC, pp. 39–32 (2012)
8. Del Corro, L., Gemulla, R.: Clausie: clause-based open information extraction. In: International Conference on World Wide Web, WWW 2013 (2013)
9. Fader, A., Soderland, S., Etzioni, O.: Identifying relations for open information extraction. In: EMNLP 2011 (2011)
10. Ferret, O.: Typing relations in distributional thesauri. In: Gala, N., Rapp, R., Bel-Enguix, G. (eds.) Language Production, Cognition, and the Lexicon. Text, Speech and Language Technology, vol. 48, pp. 113–134. Springer, Heidelberg (2015)
11. Gábor, K., Zargayouna, H., Buscaldi, D., Tellier, I., Charnois, T.: Semantic annotation of the acl anthology corpus for the automatic analysis of scientific literature. In: LREC 2016, Portoroz, Slovenia (2016, in press)
12. Gábor, K., Zargayouna, H., Tellier, I., Buscaldi, D., Charnois, T.: A typology of semantic relations dedicated to scientific literature analysis. In: SAVE-SD Workshop at the 25th World Wide Web Conference (2016)
13. Hearst, M.: Automatic acquisition of hyponyms from large text corpora. In: COLING 1992, pp. 539–545 (1992)
14. Hendrickx, I., Kim, S.N., Kozareva, Z., Nakov, D., Séaghdha, P.O., Padó, S., Pennacchiotti, M., Romano, L., Szpakowicz, S.: Semeval-2010 task 8: multi-way classification of semantic relations between pairs of nominals. In: Proceedings of the Workshop on Semantic Evaluations (2010)
15. Hobbs, J.R., Riloff, E.: Information extraction. In: Indurkhya, N., Damerau, F.J. (eds.) Handbook of Natural Language Processing, 2nd edn. CRC Press, Taylor and Francis Group, Boca Raton, FL (2010)
16. Kok, S., Domingos, P.: Extracting semantic networks from text via relational clustering. In: Daelemans, W., Goethals, B., Morik, K. (eds.) ECML PKDD 2008. LNCS (LNAI), pp. 624–639. Springer, Heidelberg (2008). doi:10.1007/978-3-540-87479-9_59
17. Korhonen, A., Krymolowski, Y., Collier, N.: The choice of features for classification of verbs in biomedical texts. In: COLING (2008)
18. Levy, O., Goldberg, Y., Dagan, I.: Improving distributional similarity with lessons learned from word embeddings. Trans. ACL 3, 211–225 (2015)
19. Levy, O., Remus, S., Biemannm, C., Dagan, I.: Do supervised distributional methods really learn lexical inference relations? In: ACL 2015 (2015)
20. Lin, D., Pantel, P.: Dirt: discovery of inference rules from text. In: ACM SIGKDD Conference on Knowledge Discovery and Data Mining (2001)
21. Mikolov, T., Chen, K., Corrado, G., Dean, J.: Efficient estimation of word representations in vector space. In: Proceedings of Workshop at ICLR (2013)

22. Mikolov, T., Sutskever, I., Chen, K., Corrado, G.S., Dean, J.: Distributed representations of words and phrases and their compositionality. Advances in Neural Information Processing Systems (2013)

23. Mikolov, T., Yih, W., Zweig, G.: Linguistic regularities in continuous space word representations. In: NAACL (2013)

24. Min, B., Shi, S., Grishman, R., Lin, C.-Y.: Ensemble semantics for large-scale unsupervised relation extraction. In: EMNLP 2012 (2012)

25. Navigli, R., Ponzetto, S.P.: BabelNet: the automatic construction, evaluation and application of a wide-coverage multilingual semantic network. Artif. Intell. **193**, 217–250 (2012)

26. Omodei, E., Cointet, J.-P., Poibeau, T.: Mapping the natural language processing domain: experiments using the acl anthology. In: LREC 2014 (2014)

27. Petasis, G., Karkaletsis, V., Paliouras, G., Krithara, A., Zavitsanos, E.: Ontology population and enrichment: state of the art. In: Paliouras, G., Spyropoulos, C.D., Tsatsaronis, G. (eds.) Knowledge-Driven Multimedia Information Extraction and Ontology Evolution. LNCS (LNAI), pp. 134–166. Springer, Heidelberg (2011). doi:10.1007/978-3-642-20795-2_6

28. Presutti, V., Consoli, S., Nuzzolese, A.G., Recupero, D.R., Gangemi, A., Bannour, I., Zargayouna, H.: Uncovering the semantics of wikipedia pagelinks. In: Janowicz, K., Schlobach, S., Lambrix, P., Hyvönen, E. (eds.) EKAW 2014. LNCS (LNAI), pp. 413–428. Springer, Heidelberg (2014). doi:10.1007/978-3-319-13704-9_32

29. Radev, D.R., Muthukrishnan, P., Qazvinian, V.: The ACL anthology network corpus. In: ACL Workshop on Text and Citation Analysis for Scholarly Digital Libraries (2009)

30. Sateli, B., Witte, R.: What's in this paper? Combining rhetorical entities with linked open data for semantic literature querying. In: Proceedings of the 24th International Conference on World Wide Web (2015)

31. Srikant, R., Agrawal, R.: Mining sequential patterns: generalizations and performance improvements. In: Apers, P., Bouzeghoub, M., Gardarin, G. (eds.) EDBT 1996. LNCS, pp. 1–17. Springer, Heidelberg (1996). doi:10.1007/BFb0014140

32. Turney, P.D.: Similarity of semantic relations. CoRR, abs/cs/0608100 (2006)

33. Weeds, J., Clarke, D., Reffin, J., Weir, D., Keller, B.: Learning to distinguish hypernyms and co-hyponyms. In: COLING 2014 (2014)

34. Yangarber, R., Lin, W., Grishman, R.: Unsupervised learning of generalized names. In: COLING 2002 (2002)

35. Yao, L., Haghighi, A., Riedel, S., McCallum, A.: Structured relation discovery using generative models. In: EMNLP 2011 (2011)

36. Zhao, Y., Karypis, G.: Evaluation of hierarchical clustering algorithms for document datasets. In: CIKM (2002)

37. Zhao, Y., Karypis, G., Fayyad, U.: Hierarchical clustering algorithms for document datasets. Data Min. Knowl. Discov. **10**, 141–168 (2005)

38. Zhou, G., Su, J., Zhang, J., Zhang, M.: Exploring various knowledge in relation extraction. In: ACL 2005 (2005)

A Framework for Interpolating Scattered Data Using Space-Filling Curves

David J. Weston[(✉)]

Department of Computer Science and Information Systems, Birkbeck College,
University of London, London, UK
dweston@dcs.bbk.ac.uk

Abstract. The analysis of spatial data occurs in many disciplines and covers a wide variety activities. Available techniques for such analysis include spatial interpolation which is useful for tasks such as visualization and imputation. This paper proposes a novel approach to interpolation using space-filling curves. Two simple interpolation methods are described and their ability to interpolate is compared to several interpolation techniques including natural neighbour interpolation. The proposed approach requires a Monte-Carlo step that requires a large number of iterations. However experiments demonstrate that the number of iterations will not change appreciably with larger datasets.

1 Introduction

Spatial interpolation is one of the many tools available for spatial data mining [10]. It is particularly useful in spatial analysis since it is often the case that data cannot be collected at every desired location due to practical issues such as cost. In addition the data may have missing values [11], that may require imputation. The literature for spatial interpolation is large and the interested reader is referred to [8] for an overview of available approaches in the practical context of environmental sciences.

Space-filling curves have been successfully used in a broad range of computational problems, for example in calculating efficiently all nearest neighbours [4] and image segmentation [9], see [1] for a comprehensive review. The primary reason for this is the fact that space-filling curves can be used to map multi-dimensional Euclidean data onto one dimension which partially preserves local spatial correlations, i.e. points that are close in the multidimensional space are likely to be close in the one dimensional ordering of the data.

[16] investigated the orderings of data along space-filling curves where the data has been repeatedly transformed using shape preserving transformations only. It was shown that the probability of an ordering is dependent on the spatial locations of the data in the higher dimensional space. This property was then used to construct novel shape descriptors for shape matching. The concern of this paper is to use the approach detailed in [16] for spatial interpolation. The essential idea is to transform the data-sites and query points using a shape

H. Boström et al. (Eds.): IDA 2016, LNCS 9897, pp. 249–260, 2016.
DOI: 10.1007/978-3-319-46349-0_22

preserving transformation and then map them onto a space-filling curve, where a simple interpolation scheme is used to impute the value at each query point. This process is repeated using different shape preserving transformations and the resulting interpolations are then aggregated. The main motivation for this work is to produce a conceptually simple approach for interpolation in two or higher dimension that is also numerically robust and simple to implement.

In the next section, relevant methods for interpolating scattered data are discussed. After which space-filling curves are introduced with a brief overview of the construction of the Hilbert curve. Following this, the framework for performing spatial interpolation using space-filling curves is introduced. Experiments to demonstrate the utility of the approach are provided. The final section concludes with ideas for future research.

2   Scattered Data Interpolation Methods

In the following discussion it will be assumed that there are n data-sites $x_1, \ldots, x_n$ with respective locations $\mathbf{x}_1 \ldots \mathbf{x}_n$ and each data-site has a value denoted $z_1 \ldots z_n$. In addition there are m query sites, $q_1 \ldots q_m$ with location $\mathbf{q}_1 \ldots \mathbf{q}_m$. It is at these locations that an imputed value is desired, i.e. we wish to estimate $\hat{f}(\mathbf{q}_j)$, for query site q_j.

There are a large array of methods available for spatial interpolation. The focus in this section will be on three methods for spatial interpolation. They have been chosen specifically because they are the higher dimensional analogues of the interpolation we do in one dimension and hence provide a clear comparison. They are *piecewise constant*, *piecewise linear* and *natural neighbour* interpolation.

Piecewise constant interpolation is a very simple approach to scattered data interpolation. The interpolated value, $\hat{f}(\mathbf{q}_j)$, for query site q_j is the value associated with the closest (in the Euclidean sense) data-site.

Piecewise linear interpolation for scattered data uses the Delauney triangulation, see e.g. [12], induced from the data-sites. The vertices in this triangulation are the data-sites. In, for example 2D, a query point will reside in one triangle. Let us assume that the vertices are the data-sites with indicies p_{1j}, p_{2j}, p_{3j}, then

$$\hat{f}(\mathbf{q}_j) = \frac{\sum_{i=1}^3 a_{p_i} z_{p_{ij}}}{\sum_{i=1}^3 a_{p_i}},$$

where a_{p_i} is the Euclidean distance between the query point q_j and the vertex location $\mathbf{x}_{p_{ij}}$

Natural neighbour interpolation is a well known approach to interpolating scattered data, see for example [3]. This interpolation scheme involves calculating a weighted sum of data-site values that are natural neighbours (definition to follow) to the query point. There exist several approaches to calculating these weights, the most well known is *Sibson* interpolation and is calculated as follows. First the Voronoi tessellation is induced from the data-site locations. This consists of partitioning the region of interest into non-overlapping tiles.

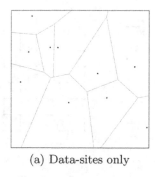

(a) Data-sites only

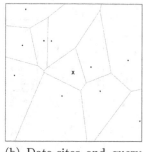

(b) Data-sites and query point

Fig. 1. Voronoi diagrams used for natural neighbour interpolation weight calculation. The query point q_j is denoted by a cross located in the centre of (b).

Each tile contains only one data-site and contains all locations that are closest to this data-site, Fig. 1(a).

The location of the query point is added to the list of data-sites and a new Voronoi diagram is produced. This is shown in Fig. 1(b) where the query point in this figure is denoted with a cross. This query point has its own Voronoi tile that contains regions taken from the original Voronoi tiles shown in Fig. 1(a). Data-sites that have had their Voronoi tile changed by the inclusion of the query point are called its natural neighbours. For a particular query point q_j

$$\hat{f}(\mathbf{q}_j) = \frac{\sum_i w_{ij} z_i}{\sum_i w_{ij}},$$

where w_{ij} is the area from the query point tile that was originally part of the Voronoi tile for the ith data-site.

Conceptually natural neighbour interpolation is relatively straightforward however computing it efficiently is rather involved [7]. Indeed approximations that rely on discretising the region of interest have been proposed to produce more efficient algorithm [13]. Natural neighbour interpolation is defined for two or more dimensions, since in 1D, the procedure for natural neighbour interpolation reduces to piecewise linear interpolation. Briefly, in 1D the Voronoi tiles are simply intervals and the natural neighbours of a query point are its predecessor and successor data-sites. Since we shall be using linear interpolation in 1D in our proposed approach, we consider natural neighbour interpolation to be a relevant method to compare against.

3   Space-Filling curves

A space-filling curve is typically defined as a continuous mapping from the unit interval $[0, 1]$, onto d-dimensional Euclidean space where the image consists of all points within the compact region $[0, 1]^d$.

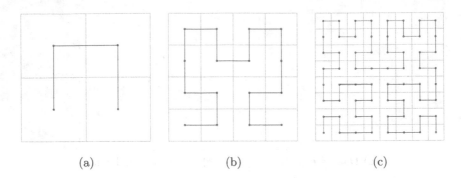

(a)   (b)   (c)

Fig. 2. First three iterations for Hilbert curve construction, the circles denote the centres of the sub-squares.

Figure 2 shows the construction of the Hilbert space-filling curve. A square is sub-divided into four sub-squares which are given a specific order and orientation. Joining the centres of these sub-squares by following their order produces a polygon approximation to the Hilbert curve, Fig. 2(a). These sub-squares are themselves recursively subdivided. Figures 2(b, c) show the polygon curve for second and third iteration respectively. In the limit as the number of iterations tends to infinity, the polygon curve tends to the Hilbert curve.

A detailed explanation regarding space-filling curves and their construction can be found in [1,14]. The code used in the paper is based on [15].

4   Framework for Interpolation

There are three stages in the proposed framework for interpolation, denoted *shape preserving embedding*, *one-dimensional interpolation* and *aggregation*. Each stage is described separately in the following sections.

4.1   Shape Preserving Embedding

The first stage involves ordering the data-sites (and query locations) in multidimensional space along a space-filling curve. The entire process is demonstrated graphically for the 2D case in Fig. 3. For simplicity it is assumed that interpolation is required over a square region of interest containing all the data-sites, Fig. 3(a). A shape preserving transformation is applied that maps the region of interest onto the unit square denoted by the grey region in Fig. 3(b). Each data-site (and query point) can then be ordered along a Hilbert curve. Figure 3(c) shows each data-site joined to their predecessor and successor along the space-filling curve.

Let e denote a shape preserving transformation that embeds the query points and data-sites successfully into the unit square. The transformation is in fact a composite, comprising a translation, a rotation, a reflection (with probability 0.5)

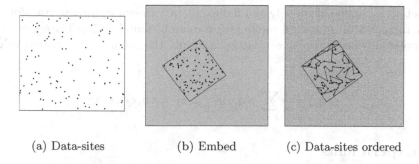

(a) Data-sites   (b) Embed   (c) Data-sites ordered

Fig. 3. Embedding data-sites onto Hilbert curve.

and a scaling. Details of the embedding can found in [16], the maximum scale factor in this study is 10. For reasons of computational simplicity the region of interest, i.e. domain over which the interpolation function $\hat{f}(\cdot)$ is to be estimated is assumed to be a discretised square with resolution 2000×2000.

Let h denote a function that maps a point in the unit square onto the unit interval using a Hilbert curve. The Hilbert index, t_i, for the ith data-site is

$$t_i = h(e(\mathbf{x}_i))$$

and similarly for query sites. The data is then sorted in ascending Hilbert index order. Let this ordering function be denoted by π, then $t_{\pi(d_i)}$ and $t_{\pi(q_j)}$ are both non-decreasing for $i = 1, \cdots, n$ and $j = 1, \cdots, m$ respectively.

4.2 One-Dimensional Interpolation

Once a Hilbert index has been associated with each datum, interpolation can proceed using any 1D interpolation method. For this study 1D piecewise constant and 1D piecewise linear interpolation (described in Sect. 2), denoted *Hilbert-Const* and *Hilbert-Linear* respectively are used. Both these methods are trivially simple to implement and due to their simplicity they are amenable to further analysis which can be achieved without the need of a ground truth function to interpolate, see Sect. 6.

4.3 Aggregation

Let g denote a function that encapsulates the two stages described above, such that interpolated value for the query site q_j is

$$\hat{f}(\mathbf{q}_j) = g(\mathbf{q}_j, \mathbf{x}_{1...m}, z_{1...m}, h, Q_1, e).$$

where Q_1 denotes a plug-in one-dimensional interpolation function and e a shape preserving transformation. Recalling that $\mathbf{x}_{1...m}, z_{1...m}$ are the data-site locations and data-site values respectively; h is the Hilbert mapping, note that this can be replaced with other space-filling curve mappings.

Let $e_1, \ldots, e_\eta$ be identically and independently drawn legitimate transformations, further details for the sampling regime can be found in [16]. The aggregated interpolated value for the query site q_j is simply the average interpolated value, i.e.

$$\hat{f}(\mathbf{q}_j) = \frac{1}{\eta} \sum_{k=1}^{\eta} g(\mathbf{q}_j, \mathbf{x}_{1\ldots m}, z_{1\ldots m}, h, Q_1, e_k).$$

5   Experiments

The following experimental design has been motivated by [6]. The interpolation schemes are tested by evaluating the mean squared error (MSE) and the maximum absolute error (Max Error) between interpolated values and a ground truth function. For the ground truth Franke's function [6] has been selected, see Fig. 6(a).

It is also instructive to visualise the interpolation, for this task a bivariate Gaussian is used see Fig. 4(a). The experiments are organised as follows. First the focus is on visualising the resulting interpolations, then a more formal approach to evaluating the interpolation methods is performed. Henceforth piecewise linear and piecewise constant shall be referred to as linear and constant respectively.

5.1   Visualising the Interpolated Function

This experiment uses data-sites located on a 21×21 regular grid spanning the entire region of interest. It should be noted that for the special case of data-sites located on regular grids there exist specific interpolation schemes that should do better than the general scattered data interpolation approaches shown here. However a regular grid generates easily interpretable images. Figure 4 shows the resulting interpolation when using the spatial interpolation techniques described in Sect. 2.

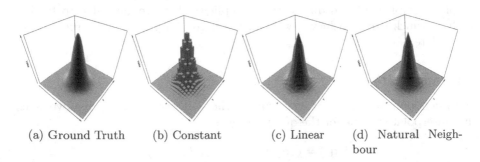

(a) Ground Truth   (b) Constant   (c) Linear   (d) Natural Neighbour

Fig. 4. Bivariate Gaussian test function and interpolated functions using standard 2D approaches based on data-sites located on a regular 21×21 grid.

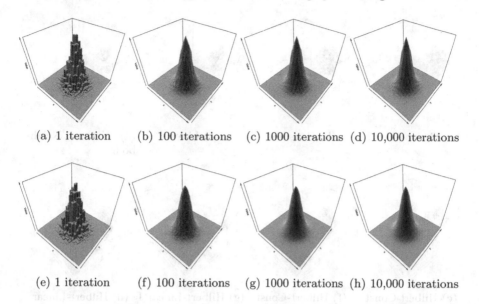

(a) 1 iteration   (b) 100 iterations (c) 1000 iterations (d) 10,000 iterations

(e) 1 iteration   (f) 100 iterations (g) 1000 iterations (h) 10,000 iterations

Fig. 5. Hilbert-Const Interpolation of a Bivariate Gaussian based on data-sites located on a regular 21×21 grid. Top row Hilbert-Const Interpolation, bottom row Hilbert-Linear Interpolation.

It is clear that constant interpolation, Fig. 4(b), does a poor job of reconstructing the ground truth. Linear interpolation, Fig. 4(c) does much better but with strong visible linear artifacts. Finally in Fig. 4(d) natural neighbour interpolation is much smoother. Although artifacts arising from the data-site locations are visible.

Figure 5(a–d) shows the interpolated function using Hilbert-Const with increasing number of iterations. Figure 5(a) shows the interpolated function after one iteration is similar to constant interpolation shown in Fig. 4(b). As the number of iterations increases, the interpolated function becomes less noisy and at 10,000 iterations the function is visibly similar to natural neighbour interpolation but with more pronounced bumps.

Figure 5(e–h) shows the interpolated function Hilbert-Linear. The first iteration may look similar Hilbert-Const however closer inspection should reveal that there are no flat regions on the tall peaks. At 10,000 iterations the interpolated function appears smoother than both natural neighbour and Hilbert-Const. In contrast to both natural neighbour and Hilbert-Const an artifact due to the data-site locations manifests itself as small dimples.

Franke's bivariate function, shown in Fig. 6(a), is a more challenging surface to interpolate. Figure 6 also shows the interpolated functions using the interpolation methods described in Sect. 2 and the resulting interpolated function after the first and $10,000th$ iteration for both Hilbert-Const and Hilbert-Linear. Visually these results are consistent with those of the bivariate Gaussian.

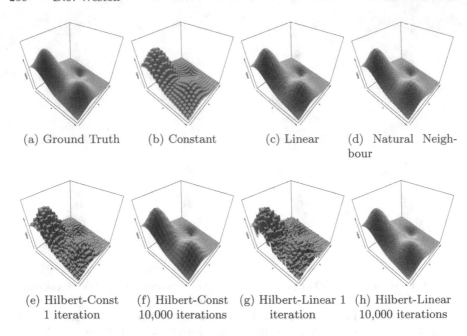

(a) Ground Truth   (b) Constant   (c) Linear   (d) Natural Neighbour

(e) Hilbert-Const   (f) Hilbert-Const   (g) Hilbert-Linear 1   (h) Hilbert-Linear
1 iteration   10,000 iterations   iteration   10,000 iterations

Fig. 6. Franke's bivariate test function and interpolated functions using standard 2D approaches based on data-sites located on a regular 21×21 grid.

5.2   Scattered Data Interpolation

The following experiments focus on more quantitative measures of the quality of an interpolation. Scattered data-sites are generated by selecting uniformly at random a location on a 2000×2000 grid without replacement, this is to ensure that all the data-site locations are unique. The number of data-sites used in the experiments are 100, 300 and 500. Finally, for completeness, the data-sites using regular locations used in the first experiment will also be used (which has 441 data-sites). The number of aggregation iterations η is set to 50,000. Natural neighbour interpolation is not defined outside the convex hull of data-sites, so to make all the results commensurate only query sites within the convex hull are included in the analysis.

Table 1 shows the evaluations for the three standard interpolation methods (natural neighbor interpolation is denoted NN in this table) and the two proposed approaches for different sets of data-site locations and number. Notable observations include the following. Constant Interpolation in 2D is consistently poor; Linear interpolation in 2D in most cases performs better than natural neighbour.

Hilbert-Const has lower Max Error than natural neighbour for all scattered data sets, but natural neighbour performs consistently better with respect to MSE. The performance difference between Hilbert-Linear and Hilbert-Const is somewhat inconclusive but it appears that Hilbert-Const performs better.

Table 1. Performance of interpolation schemes with respect to maximum absolute error and for mean squared error for the Franke function.

# Data-sites	100		300		500		441 Regular	
	Max Err	MSE	Max Err	MSE	Max Err	MSE	Max Err	MSE
Constant 2D	0.421	0.00463	0.219	0.00147	0.173	0.000821	0.119	0.000529
Linear 2D	0.229	0.00181	0.139	0.000313	0.109	8.04e−05	0.0185	1.24e−05
NN 2D	0.249	0.00202	0.147	0.000332	0.0965	7.9e−05	0.0188	1.2e−05
Hilbert-Const	0.234	0.00239	0.117	0.000407	0.0635	0.000151	0.0217	2.03e−05
Hilbert-Linear	0.308	0.00381	0.0975	0.000516	0.0704	0.000135	0.0279	4.85e−05

A key issue with the proposed approach is the number of aggregation iterations required to produce a reasonable interpolation. Figure 7 shows the number of aggregation iterations against MSE and Max Error for the Hilbert-Const interpolation of Franke's function. The convergence is largely independent of the number of data-sites and there is little to gain after around 1000 iterations. A similar result has been obtained for Hilbert-Linear interpolation.

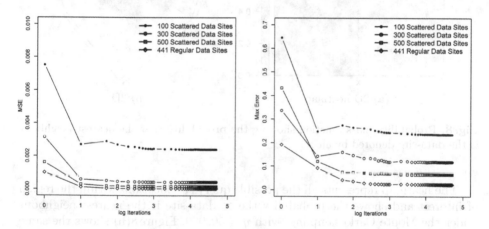

Fig. 7. Number of aggregation iterations versus MSE (leftmost graph) and Max Error (rightmost graph). Note the x-axis has a base-10 log scale.

In this section, analysis of interpolating specific functions was considered. In the next section the proposed approach is re-interpreted so that we can reason about it by considering only the location of the data-sites, i.e. without the need of a ground truth function to interpolate.

6   Further Analysis

Hilbert-Const interpolation has a particularly simple interpretation. It can be viewed as a simple weighted sum of data-site values. The weight denoted p_i is the probability that data-site x_i is the nearest neighbour to q_j along the Hilbert curve under the Monte Carlo sampling described in the Aggregation Section, i.e.

$$\hat{f}(\mathbf{q}_j) = \sum_i^n p_i z_i \qquad (1)$$

Consider the case where there are only two data-sites, denoted $\times$ and $\cdot$, located within a region of interest shown in Fig. 8(a) (ignoring the heatmap for the moment). Under Euclidean distance, locations that are closest to the data-site denoted by an $\times$ are to the left of the dashed line.

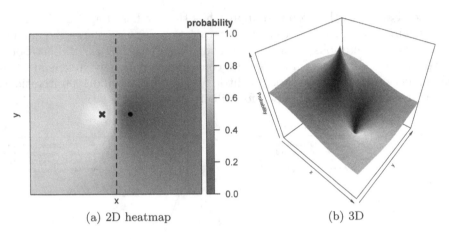

(a) 2D heatmap   (b) 3D

Fig. 8. Probability mass function showing the probability that the nearest neighbour is the data-site denoted by an $\times$.

The heatmap represents all the possible query site locations within the region of interest and shows the probability the $\times$ data-site is the nearest neighbour under the Monte Carlo sampling, with $\eta = 50,000$. Figure 8(b) shows the same probability mass function but in 3D. It is clear that there is a discontinuity at the peak (corresponding to the $\times$ data-site).

Referring back to Eq. 1, for the interpolated function $\hat{f}(\cdot)$ to be continuous, both p_i and z_i are required to spatially continuous over the region of interest. Assuming that the function to be interpolated is indeed continuous, i.e. z_i is continuous, then p_i needs to be continuous. As has been noted, p_i is not continuous at data-sites. Hence the interpolated function will not be continuous at data-sites. Note that natural neighbour interpolation also has this issue. The smoothness of probability mass function elsewhere in Fig. 8(b) is consistent with p_i being continuous everywhere apart from at data-sites.

A final observation from the heatmap in Fig. 8(a) is the probability mass function behaves sensibly since in general the closer to data-site $\times$ the higher the probability. Figure 9 shows the average number of data-sites that contribute

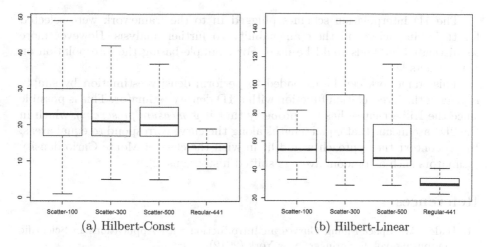

(a) Hilbert-Const   (b) Hilbert-Linear

Fig. 9. Number of non zero weighted data-sites for each query location. Hilbert-Linear has more than Hilbert-Const since for linear interpolation uses two data-sites for interpolation whereas constant only requires one.

to a query location. For the scattered data it is clear the average number is largely independent of the increasing number of data-sites. This is a crucial observation for the utility of the proposed approach for large datasets. It suggests that the interpolation is local to the query data-site.

There is one exception, query points near the boundary of the region of interest. Figure 9 show there is a large variance in the number contributing data-sites per query site. This phenomena is due to the space-filling curve exiting region of interest and entering at some other location along the boundary. This behaviour is not necessarily wrong, it is making the assumption that the function is homogeneous around the boundary. One way to remove this edge effect is to introduce a post-processing step that keeps the closest w contributing data-sites for each query site near the boundary, where w is the overall mean.

7   Conclusion and Future Work

This paper proposed a novel framework to interpolating scattered data. The approach is conceptually easy to understand and straightforward to code and delivers results that are in some ways commensurate with natural neighbour interpolation. If radix sort is used, then the computational complexity of the algorithm is linear with respect to input size. More specifically it is $O(\eta m d k)$ for d-dimensional data where each coordinate is represented using k bits and the combined total of query and data points is m. The number of iterations η is typically large but remains constant with increasing size of the data set. Hence it is likely that this approach will be useful in circumstances where interpolation is required for very large datasets. Note that η is likely to grow exponentially with respect to d, its precise relationship is left for future work.

The 1D interpolation schemes plugged in to the framework were selected for their simplicity and their amenability to further analysis. However more sophisticated methods could be used. For example basing the interpolation on 1D wavelets [5].

This framework can be extended to perform density estimation by simply replacing the interpolation function with a 1D density estimator. This is possible since the Hilbert curve has the property that it is *measure preserving*, which in the 2D case means that equal lengths along the curve correspond to equal areas. In his context the approach would fit in with the class of Monte Carlo density estimators such as random average shifted histograms [2].

References

1. Bader, M.: Space-Filling Curves an Introduction with Applications in Scientific Computing, vol. 9. Springer, New York (2012)
2. Bourel, M., Fraiman, R., Ghattas, B.: Random average shifted histograms. Comput. Stat. Data Anal. **79**, 149–164 (2014)
3. Braun, J., Sambridge, M.: A numerical method for solving partial differential equations on highly irregular evolving grids. Nature **376**, 655–660 (1995)
4. Chen, H.L., Chang, Y.I.: All-nearest-neighbors finding based on the Hilbert curve. Expert Syst. Appl. **38**(6), 7462–7475 (2011)
5. Lamarque, C.H., Robert, F.: Image analysis using space-filling curves and 1D wavelet bases. Pattern Recogn. **29**(8), 1309–1322 (1996)
6. Lazzaro, D., Montefusco, L.B.: Radial basis functions for the multivariate interpolation of large scattered data sets. J. Comput. Appl. Math. **140**(1), 521–536 (2002)
7. Ledoux, H., Gold, C.: An efficient natural neighbour interpolation algorithm for geoscientific modelling. In: Fisher, P.F. (ed.) Developments in Spatial Data Handling, pp. 97–108. Springer, Heidelberg (2005)
8. Li, J., Heap, A.D.: Spatial interpolation methods applied in the environmental sciences: a review. Environ. Model. Softw. **53**, 173–189 (2014)
9. Mari, J.F., Le Ber, F.: Temporal and spatial data mining with second-order hidden Markov models. Soft. Comput. **10**(5), 406–414 (2006)
10. Mennis, J., Guo, D.: Spatial data mining and geographic knowledge discovery. An introduction. Comput. Environ. Urban Syst. **33**(6), 403–408 (2009)
11. Ohashi, O., Torgo, L.: Spatial interpolation using multiple regression. In: 2012 IEEE 12th International Conference on Data Mining, pp. 1044–1049. IEEE (2012)
12. Okabe, A., Boots, B., Sugihara, K., Chiu, S.N.: Spatial Tessellations: Concepts and Applications of Voronoi Diagrams, vol. 501. Wiley, New York (2009)
13. Park, S., Linsen, L., Kreylos, O., Owens, J., Hamann, B.: Discrete Sibson interpolation. IEEE Trans. Vis. Comput. Graph. **12**(2), 243–253 (2006)
14. Sagan, H.: Space-Filling Curves. Springer, New York (1994)
15. Skilling, J.: Programming the Hilbert curve. In: Bayesian Inference and Maximum Entropy Methods in Science and Engineering: 23rd International Workshop on Bayesian Inference and Maximum Entropy Methods in Science and Engineering, vol. 707, pp. 381–387. AIP Publishing (2004)
16. Weston, D.J.: Exploiting novel properties of space-filling curves for data analysis. In: Blockeel, H., Leeuwen, M., Vinciotti, V. (eds.) IDA 2014. LNCS, vol. 8819, pp. 356–367. Springer, Heidelberg (2014). doi:10.1007/978-3-319-12571-8_31

Privacy-Awareness of Distributed Data Clustering Algorithms Revisited

Josenildo C. da Silva[1(✉)], Matthias Klusch[2], and Stefano Lodi[3]

[1] Depto. de Informática, Instituto Federal do Maranhão (IFMA),
Av. Getúlio Vargas, 04, Monte Castelo, CEP, São Luís, MA 65030-005, Brazil
jcsilva@ifma.edu.br
[2] DFKI GmbH, Stuhlsatzenhausweg 3, Campus D3.2, 66123 Saarbrücken, Germany
klusch@dfki.de
[3] Dipartimento di Informatica - Scienza e Ingegneria,
Viale Risorgimento 2, Bologna, Italy
stefano.lodi@unibo.it

Abstract. Several privacy measures have been proposed in the privacy-preserving data mining literature. However, privacy measures either assume centralized data source or that no insider is going to try to infer some information. This paper presents distributed privacy measures that take into account collusion attacks and point level breaches for distributed data clustering. An analysis of representative distributed data clustering algorithms show that collusion is an important source of privacy issues and that the analyzed algorithms exhibit different vulnerabilities to collusion groups.

1 Introduction

The goal of Distributed Data Mining (DDM) is to find patterns or models from a collection of distributed datasets, that is, datasets residing on the nodes of a communication network, possibly under constraints of limited bandwidth and data privacy [16]. In DDM, Distributed Data Clustering (DDC) is the problem of finding groups of similar objects in distributed datasets [4].

Privacy and data ownership play an important role in DDM and in DDC in particular, a role which calls for a privacy preserving solution [6,13]. Two main approaches have emerged to address this problem: *secure multiparty computation* and *model-based data mining*. With the secure multi-party computation (SMC) approach, all computations are performed by a group of mining parties following a given protocol and using cryptographic techniques to ensure that only the final results will be revealed to the participant, e.g. secure sum, secure comparison [5], secure set union [3]. In the model-based approach, each site computes a partial local model from the local dataset and all local models are aggregated to produce a global model, which is shared with the participants.

Each approach proposes a different privacy measure. These privacy measures, however, either assume that no insider is going to try to infer some sensitive information, or do not account for particularities of specific data mining tasks.

© Springer International Publishing AG 2016
H. Boström et al. (Eds.): IDA 2016, LNCS 9897, pp. 261–272, 2016.
DOI: 10.1007/978-3-319-46349-0_23

The privacy definition in SMC considers only threats from the outside and does not care about how much an inside party can learn from the protocol output. For example, in a protocol where three parties compute the sum of numbers in a secure multiparty protocol, e.g. secure sum, the process does not leak any input information. However, when two parties collude they can subtract its own input and learn the input of remaining party.

Model-based approaches, on the other hand, define privacy from the perspective of the whole dataset, and not of single points. For example, in [11] dataset privacy is based on the average privacy of all points. Some points will, of course, have privacy level much lower than the average privacy level. Thus, even if a single point has a very low privacy level, this privacy breach may go unnoticed.

In this paper, we present distributed privacy measures that take into account inference attacks from insiders and are able to detect point-level privacy breaches. We follow an information theoretic approach and define a set of properties from which our privacy measures are derived. We also apply the proposed measures to representative algorithms to demonstrate previous undetected privacy issues.

As main contribution, this paper: (i) introduces the first privacy measures for DDM in general and DDC in particular with respect to insider and collusion attacks, and single data point privacy; (ii) re-evaluates the privacy preservation of representative DDC algorithms with these measures, and reveals that insider collusion is an important source of privacy breach and (iii) exemplifies the respective privacy analysis for selected algorithms.

2   Privacy Measures for Distributed Data Clustering

Major threats in a distributed mining session come from malicious insiders. Therefore, privacy measures should take into account the presence of collusion groups of malicious peers. Moreover, privacy measures should detect the privacy level of single data points.

In this paper, we will regard privacy measures as functions which, for a given distributed data mining algorithm, map a dataset subset and the maximum size of collusion groups of parties to a real number, and satisfy certain properties. We will call the value of such a measure a *privacy level*.

Let $L_1, \ldots, L_p$ be sites hosting one element of a partition of a dataset D each, and $\mathcal{A}$ be any distributed data mining algorithm running on $L_1, \ldots, L_p$. We will assume that up to $p - 1$ sites among $L_1, \ldots, L_p$ are malicious, in that they seek to infer objects of D, or parts thereof, possibly in collusion groups of at most $c < p$ members, by either exchanging information or violating the protocol of $\mathcal{A}$, or both. By *privacy measure* for $\mathcal{A}$ we mean a computable partial function

$$\mathbf{PR}_{\mathcal{A}} \colon (X, c) \in 2^D \times \{0, 1, \ldots, p - 1\} \to \mathbf{PR}_{\mathcal{A}[c]}(X) \in [0, \infty) \tag{1}$$

which satisfies one or multiple of the following properties:

P1 (collusion) $\mathbf{PR}_{\mathcal{A}[1]}(X) \geq \mathbf{PR}_{\mathcal{A}[c]}(X)$ when there are at most c malicious peers colluding, for all $c \in \{1, \ldots, p - 1\}$ and for all $X \subseteq D$;

P2 (point monotonicity) it is non-increasing from singletons to dataset, i.e., $\mathbf{PR}_{\mathcal{A}[c]}(\{x\}) \geq \mathbf{PR}_{\mathcal{A}[c]}(D)$ for all $c \in \{0, 1, \dots, p - 1\}$.

Property **P1** expresses the decrease of privacy level in scenarios with inference attacks and collusion from malicious parties. Note that $c = 1$ expresses the absence of collusion groups of size two or greater. Therefore, parties are semi-honest, i.e., they adhere to the protocol of $\mathcal{A}$, but may exploit information gathered during execution for inference purposes. When no parties attempt to infer objects, i.e., are honest, there are no inside threats and $c = 0$. Property **P2** constrains $\mathbf{PR}_{\mathcal{A}}$ to behave as a worst-case measure: a greater privacy level than the one at singletons is not attainable for the dataset. We call this property *point-level* awareness.

Throughout this paper, we use the following notation. To indicate explicitly a privacy measure m in the evaluation of a given algorithm $\mathcal{A}$ we use the notation $\mathbf{PR}_{\mathcal{A}}^m$. We indicate the privacy of a given point x, given an algorithm $\mathcal{A}$ and measure m, as $\mathbf{PR}_{\mathcal{A}}^m(X)$; for a dataset D we use $\mathbf{PR}_{\mathcal{A}}^m(D)$. For the sake of simplicity, we omit algorithm, measure, dataset or data point, when they are implicit in a given context.

2.1 Existing Privacy Measures for DDC

In SMC, privacy is equivalent to having a trusted third party perform the computation and erasing all of the input data after the computation [7]. An SMC protocol is said to preserve privacy if one can prove that after the computation no party learns anything but the final results, as it would be the case if there were a trusted third party in the setting. This notion of privacy is known as the *simulation paradigm* [5], and is used to formally define privacy for SMC protocols. For a discussion on proofs for SMC protocols, the reader may refer to [5,9]. The privacy measure which is used in SMC protocols will be called *private computation*, and denoted in this paper as $\mathbf{PR}_{\mathcal{A}}^{PC}(D)$.

Model-based approaches work by producing partial local models, which are then aggregated into a global model [8,10,11,14]. Finally, using the global model each party computes the mining results. Examples of models include wavelets coefficients, parametric models, like a mixture of Gaussians, or non-parametric models, like kernel density estimates.

In the model-based approach the *likelihood-based* measure is used in the context of clustering and classification [10]. Let D be a given dataset and $f_\lambda(x)$ be the probability density function associated with a given probabilistic model λ. The privacy $\mathbf{PR}_{\mathcal{A}}^{like}$ of data set D given model λ is defined as the reciprocal of the geometrical mean likelihood of the dataset being generated under model λ and can be expressed as:

$$\mathbf{PR}_{\mathcal{A}}^{like}(D) = 2^{\left(-\frac{1}{|D|}\sum_{x \in D}\log f_\lambda(x)\right)} \tag{2}$$

This measure indicates how likely a sampled data set is to occur given a probability model [10]. If the likelihood is high, then the privacy is low and vice-versa.

2.2   Limitations of Current Measures

In the presence of collusion groups a secure multiparty protocol (SMC) may fail [9] because its privacy definition gives only the privacy level from the outsiders' point of view. Any malicious insiders will receive the correct output, from which they may try to reconstruct sensitive inputs from other parties.

$\mathbf{PR}_{\mathcal{A}}^{PC}(D)$ **Does Not Address Inference or Collusion (¬P1).** The private computation measure was designed to detect leaks from the protocol and not from outputs. Consider a protocol where three parties compute the set union in a secure multiparty protocol. The process does not leak any input information, but when two parties collude they can remove its own input sets and learn the input set of remaining party. However, privacy computation does not address this inference attack situation and $\mathbf{PR}_{SMCSum[0]}^{PC}(x) = \mathbf{PR}_{SMCSum[2]}^{PC}(x)$ when they should indicate the decrease in privacy in the presence of 2 malicious parties working in collusion.

$\mathbf{PR}_{\mathcal{A}}^{PC}(D)$ **is Point-Level (P2).** By definition, if any point $x \in D$, the dataset of inputs of a given party, is leaked, the protocol is considered not private, i.e. $\forall x \in D : \mathbf{PR}_{\mathcal{A}}^{PC}(x) = 0 \rightarrow \mathbf{PR}_{\mathcal{A}}^{PC}(D) = 0$. Therefore, $\forall x \in D : \mathbf{PR}_{\mathcal{A}}^{PC}(x) \geq \mathbf{PR}_{\mathcal{A}}^{PC}(D)$.

The likelihood-based measure is discussed in the following. Let $D = \{1, 4, 6, 9\}$, a dataset, and a mixture of two Gaussian with the first model be centered at $x_0 = 1$, i.e. it has mean $\mu_1 = 1$ with variance $\sigma_1^2 = 0.1$. The second model models the three remaining points, i.e., it has mean $\mu_2 = 6.3$ and variance $\sigma_2^2 = 1.0$. With probability density function of model denoted by $f(x)$, using Eq. (2) we have: $\mathbf{PR}_{\mathcal{A}}^{like}(D) = 2^{\left(-\frac{1}{|4|}(\log_2(f(1))+log_2(f(4))+log_2(f(6))+log_2(f(9)))\right)} = 13.7326$.

$\mathbf{PR}_{\mathcal{A}}^{like}(D)$ **Does Not Address Inference or Collusion (¬P1).** If the mixture of local models represents datasets from participants and a malicious insider has access to all local models, it can try to reconstruct other participants' datasets. In the above example, the attacker could reconstruct the first point with high precision with the first model, which is centered at x_0 with small variance. $\mathbf{PR}_{\mathcal{A}[1]}^{like}(x_0) = 2^{-\log_2 f(x_0)} = 0.5013$. However, even in this case $\mathbf{PR}_{\mathcal{A}[1]}^{like}(D) = 13.7326$, i.e. the drop in privacy due to a insider attack ($c \geq 1$) is not reflected in $\mathbf{PR}^{like}$. Therefore, $\mathbf{PR}_{\mathcal{A}}^{like}(D)$ does not fulfill property P1.

$\mathbf{PR}_{\mathcal{A}}^{like}(D)$ **is Not Point-Level (¬P2).** When only a few points have a high likelihood of being reconstructed with a high precision, $\mathbf{PR}_{\mathcal{A}}^{like}(D)$ measure will still indicate a high privacy protection. Consider a dataset D above. The geometrical mean in the privacy measure smoothed out the measure for $x_0 = 1$, masking a possible privacy breach. Thus,

$$\mathbf{PR}_{\mathcal{A}}^{like}(x_0) = 2^{-\log_2 f(x_0)} = 0.5013 < \mathbf{PR}_{\mathcal{A}}^{like}(D) = 13.7326$$

Therefore, $\mathbf{PR}_{\mathcal{A}}^{like}(D)$ does not fulfill property P2. Table 1 presents a summary of all studied privacy measures and their properties.

Table 1. Summary of privacy measures and properties

	Reference	Approach	Collusion (P1)	Point-level (P2)
PR^{PC}	[5]	Simulation	no	yes
PR^{like}	[10]	Probability	no	no
PR^{range}	(Def. 1)	Info. theory	yes	yes
PR^{rec}	(Def. 2)	Inference analysis	yes	yes
PR^{BK}	(Def. 3)	Info. theory	yes	yes

3   New Privacy Measures for DDC

In this section, we propose new privacy measures to analyze distributed data clustering algorithms. We assume that the attackers are members of the mining group and that they have access to the resulting cluster map and other information defined by the mining protocol being analyzed.

Our first measure defines the privacy of a cluster as the size of the interval between its maximal and minimal values.

Definition 1 (Cluster range measure). *Given a dataset D and a cluster map $\mathcal{C} = \{C_k\} \subseteq 2^D$, whose elements C_k are pairwise disjoint. We define the cluster privacy of a given point x in a given cluster $C_k \in \mathcal{C}$ as:*

$$PR^{range}(x) = max\{C_k\} - min\{C_k\}. \tag{3}$$

Extending to the whole dataset:

$$PR^{range}(D) = min\{PR^{range}(x) : x \in D\}. \tag{4}$$

As an example, consider a cluster of data points over the dimension "annual income" ranging from US\$ 100 000 to US\$ 150 000 reveals the value of each data point with a maximal error of US\$ 50 000 and maximal mean error of US\$ 25 000 (assuming uniform distribution). Consequently, each point in this cluster is said to have a privacy level of 50 000 dimension units, US\$ in this case.

If a reconstruction method is known, it is possible to measure how close the reconstructed data gets to the original sensitive data.

Definition 2 (Reconstruction based measure). *Let $R \subset \mathbb{R}$ denote a set of reconstructed data objects such that each $r_i \in R$ is a reconstructed version of $x_i \in D$. We define the privacy level, given a reconstruction method, by:*

$$PR^{rec}(x_i) = \mid x_i - r_i \mid. \tag{5}$$

Extending to the whole dataset:

$$PR^{rec}(D) = min\{PR^{rec}(x_i) : x_i \in D, r_i \in R, 1 \le i \le N\} \tag{6}$$

where N is the size of the dataset D.

Consider secure k-means algorithm [15]. In this algorithm parties L_1 and L_p hold together the information on the distance $d = |x - \mu_i|$ between a given centroid μ_i and other parties data points x. Thus, attackers can use the inverse of the distance as a reconstruction method to infer data points x. $\mathbf{PR}^{rec}(x)$ will denote the precision of this specific reconstruction method.

A general definition of privacy proposed in the centralized data mining setting is the *bounded knowledge* measure [1], which defines privacy as the length of the interval from which a random variable X is generated. This measure can be expressed in terms of the entropy of X, as follows.

Definition 3 (Bounded Knowledge). *Given a random variable X with probability density function f_X and domain Ω_X, the privacy of X given by its bounded knowledge is:*

$$\mathbf{PR}^{BK}(X) = 2^{h(X)} \tag{7}$$

where $h(X) = -\int_{\Omega_X} f_X(x) log_2 [f_X(x)]\, dx$ is the differential entropy.

As an example, consider a random variable X uniformly distributed between 20 and 70, abbreviated $X \sim U(20, 70)$, has probability density function $f(x) = \frac{1}{50}$, for $20 \leq x \leq 70$, and 0 otherwise. The entropy of X is $h(X) = log_2(50)$. Thus, the privacy provided by X according to bounded knowledge measure is $\mathbf{PR}^{BK}(X) = 2^{log_2(50)} = 50$. This definition is general enough to be used in different data mining contexts, e.g. cluster analysis, association rules, etc. [2].

For a given point $x \in C_i$, a cluster in cluster map $\mathcal{C}$ induced from D, X_i a random variable for values of C_i and a probability density function $f_{X_i}(x)$, let:

$$\mathbf{PR}^{BK}(x) = \mathbf{PR}^{BK}(X_i) = 2^{h(X_i)} \tag{8}$$

with $f_{X_i}(x)$ being zero outside C_i.

In the case of a cluster map, we are interested in the smallest interval size in the said map[1]. Therefore,

$$\mathbf{PR}^{BK}(D) - \min\{\mathbf{PR}^{BK}(x)\} = \min\{2^{h(X_i)}\}. \tag{9}$$

The next definition extends each of the previously defined measures to include collusion groups.

Definition 4. *Let $\mathcal{A}$ be a distributed data clustering algorithm, D be a dataset, and measure $m \in \{rec, range, BK\}$, with collusion groups containing at most c attackers. We define:*

$$\mathbf{PR}^m_{\mathcal{A}[c]}(D) = \min\{\mathbf{PR}^m_{\mathcal{A}[i]}(D) : 1 < i \leq c\}. \tag{10}$$

---

[1] This notion comes from the well-known idea in computer security that defines the security level of a system as the level of its *weakest link*.

$\mathbf{PR}^m_{\mathcal{A}[c]}(D)$ represents the minimum privacy level provided to dataset D when the collusion groups have at most c peers, using any privacy measure m. For example, $\mathbf{PR}^{BK}_{\mathcal{A}[2]}(D)$ denotes the privacy level provided by algorithm $\mathcal{A}$ to dataset D when collusion groups are formed with at most 2 malicious peers analyzed with BK measure.

Properties Analysis of $\mathbf{PR}^m_{\mathcal{A}[c]}(D)$

Lemma 1 (Collusion). *Given an algorithm $\mathcal{A}$, for all dataset D and privacy measures $m \in \{range, BK, rec\}$, and $c > 1$ (presence of non-singleton collusion groups), if there is a collusion scenario decreasing the privacy level of dataset D, then $\mathbf{PR}^m_{\mathcal{A}[1]}(D) \geq \mathbf{PR}^m_{\mathcal{A}[c]}(D)$.*

Proof. Let $a = \mathbf{PR}^m_{\mathcal{A}[1]}(D)$ be the privacy level of dataset D with algorithm $\mathcal{A}$ with no collusion (i.e., $c = 1$), and $b = \mathbf{PR}^m_{\mathcal{A}[c]}(D)$ be the privacy level in a collusion scenario with $c > 1$ malicious peers. By definition $\mathbf{PR}^m_{\mathcal{A}[c]}(D)$ is the smallest privacy level considering all collusion scenarios. Thus, $\mathbf{PR}^m_{\mathcal{A}[c]}(D) = \min\{a, b\}$. Therefore, if the collusion group decreases the privacy level of the $c = 1$ scenario, then $a \geq \min\{a, b\}$. $\square$

Lemma 2 (Point level privacy). *$\forall x \in D : \mathbf{PR}^m_{\mathcal{A}[c]}(x) \geq \mathbf{PR}^m_{\mathcal{A}[c]}(D)$, for all dataset D and privacy measures $m \in \{range, BK, rec\}$.*

Proof. (Range) Consider a cluster map $\mathcal{C}$ from D, with only two clusters C_a and C_b. Let r_a and r_b denote $r_a = max\{C_a\} - min\{C_a\}$ and $r_b = max\{C_b\} - min\{C_b\}$, the cluster range of C_a and C_b respectively. For a given point $x_a \in C_a$, by definition, $\mathbf{PR}^{range}(x_a) = r_a$ and $\mathbf{PR}^{range}(D) = \min\{r_a, r_b\}$. Therefore, $r_a \geq \min\{r_a, r_b\}$.

(Rec) Consider a dataset D and a reconstructed set R. Let x_a be any given point in D and r_a its reconstructed counterpart in R. By definition, $\mathbf{PR}^{rec}_{\mathcal{A}[c]}(x_a)$ is $|x_a - r_a|$ and $\mathbf{PR}^{rec}_{\mathcal{A}[c]}(D) = \min\{|x_i - r_i| : x_i \in D, r_i \in R, 1 \leq i \leq N\}$. Therefore, $|x_a - r_a| \geq \min\{|x_i - r_i| : x_i \in D, r_i \in R\}$.

(BK) Consider a cluster map $\mathcal{C}$ from D, with only two clusters C_a and C_b. Let X_a be a random variable modeling a data point $x_a \in C_a$, and X_b a random variable modeling data points $x_b \in C_b$. By definition, $\mathbf{PR}^{BK}_{\mathcal{A}[c]}(x_a)$ is $2^{h(X_a)}$ and $\mathbf{PR}^{BK}_{\mathcal{A}[c]}(D) = \min\{2^{h(X_a)}, 2^{h(X_b)}\}$.

Therefore, $\mathbf{PR}^{BK}_{\mathcal{A}[c]}(x_a) = 2^{h(X_a)} \geq \min\{2^{h(X_a)}, 2^{h(X_b)}\}$. Similarly, we have that $\mathbf{PR}^{BK}_{\mathcal{A}[c]}(x_b) = 2^{h(X_b)} \geq \min\{2^{h(X_a)}, 2^{h(X_b)}\}$. $\square$

We have thus derived three privacy measures, Cluster Range, Reconstruction, and Bounded Knowledge, that are inspired by different abstractions of privacy, and satisfy the natural properties of collusion and point-level awareness. In contrast, the Private Computation and Likelihood privacy measures fail to capture at least one of such properties. We will now revisit prominent DDC algorithms to examine if and how applying the new measures changes their evaluation, as to the amount of privacy that is guaranteed by each of them.

4 Application to DDC Algorithms

To apply our measures to DDC algorithms, we need to analyze which information is available to each party during the mining session, which collusion groups can be formed and how they can reconstruct information from available information (including single malicious attacks). In the following, a few algorithms for distributed data clustering are briefly reviewed and their privacy properties are then analyzed in light of our privacy definitions. We selected these algorithms because they are based on prominent methods for distributed data clustering.

4.1 Secure Multiparty k-means

Vaidya and Clifton [15] proposed an extension of the classic k-means algorithm to the distributed setting, using cryptographic protocols to achieve privacy (VC-kmeans). Data is assumed to be vertically partitioned. The solution is based on a secure sum protocol to find the closest cluster for any given point. It also uses secure permutation and secure comparison. VC-kmeans assumes three trusted parties L_1, L_2 and L_p. Additionally, let L_j be any other non-trusted party in the mining group. It was originally evaluated with $\mathbf{PR}^{PC}$ as private with three trusted parties, but no analysis is presented on how much privacy is preserved under collusion.

Single Insider Attacks. A given party L_j knows only: (i) $\boldsymbol{\mu}_j$, a share of the centroid; (ii) d_{ij}, the distance from the cluster centroid $\boldsymbol{\mu}_i$ to the view of point x_j; (iii) and a random vector $\boldsymbol{v}_j$. L_1 is the party which starts the protocol and knows: (i) a partial view of the cluster centroids, $\boldsymbol{\mu}_1$; (ii) the cluster assignment for each data point x; (iii) a random vector $\boldsymbol{v}$; and (iv) a permutation π of 1 to k, used to preserve the privacy of information in the SMC protocol. L_2 knows $\boldsymbol{T}_2 = \pi(\boldsymbol{v}_2 + \boldsymbol{d}_2)$, the permuted sum of $\boldsymbol{v}_2$ with $\boldsymbol{d}_2$, which is hidden from the other parties but L_p. L_p knows its share of the centroid $\boldsymbol{\mu}_p$, and $\boldsymbol{T}_i = \pi(\boldsymbol{v}_i + \boldsymbol{d}_i)$, $i = 1, 3, 4, \ldots, p$, the permuted sum of $\boldsymbol{v}_i$ with $\boldsymbol{d}_i$ of each party but L_2. Moreover, L_p knows the combined sum of $\boldsymbol{T}_i$ from all parties but L_2, i.e. $\boldsymbol{Y} = \boldsymbol{T}_1 + \sum_{i=3}^{p} \boldsymbol{T}_i$. L_1 is the party holding the most important information, which can be used to reconstruct sensitive data, including the random vector $\boldsymbol{v}$ and the permutation π. However, without the permuted sum of distances $\boldsymbol{Y}_i$ from other parties ($i = 1, 3, 4, \ldots, p$) L_1 will not learn anything, because it cannot reconstruct data points from other parties. Similarly, L_2 and L_p will not learn anything from the information they hold alone.

Let D be a n-dimensional dataset distributed over a network of peers. When there are only single insider attacks, algorithm VC-kmeans produces a cluster map of C from D with a privacy level given by:

$$\mathbf{PR}^{range}_{VCkmeans[1]}(D) = \min\{\max(C_i) - \min(C_i)\} \tag{11}$$

with $\forall C_i \in C$. Any insider attacker working solo can only learn what is disclosed by the cluster map itself – namely, that each point ranges in the interval min, max, for a given cluster. Contrast this information with the result of an

SMC analysis, which only tells us that the protocol is private, but does not quantify it in terms of original data space units.

Attack with Collusion of Insiders L_1 and L_p. Together, L_1 and L_p hold information on the permuted sum of all parties except for L_2. Moreover, they hold information on the permutation π and the random vector v. Therefore, this collusion group may compute the vector d_i using inverse of permutation π:

$$d_i = \pi^{-1}(Y_i) - v_i \qquad (12)$$

with $i = 1, 3, 4, \ldots, p$. The vector d_i represents the distance between a given point x and the cluster centroid i with mean μ_i, therefore, with the true distance, every point x can be located with an arbitrary error. Using Eq. (12) as reconstruction method, we apply $\mathbf{PR}^{rec}(D)$.

$$\mathbf{PR}^{rec}_{VCkmeans[2]}(D) = \min\{|x - r| : x \in D, r \in R\} \approx 0 \qquad (13)$$

where D is the original dataset and R is a reconstructed dataset. Original evaluation with $\mathbf{PR}^{PC}_{VCkmeans}(D)$ does not inform how much privacy is lost with only one attacker. However, we find in our analysis that a malicious alone can learn no more than the size of each cluster.

4.2 Distributed Data Clustering with Generative Models

Merugu and Ghosh [10] present an algorithm for distributed clustering and classification based on generative models approach (DDCGM). Their algorithm outputs an approximate model $\hat{\lambda}_c$ of a true global model λ_c from a predefined fixed family of models F, *e.g.* multivariate 10-component Gaussian mixtures. DDCGM first computes local models λ_i, from which the average global model $\bar{\lambda}$ is generated by $p_{\bar{\lambda}}(x) = \sum_{i=1}^n \nu_i p_{\lambda_i}(x)$ where $p_\lambda(x)$ is the probability density function of a given model λ. The algorithm uses $\bar{\lambda}$ to find a good approximation $\hat{\lambda}_c$ of the true (and unknown) global model λ_c. The model $\hat{\lambda}_c$ is used as cluster map. Original privacy evaluation was based on $\mathbf{PR}^{like}(D)$, with all models in a mixture, regardless of the possible weakness of any component. The new evaluation reflects the weakest model in the mixture.

Single Insider Attacks. In the DDCGM scheme, a central entity receives local generative models and combines them into an average generative model. This entity knows individual generative model from each party. Arbitrary parties know only the global model. Since the models represent clusters, we can apply $\mathbf{PR}^{range}(D)$. Let $p_\lambda(x)$ be a mixture model with k elements. The privacy level provided by DDCGM using $p_\lambda(x)$ and with no collusion is:

$$\mathbf{PR}^{range}_{DDCGM[1]}(D) = \min\{x_{max} - x_{min}\} \qquad (14)$$

where x_{max} and x_{min} are inferior and superior elements at the each cluster, according to the model $p_\lambda(x)$.

Assuming that each component model λ_i in the mixture is a Gaussian in a n dimensional data space with covariance matrix Σ_i, the entropy is $h_i(x) =$

$\ln \sqrt{(2\pi e)^n |\Sigma_i|}$, where $|\Sigma_i|$ is the determinant of the covariance matrix of the given model, and consequently, a cluster. Therefore, we can compute:

$$\mathbf{PR}_{DDCGM[1]}^{BK}(D) = \min\left\{2^{h_i(x)}\right\} = \min\left\{2^{\ln\sqrt{(2\pi e)^n |\Sigma_i|}}\right\} \tag{15}$$

Collusion Attack. Any collusion group must include the central party since there is little information for arbitrary parties, and collusion attacks reduce to single aggregator attack. Thus, $\mathbf{PR}_{DDCGM[1]}^{BK}(D) = \mathbf{PR}_{DDCGM[c]}^{BK}(D)$ with $c \geq 2$.

4.3   Information Theoretical Approach to Distributed Clustering

Shen and Li [14] proposed an information theoretical approach to distributed clustering (ITDDC). They assume a peer-to-peer network where each node solves a local clustering problem and updates its neighbors. The clustering problem is to fit a discriminative model to cluster boundaries that maximize the mutual information between cluster labels and data points. With low communication, local clusters are formed based on global information spread through the network. The algorithm needs several rounds of iterations to converge. When it comes to privacy, the authors do not investigate how the algorithm would behave under inference attacks and do not investigate how much privacy this approach does provide.

Single Insider Attack. Each party in ITDDC knows a set of discriminative models defining the clusters boundaries of points on data sets and from all its direct neighbors. We can apply $\mathbf{PR}^{range}(D_j)$ to compute how much privacy is preserved at local dataset D_j for a given model. Each party estimates $\hat{p}_j(k|x)$, a class label distribution defined by a local discriminative model (for instance, logistic regression). The distribution of x in a given cluster is not disclosed. Thus, each point can only be located in the interval corresponding to its cluster boundaries. The privacy provided by DDCGM using $\hat{p}_j(k|x)$ and with no collusion is:

$$\mathbf{PR}_{ITDDC[1]}^{range}(D) = \min\{x_{max} - x_{min}\} \tag{16}$$

where x_{max} and x_{min} are inferior and superior elements at the each cluster, according to the boundaries defined by model $\hat{p}_j(k|x)$.

Collusion Attack. The only information being exchanged among the parties is the local models. Moreover, there is no special central entity holding extra information on data distribution at local datasets. Therefore, even if malicious parties collude against another party, they cannot improve on the single insider attack. Therefore, $\mathbf{PR}_{ITDDC[c]}^{range}(D) = \mathbf{PR}_{ITDDC[1]}^{range}(D)$, with $c \geq 1$ colluding parties.

4.4   Elliptic Curves for Multiparty k-means

Patel and colleagues [12] present a privacy-preserving distributed k-means algorithm based on elliptic curves (EC-kmeans). They assume no trusted party and use elliptic curves to achieve low overhead cryptography. No analysis on inference attack or collusion is presented by the authors.

Single Insider Attack. Each peer knows its own centroids, its own cluster boundaries and the encrypted version of the global centroids and the number of points in a global cluster. Without collusion, a given malicious party does not even know the boundaries of clusters residing on other parties.

Collusion. The initiator knows the information necessary to decrypt data in the mining session. Therefore, a collusion group with the initiator and any party L_i can learn about the centroids and number of points in each cluster on the party L_{i-1}. With the centroids, cluster boundaries of dataset D_j at L_j could be estimated and $\mathbf{PR}_{ECkmeans[2]}^{range}(D_j) = \min\{x_{max} - x_{min}\}$.

4.5 Discussion

Table 2 presents an overview of the studied algorithms. The analysis above shows that collusion is indeed a chief source of privacy breach, and that algorithms can be separated according to their vulnerability to collusion groups and to the malicious behavior of a site with a special role in the protocol, e.g., a central site, or an aggregator, or a protocol initiator. VC-kmeans is almost completely not private if the central site colludes, whereas DDCGM has limited vulnerability to the central site and not to collusion; ITDDC does not use a central site and only disclose cluster ranges, irrespective of collusions. EC-kmeans, finally, is secure and only discloses range information under a collusion attack that involves the initiator.

Table 2. Summary of privacy preserving distributed data clustering algorithms.

	Original assessment	Single attacks	Collusion attacks
VC-kmeans [15]	private (3 trusted)	$\min\{x_{max} - x_{min}\}$	decrease to ≈ 0, $c \geq 2$
DDCGM [10]	$2^{\left(-\frac{1}{|D|}\sum_{x \in D} \log f_\lambda(x)\right)}$	$\min\{2^{\ln\sqrt{(2\pi e)^n |\Sigma_i|}}\}$	same level, $c \geq 1$
ITDDC [14]	N/A	$\min\{x_{max} - x_{min}\}$	same level, $c \geq 1$
EC-kmeans [12]	private (0 trusted)	$\min\{x_{max} - x_{min}\}$	$\min\{x_{max} - x_{min}\}$, $c \geq 2$

5 Conclusions

We presented new privacy measures for distributed data clustering, in order to overcome the limitations of existing measures. Starting from a set of formal properties, it was shown that the new measures satisfy the properties and, therefore, improve over previous ones. The new measures were applied to selected representative of privacy-preserving distributed data clustering algorithms. Some identified benefits from the new measures are the ability to detect the vulnerabilities of the representative algorithm to collusion in different scenarios and detect point level privacy breach. In fact, it was shown that collusion is indeed an important source of privacy breach, and that algorithms can be separated according to their vulnerability to collusion groups and to the potential malicious behavior of a site with a special role in the protocol, e.g., a central site, or an aggregator, or a protocol initiator.

Acknowledgment. This work was partly supported by the EU-funded project TOREADOR (contract n. H2020-688797)

References

1. Agrawal, D., Aggarwal, C.C.: On the design and quantification of privacy preserving data mining algorithms. In Proceedings of the 20th Symposium on Principles of Database Systems (PODS), pp. 247–255. ACM, May 2001
2. Bertino, E., Fovino, I., Provenza, L.: A framework for evaluating privacy preserving data mining algorithms. Data Min. Knowl. Discov. **11**(2), 121–154 (2005)
3. Clifton, C., Kantarcioglu, M., Vaidya, J., Lin, X., Zhu, M.: Tools for privacy preserving data mining. ACM SIGKDD Explor. Newsl. **4**(2), 28–34 (2002)
4. Forman, G., Zhang, B.: Distributed data clustering can be efficient and exact. SIGKDD Explor. Newsl. **2**(2), 34–38 (2000)
5. Goldreich, O.: Foundations of Cryptography: Basic Applications, vol. 2. Cambridge University Press, Cambridge (2004)
6. Jones, C., Hall, J., Hale, J.: Secure distributed database mining: principle of design. In: Advances in Distributed and Parallel Knowledge Discovery, Chap. 10, pp. 277–294. AAAI Press/MIT Press, Menlo Park (2000)
7. Kantarcioglu, M.: A survey of privacy-preserving methods across horizontally partitioned data. In: Aggarwal, C.C., Yu, P.S. (ed.) Privacy-Preserving Data Mining. The Kluwer International Series on Advances in Database Systems, vol. 34, pp. 313–335. Springer, New York (2008)
8. Klusch, M., Lodi, S., Moro, G.: Agent-based distributed data mining: the KDEC scheme. In: Klusch, M., Bergamaschi, S., Edwards, P., Petta, P. (eds.) Intelligent Information Agents. LNCS (LNAI), vol. 2586, pp. 104–122. Springer, Heidelberg (2003). doi:10.1007/3-540-36561-3_5
9. Lindell, Y., Pinkas, B.: Secure multiparty computation for privacy-preserving data mining. J. Priv. Confidentiality **1**(1), 5 (2009)
10. Merugu, S., Ghosh, J.: Privacy-preserving distributed clustering using generative models. In: Proceedings of the 3rd International Conference on Data Mining (ICDM). IEEE (2003)
11. Merugu, S., Ghosh, J.: A privacy-sensitive approach to distributed clustering. Pattern Recogn. Lett. **26**, 399–410 (2005)
12. Patel, S.J., Punjani, D., Jinwala, D.C.: An efficient approach for privacy preserving distributed clustering in semi-honest model using elliptic curve cryptography. Int. J. Netw. Secur. **17**(3), 328–339 (2015)
13. Provost, F.: Distributed data mining: scaling up and beyond. In: Advances in Distributed and Parallel Knowledge Discovery, pp. 3–27. AAAI Press, Palo Alto (2000)
14. Shen, P., Li, C.: Distributed information theoretic clustering. IEEE Trans. Signal Process. **62**(13), 3442–3453 (2014)
15. Vaidya, J., Clifton, C.: Privacy-preserving k-means clustering over vertically partitioned data. In: Proceedings of the 9th International Confernce on Knowledge Discovery and Data Mining (KDD), pp. 206–215. ACM (2003)
16. Zaki, M.J.: Parallel and distributed data mining: an introduction. In: Zaki, M.J., Ho, C.-T. (eds.) LSPDM 1999. LNCS (LNAI), vol. 1759, pp. 1–23. Springer, Heidelberg (2000). doi:10.1007/3-540-46502-2_1

Bi-stochastic Matrix Approximation Framework for Data Co-clustering

Lazhar Labiod$^{(\boxtimes)}$ and Mohamed Nadif

LIPADE, University Paris Descartes, 75006 Paris, France
{lazhar.labiod,mohamed.nadif}@parisdescartes.fr

Abstract. The matrix approximation approaches like *Singular Value Decomposition* SVD and *Non-negative Matrix Tri-Factorization* (NMTF) have recently been shown to be useful and effective to tackle the co-clustering problem. In this work, we embed the co-clustering in a *Bistochastic Matrix Approximation* (BMA) framework and we derive from the double kmeans objective function a new formulation of the criterion to optimize. First, we show that the double k-means is equivalent to algebraic problem of BMA under some suitable constraints. Secondly, we propose an iterative process seeking for the optimal simultaneous partitions of rows and columns data, the solution is given as the steady state of a markov chain process. We develop two iterative algorithms; the first consists in learning rows and columns similarities matrices and the second consists in obtaining the simultaneous rows and columns partitions. Numerical experiments on simulated and real datasets demonstrate the interest of our approach which does not require the knowledge of the number of co-clusters.

Keywords: Bi-stochastic matrix approximation · Co-clustering · Power method

1 Introduction

Clustering has received a significant amount of attention as an important problem with many applications, and a number of different methods have emerged over the years. For datasets arising in text mining and bioinformatics where the data is represented in a very high dimensional space, clustering both dimensions of data matrix simultaneously is often more desirable than traditional one side clustering. Co-clustering which is a simultaneous clustering of rows and columns of data matrix consists in interlacing row clustering with column clustering at each iteration [1, 4, 12]. It exploits the duality between rows and columns and therefore allows to effectively deal with high dimensional data.

The earliest co-clustering formulation called *Direct clustering* has been introduced by Hartigan [10], who has proposed a greedy algorithm for hierarchical co-clustering. In [3], the author developed a spectral co-clustering algorithm on word-document data, the largest several left and right singular vectors of the normalized word-document matrix are computed and then a final clustering step

© Springer International Publishing AG 2016
H. Boström et al. (Eds.): IDA 2016, LNCS 9897, pp. 273–283, 2016.
DOI: 10.1007/978-3-319-46349-0_24

using $k$means is applied to the data projected to the topmost singular vectors. Based on the information-theoretic, the authors [4] have proposed co-clustering algorithm that presents a non-negative matrix as an empirical joint probability distribution of two discrete random variables and set the co-clustering problem as an optimization problem. Model-based co-clustering techniques have also shown promising results in several situations, the co-clustering of binary and contingency data have been treated by using latent block Bernoulli and Poisson models [7,8]. Note that all co-clustering methods implicitly perform an adaptive dimensionality reduction at each iteration, leading to better document clustering accuracy compared to one side clustering methods. Even if the co-clustering problem is not the main objective of low rank matrix approximation approaches, these ones have attracted many authors for document clustering. Then, different algorithms based on low rank matrix approximation or more precisely on the non negative matrix factorization (NMF) are proposed [5,11,15].

In this paper, we propose a new co-clustering framework based on a *Bistochastic Matrix Approximation* (BMA) formulation. To this end and contrary to classical approaches [5,15], we first embed the co-clustering aim under NMF at the beginning. The key idea is that the latent block structure, in a rectangular non-negative data matrix, is factorized into two factors, the row-coefficient matrix $\mathbf{R}$ and column-coefficient matrix $\mathbf{C}$ indicating respectively the degree in which a row and a column belongs to a cluster. Let A be $N \times M$ data matrix, the proposed approach optimizes a relaxed formulation of the double $k$means criterion in an NMF style. Then the optimization procedure looks for the best approximation $A \approx \mathbf{R}\mathbf{R}^T A \mathbf{C}\mathbf{C}^T$ meaning the Frobenius norm with respect to some suitable constraints generated by the properties of the matrices $\mathbf{R}$ and $\mathbf{C}$. Knowing that $\mathbf{\Pi}_r = \mathbf{R}\mathbf{R}^T$ and $\mathbf{\Pi}_c = \mathbf{C}\mathbf{C}^T$ are both doubly stochastic, the theory of Markov chains says that the following iterative process

$$\hat{A}^{(t+1)} = \mathbf{\Pi}_r \hat{A}^{(t)} \mathbf{\Pi}_c \text{ where } \hat{A}^{(0)} = A. \tag{1}$$

converges to an equilibrium (steady) state. In this framework, we develop a novel co-clustering algorithm for non-negative data matrix, which iteratively computes the approximation $\hat{A}$ from which we derive a co-clustering by the calculation of its first left and right eigenvectors. Numerical experiments demonstrate the potential of the used framework in proposing an efficient algorithm to co-clustering without requiring the numbers of row and column clusters.

The rest of paper is organized as follows. Section 2 introduces the co-clustering problem. Section 3 provides a sound BMA framework for co-clustering. Section 4 focus on some details on the proposed BMA algorithms. Section 5 establish the connection between BMA algorithm and spectral co-clustering. Section 6 is devoted to numerical experiments. Finally, the conclusion summarizes the advantages of our contribution.

2   Co-clustering: Criterion and Algorithms

Given a data matrix $A = (a_{ij}) \in \mathcal{R}^{N \times M}$, the aim of co-clustering can be reached by optimizing the difference between $A = (a_{ij})$ and the clustered matrix revealing significant block structure. More formally, we seek to partition the set of rows $I = \{1, \ldots, N\}$ into K clusters $P = \{P_1, \ldots, P_K\}$ and the set of columns $J = \{1, \ldots, M\}$ into L clusters $Q = \{Q_1, \ldots, Q_L\}$. The two partitioning naturally induce clustering index matrices $R = (r_{ik}) \in \mathcal{R}^{N \times K}$ and $C = (c_{j\ell}) \in \mathcal{R}^{M \times L}$, defined as binary classification matrices such as $\sum_{k=1}^{K} r_{ik} = 1$ and $\sum_{\ell=1}^{L} c_{j\ell} = 1$. Specifically, we have $r_{ik} = 1$, if the row $\mathbf{a}_i \in P_k$, and 0 otherwise. The matrix C is defined similarly by $c_{j\ell} = 1$, if the column $\mathbf{a}_j \in Q_\ell$, and 0 otherwise. Thanks to r_{ik} and $c_{j\ell}$, a submatrix or block $A_{k\ell}$ is therefore defined by $\{(a_{ij}) | r_{ik} c_{j\ell} = 1\}$. On the other hand, we note $S = (s_{k\ell}) \in \mathcal{R}^{K \times L}$ a reduced matrix specifying the cluster representation. The detection of homogeneous blocks in A can be reached by looking for the three matrices R, C and S minimizing the total squared residue measure

$$J(A, RSC^T) = ||A - RSC^T||^2. \tag{2}$$

The term RSC^T characterizes the information of A that can be described by the cluster structures. The clustering problem can be formulated as a matrix approximation problem where the clustering aim is to minimize the approximation error between the original data A and the reconstructed matrix based on the cluster structures. Note that this matricial formulation can take the following form

$$J(A, RSC^T) = \sum_{i,j,k,\ell} r_{ik} c_{j\ell} (a_{ij} - s_{k\ell})^2.$$

With fixed P_k and Q_ℓ, it is easy to check that the optimum S is obtained by $s_{k\ell} = \frac{\sum_{i,j,k,\ell} r_{ik} c_{j\ell} a_{ij}}{r_k c_\ell}$ where, $r_k = |P_k|$ and $c_\ell = |Q_\ell|$ ($|.|$ denotes the cardinality of a cluster). In other words, each $s_{k\ell}$ is the centroid of $A_{k\ell}$. The approximation of A can be solved by an iterative alternating least-squares optimization procedure. When A is not necessarily non negative, different algorithms have been proposed to minimize this criterion (see for instance, [9]). These algorithms are equivalent and consist in using the principle of a double $k$means; simplicity and scalability are its advantages.

3   BMA Framework for Co-clustering

3.1   NMF Formulation

By considering double $k$means as a lower rank matrix factorization, with constraints, rather than a co-clustering method, we can formulate constraints to impose on NMF formulation. Let $D_r^{-1} \in \mathcal{R}^{K \times K}$ and $D_c^{-1} \in \mathcal{R}^{L \times L}$ be diagonals matrices defined as follow $D_r^{-1} = Diag(r_1^{-1}, \ldots, r_K^{-1})$ and $D_c^{-1} = Diag(c_1^{-1}, \ldots, c_L^{-1})$. Using the matrices D_r, D_c, A, R and C, the matrix summary S can be expressed as $S = D_r^{-1} R^T A C D_c^{-1}$. Plugging S into the objective function Eq. (2) leads to optimize $||A - R(D_r^{-1} R^T A C D_c^{-1}) C^T||^2$ equal to

$$||A - \mathbf{R}\mathbf{R}^T A \mathbf{C}\mathbf{C}^T||^2, \tag{3}$$

where $\mathbf{R} = RD_r^{-0.5}$ and $\mathbf{C} = CD_c^{-0.5}$. Note that this formulation holds even if A is not nonnegative, i.e., A has mixed signs entries. On the other hand, it is easy to verify that the approximation $\mathbf{R}\mathbf{R}^T A \mathbf{C}\mathbf{C}^T$ of A is formed by the same value in each block $A_{k\ell}$. Specifically, the matrix $\mathbf{R}^T A \mathbf{C}$, equal to S, plays the role of a summary of A and absorbs the different scales of A, $\mathbf{R}$ and $\mathbf{C}$. Finally the matrices $\mathbf{R}\mathbf{R}^T A$, $A\mathbf{C}\mathbf{C}^T$ give respectively the row and column clusters mean vectors. Note that it is easy to show that $\mathbf{R}$ and $\mathbf{C}$ verify the following properties

$$\begin{cases} \mathbf{R} \geq 0, \mathbf{R}^T\mathbf{R} = I_K, \mathbf{R}\mathbf{R}^T\mathbb{1} = \mathbb{1}, Trace(\mathbf{R}\mathbf{R}^T) = K, (\mathbf{R}\mathbf{R}^T)^2 = \mathbf{R}\mathbf{R}^T \\ \text{and} \\ \mathbf{C} \geq 0, \mathbf{C}^T\mathbf{C} = I_L, \mathbf{C}\mathbf{C}^T\mathbb{1} = \mathbb{1}, Trace(\mathbf{C}\mathbf{C}^T) = L, (\mathbf{C}\mathbf{C}^T)^2 = \mathbf{C}\mathbf{C}^T. \end{cases} \tag{4}$$

The formulation (3) has been also proposed in [2,11] but in this paper, the optimization process and the derived algorithm are different. Next, we consider co-clustering from the perspective of bi-stochastic matrix approximation.

3.2 BMA Formulation

Given a data matrix A which derives from K rows clusters and L columns clusters. Let us define $\mathbf{\Pi}_r = \mathbf{R}\mathbf{R}^T$ and $\mathbf{\Pi}_c = \mathbf{C}\mathbf{C}^T$, we can hope to discover the co-cluster structure of A from $\mathbf{\Pi}_r$ and $\mathbf{\Pi}_c$. Notice that from (4) $\mathbf{\Pi}_r$ and $\mathbf{\Pi}_c$ are both nonnegative, symmetric, bi-stochastic (doubly stochastic) and idempotent. Setting the double $k$means in the BMA framework, the problem of clustering can be reformulated as the seek of $\mathbf{\Pi}_r$ and $\mathbf{\Pi}_c$ minimizing $||A - \mathbf{\Pi}_r A \mathbf{\Pi}_c||^2$. The computation of $\mathbf{\Pi}_r$ and $\mathbf{\Pi}_c$ requires an iterative algorithm. We derive a general algebraic model for co-clustering by considering the double k means as a BMA problem given by

$$\arg \min_{\{\mathbf{\Pi}_r, \mathbf{\Pi}_c\}} ||A - \mathbf{\Pi}_r A \mathbf{\Pi}_c||^2. \tag{5}$$

In the rest of this paper, we will consider only nonnegativity, symmetry and bi-stochastic constraints. The Eq. 6 below establishes the equivalence between double $k$means and the BMA formulation. Then, solving the BMA objective function (5) is equivalent to finding a global solution of the double $k$means criterion (2).

$$\arg \min_{\{R,S,C\}} ||A - RSC^T||^2 \Leftrightarrow \arg \min_{\{\mathbf{\Pi}_r, \mathbf{\Pi}_c\}} ||A - \mathbf{\Pi}_r A \mathbf{\Pi}_c||^2 \tag{6}$$

Due to the limited space, the detailed proof of this equivalence is omitted here.

4   BMA Co-clustering Algorithm

The bi-stochastic matrices $\mathbf{\Pi}_r$ and $\mathbf{\Pi}_c$ constructed from the cluster memberships matrix $\mathbf{R}$ and $\mathbf{C}$ can be viewed as a special type of similarity matrices. Obviously, an arbitrary similarity matrix cannot be bi-stochastic. For a given affinity matrix, there are multiple ways to derive a bi-stochastic matrix based on different divergence functions. Next, we focus on the approximation of the nearest bi-stochastic matrix using the Frobenius norm.

4.1   Learning Similarity Matrix

First, we establish the relationship between our objective function and that used in [14,16]. From

$$||A - \mathbf{\Pi}_r A||^2 = Trace(AA^T) + Trace(\mathbf{\Pi}_r AA^T \mathbf{\Pi}_r) - 2Trace(AA^T \mathbf{\Pi}_r) \quad (7)$$

and the properties in the first line of (4), we can show that

$$\arg\min_{\mathbf{\Pi}_r}||A - \mathbf{\Pi}_r A||^2 \Leftrightarrow \arg\min_{\mathbf{\Pi}_r}||AA^T - \mathbf{\Pi}_r||^2 \Leftrightarrow \arg\max_{\mathbf{\Pi}_r} Trace(AA^T \mathbf{\Pi}_r).$$

In the same way, on the other side we have

$$\arg\min_{\mathbf{\Pi}_c}||A - A\mathbf{\Pi}_c||^2 \Leftrightarrow \arg\min_{\mathbf{\Pi}_c}||A^T A - \mathbf{\Pi}_c||^2 \Leftrightarrow \arg\max_{\mathbf{\Pi}_c} Trace(A^T A\mathbf{\Pi}_c).$$

We aim to optimize the quadratic form above with nonnegativity, symmetry and bistochastic constraint on Π. We follow the standard optimization theory to find the minima and we introduce the Lagrangian function

$$\mathcal{L} = ||AA^T - \Pi||^2 - \Lambda_1^T(\Pi - \mathbb{1}) - \Lambda_2^T(\Pi - \mathbb{1}) \quad (8)$$

where $\Lambda_1, \Lambda_2 \in R^{N \times 1}$ are the lagrangian multiplier vectors. The algorithm for learning similarity matrix is summarized in Algorithm 1 as in [14,16].

Algorithm 1. Learning similarity matrix

Input: data A
Output: similarity matrix Π
Initialize: $t = 0$ and $\Pi^{(0)} = AA^T$
repeat
$\quad \Pi^{(t+1)} \leftarrow [\Pi^{(t)} + \frac{1}{N}(I - \Pi^{(t)} + \frac{\mathbb{1}\mathbb{1}^T \Pi^{(t)}}{N})\mathbb{1}\mathbb{1}^T - \frac{1}{N}\mathbb{1}\mathbb{1}^T \Pi^{(t)}]$
until Satisfied convergence condition

4.2   BMA Algorithm

Once the bi-stochastic similarity matrices $\mathbf{\Pi}_r$ and $\mathbf{\Pi}_c$ are obtained by using Algorithm 1, the basic idea of the *Bistochastic Matrix Approximation* (BMA) algorithm is based on the following steps:

1. Estimating iteratively A by applying at each time the matrices $\mathbf{\Pi}_r$ and $\mathbf{\Pi}_c$ on the current A using the following update

$$\hat{A}^{(t+1)} = \mathbf{\Pi}_r A^{(t)} \mathbf{\Pi}_c. \quad (9)$$

This process converges to an equilibrium (steady) state. Let g be the multiplicity of the eigenvalue of matrix $\mathbf{\Pi}_r$ equal to 1 and l the multiplicity of the eigenvalue equal to 1 of $\mathbf{\Pi}_c$, $\hat{A}$ is composed of $g \ll m$ quasi-similar rows and quasi-similar columns $l \ll n$, where each row and each column is represented by its prototype.

2. Extracting the first left and right singular vectors π_r and π_c of $\hat{A}$ using the *Power Method* [6] as described in Algorithm 2. The power method is the well-known technique used for computing the largest left and right eigenvectors of data matrix. For numerical computation of the leading singular vectors of $\hat{A}$, we use a new variant of this method adapted to the case of rectangular data matrix which. It consists in starting with an arbitrary vector $\pi_r^{(0)}$, repeatedly performing alternated updates of π_c and π_r. These steps are grouped in Algorithm 2.

Algorithm 2. Modified power method

Input: data $\hat{A}$

Initialize: $\pi_r^{(0)} = \hat{A}\mathbb{1}, \pi_r^{(0)} \leftarrow \frac{\pi_r^{(0)}}{||\pi_r^{(0)}||}$

repeat

$\quad \pi_c^{(t+1)} = \hat{A}\pi_r^{(t)}$ where $\pi_c^{(t)} \leftarrow \frac{\pi_c^{(t)}}{||\pi_c^{(t)}||}$

$\quad \pi_r^{(t+1)} = \hat{A}^T\pi_c^{(t)}$ where $\pi_r^{(t)} \leftarrow \frac{\pi_r^{(t)}}{||\pi_{(r)}^{(t)}||}$

$\quad \gamma^{(t+1)} \leftarrow ||\pi_r^{(t+1)} - \pi_r^{(t)}|| + ||\pi_c^{(t+1)} - \pi_c^{(t)}||$

until stabilization of π_r, π_c, $|\gamma^{(t+1)} - \gamma^{(t)}| \simeq 0$

At first sight, this process might seem uninteresting since it eventually leads to a vector with all rows and columns coincide for any starting vector. However our practical experience shows that, first the vectors π_r and π_c very quickly collapse into rows and columns blocks and these blocks move towards each other relatively slowly. If we stop the power method iteration at this point, the algorithm would have a potential application for data visualization and co-clustering. The structure of π_r and π_c during short-run stabilization makes the discovery of rows and columns data ordering straightforward. The key is to look for values of π_r and π_c that are approximately equal and reordering rows and columns data accordingly. The BMA algorithm involves a reorganization of the rows and the columns of data matrix A according to sorted π_r and π_c. It also allows to locate the points corresponding to an abrupt change in the curve of the first left and right singular vectors π_r and π_c, and then asses the number of clusters and the rows or columns belonging to each cluster. The principal steps of the BMA algorithm are summarized in Algorithm 3. To illustrate why the BMA algorithm works well in co-clustering context, we show in the sequel, the connection between BMA co-clustering algorithm and spectral co-clustering method [3].

5   Relationship Between BMA Algorithm and Spectral Co-clustering

BMA is related to spectral co-clustering in that it finds a low dimensional embedding of data, and $k$means or another clustering techniques are used to produce

---

Algorithm 3. Approximation of A

Input: data A, $\mathbf{\Pi}_r$ and $\mathbf{\Pi}_c$
Output: data $\hat{A}$, sorted π_r and π_c
repeat
    $A^{(t+1)} = \mathbf{\Pi}_r A^{(t)} \mathbf{\Pi}_c$, $\gamma^{(t+1)} \leftarrow ||A^{(t+1)} - A^{(t)}||^2$
until $|\gamma^{(t+1)} - \gamma^{(t)}| \simeq 0$
Deduce: $\hat{A}$
Compute: π_r and π_c by using Algorithm 2.

---

the final co-clustering. But as a result, in this paper it is not necessary to find any singular vector as most co-clustering methods do. In order to find a low dimensional embedding for co-clustering, the embedding just needs to be a good linear combination of the left and the right singular vectors respectively. In this respect BMA is a different approach. In spectral co-clustering the embedding is formed by the bottom left and right eigenvectors of a normalized data matrix [3]. In BMA, embedding is defined as weighted linear combination of singular vectors, then π_r is a defined as linear combination of all left singular vectors of $\hat{A}$ and π_c as weighted linear combination of all right singular vectors of $\hat{A}$. The left and right embedding turn out to be very interesting for data reordering and co-clustering. From the start, the first largest left and right singular vectors of $\hat{A}$ are not very interesting since they move towards the uniform distribution via a long run times. However the intermediate π_r and π_c obtained by BMA after a short run time are very interesting. The experimental observation suggests that an effective reordering might run BMA for a small number of iterations.

Let us define the $(n+m)$ by $(n+m)$ data matrix $\mathbf{M}$ as follow $\mathbf{M} = \begin{pmatrix} 0 & \hat{A} \\ \hat{A}^T & 0 \end{pmatrix}$. Assuming that $\mathbf{M}$ is diagonalizable, then it exists a non singular matrix Q of eigenvectors such that $Q^{-1}\mathbf{M}Q = diag(\lambda_1, \lambda_2, \ldots, \lambda_{n+m})$. Furthermore, assuming that the eigenvalues are ordered $|\lambda_1| > |\lambda_2| \geq |\lambda_3| \geq \ldots \geq |\lambda_{n+m}|$ and expanding the initial approximation $\pi^{(0)} = \begin{bmatrix} \pi_r^{(0)} \\ \pi_c^{(0)} \end{bmatrix}$ in terms of the eigenvectors of $\mathbf{M}$: $\pi^{(0)} = c_1 q_1 + c_2 q_2 + \ldots + c_{n+m} q_{n+m}$ with $q_k = \begin{bmatrix} u_k \\ v_k \end{bmatrix}$ where the upper part u_k is for the rows of $\hat{A}$ and the lower part v_k is for the columns of $\hat{A}$, $c_k \neq 0$ is assumed. We have

$$\pi^{(t)} = \mathbf{M}^{(t)}\pi^{(0)} = \lambda_1^{(t)}\left(c_1 q_1 + \sum_{k=2}^{n+m} c_k \left(\frac{\lambda_k}{\lambda_1}\right)^{(t)} q_k\right) \tag{10}$$

for $k = 2, \ldots, n+m$, we have $|\lambda_i| < |\lambda_1|$, so the second term tends to zero, and the power method converges to the eigenvector $q_1 = \begin{bmatrix} u_1 \\ v_1 \end{bmatrix}$ corresponding to the dominant eigenvalue λ_1. The rate of convergence is determined by the ratio $|\frac{\lambda_2}{\lambda_1}|$, if this is close to one the convergence is very slow.

In Eq. (10), we expand the left eigenvector $\pi^{(t)}$ as a linear combination of the eigenvectors of $\mathbf{M}^{(t)}$. It is easy to see that $\pi^{(t)} = \begin{bmatrix} \pi_r^{(t)} \\ \pi_c^{(t)} \end{bmatrix}$ is the leading eigenvector of $\mathbf{M}^{(t)}$. Mathematically, in Eq. (11) we show that the leading eigenvector of $\mathbf{M}^{(t)}$ is closely related to the first singular vectors of $\hat{A}^{(t)}$, $\pi_r^{(t)}$ and $\pi_c^{(t)}$ are the left and right singular vector of $\hat{A}^{(t)}$ respectively.

$$\begin{bmatrix} \pi_r^{(t)} \\ \pi_c^{(t)} \end{bmatrix} = \begin{bmatrix} 0 & \hat{A}^{(t)} \\ \hat{A}^{T(t)} & 0 \end{bmatrix} \begin{bmatrix} \pi_r^{(0)} \\ \pi_c^{(0)} \end{bmatrix} = \begin{bmatrix} \hat{A}^{(t)} \pi_c^{(0)} \\ \hat{A}^{T(t)} \pi_r^{(0)} \end{bmatrix}. \tag{11}$$

More interestingly, Eq. (11) above provides the update equations given in Algorithm 2. Also, instead of constructing $\mathbf{M}$ (like the most spectral co-clustering methods do) which is bigger and sparser than the approximated data $\hat{A}$, we provide a way to cocluster, not using $\mathbf{M}$ but directly from $\hat{A}$.

Now, by exploiting the block structure of $\mathbf{M}$ in Eq. (10), we show in Eq. (12); $\pi_r^{(t)}$ as a linear combination of the left singular vectors of $\hat{A}$ and $\pi_c^{(t)}$ as a linear combination of the right singular vectors of $\hat{A}$.

$$\begin{bmatrix} \pi_r^{(t)} \\ \pi_c^{(t)} \end{bmatrix} = \begin{bmatrix} \lambda_1^{(t)}(c_1 u_1 + \sum_{i=2}^{n} c_i (\frac{\lambda_i}{\lambda_1})^{(t)} u_i) \\ \lambda_1^{(t)}(c_1 v_1 + \sum_{j=n+1}^{n+m} c_j (\frac{\lambda_j}{\lambda_1})^{(t)} v_j) \end{bmatrix}. \tag{12}$$

The BMA insight is the observation that while BMA iteration run to convergence will give us the principal singular vector for a matrix, running BMA iteration on $\hat{A}$ to what we call local convergence will give us tow vectors π_r that is a linear combination of the k right singular vectors and a vector π_c that is a linear combination of the k left singular vectors corresponding to the k largest singular values of $\hat{A}$. Assuming that the gap between λ_k and λ_{k+1} is large, and that the gap between λ_1 and $\lambda_2, \ldots, \lambda_k$ is not very large, we note that BMA iteration on $\hat{A}$ will converge locally before converging globally. We define local convergence as the point at which the kth to nth singular values are no longer influencing the term π_r and π_c in BMA iteration.

Furthermore, we assume that at this point, the vector π_r (resp π_r) has converged locally to the row clusters represented by an embedding into the space spanned $u_1, \ldots, u_k$ (resp $v_1, \ldots, v_k$) in spectral co-clustering. Local convergence is detected by recognizing the point at which $|\gamma^{(t+1)} - \gamma^{(t)}| \simeq 0$ where $\gamma^{(t+1)} \leftarrow ||\pi_r^{(t+1)} - \pi_r^{(t)}|| + ||\pi_c^{(t+1)} - \pi_c^{(t)}||$.

Now, to use the locally-converged π_r and π_c to clustering the rows and columns of data, we observe that the entries in π_r and π_c should be approximately piecewise-constant. Thus, the linear combination of these singular left and right vectors will give an approximately piecewise-constant vector, where each piece corresponds to one of the clusters.

6   Numerical Experiments

We now provide experimental results to illustrate the behavior of the BMA algorithm in data visualization and co-clustering contexts.

6.1   Simulated Data

We visualize two synthetic (500×300) data sets with 6 co-clusters, generated according to a latent block Bernoulli mixtures model [7]: data1 and data2 with three different patterns illustrated respectively in the left of Figs. 1 and 2. The co-clustering task is to recover groups of rows and columns. After the step learning, the clusters indicators are given by the vectors π_r and π_c. We then reorganize the rows and columns separately or simultaneously according the obtained co-clusters in the middle of Figs. 1 and 2. From the simulated data, the π_r and π_c plots are shown in the right of these figures. It can be seen that our method reconstructs effectively all co-clusters for balanced and unbalanced data sets. For all data, the π_r and π_c plots show that the number of abrupt changes and in each plot it correlates well with the true number of clusters.

6.2   Real Datasets

To evaluate BMA in terms of clustering, we used four datasets whose characteristics are reported in Table 1. The first three datasets, commonly used in document clustering, are word-by-document matrices whose rows correspond to documents, and columns to words; each cell denotes the frequency of a word in a document. **Classic3** [3] contains three classes denoted Medline, Cisi, Cranfield. **CSTR** consists of abstracts, which were divided into four research areas: Natural Language Processing (NLP), Robotics/Vision, Systems and Theory. **WebKB4** consists of four webpages collected from computer science departments: student,

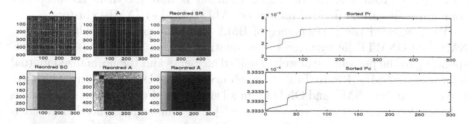

Fig. 1. Data1: A, $\hat{A}$, reordered $SR = AA^T$ according π_r, reordered $SC = A^T A$ according π_c, and finally reordered A and reordered $\hat{A}$ according π_r and π_c simultaneously.

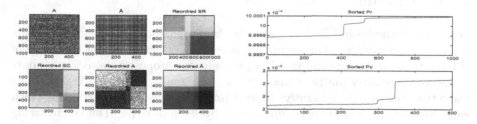

Fig. 2. Data2: A, $\hat{A}$, reordered $SR = AA^T$ according π_r, reorderd $SC = A^T A$ according π_c, and finally reordered A and reordered $\hat{A}$ according π_r and π_c simultaneously.

Table 1. Clustering accuracy and normalized mutual information (%)

Datasets	# samples	# features	# classes	per	NMF	ONMTF	SpecCO	BMA
Classic3	3891	4303	3	ACC	73.33	70.10	97.89	98.30
				NMI	51.46	51.46	91.17	91.91
CSTR	476	1000	4	ACC	75.30	77.41	80.21	90.73
				NMI	66.40	67.30	66.36	77.86
Webkb4	4199	1000	4	ACC	66. 30	67.10	61.68	68.8
				NMI	42.70	45.36	48.64	49
Leukemia	38	5000	3	ACC	89.21	90.32	94.73	97.36
				NMI	75.42	80.50	82	90.69

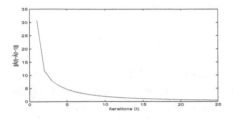 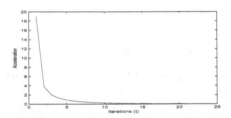

Fig. 3. Number of BMA iterations necessary to achieve: (left) - $\gamma^{(t+1)} = ||A^{(t+1)} - A^{(t)}||^2 \simeq 0$ and (right) - *Acceleration* $= |\gamma^{(t+1)} - \gamma^{(t)}| \simeq 0$.

faculty, course and project. The last dataset **Leukemia** contains expression levels of genes taken over samples. This data set is well known in the academic community, it can be divided into three (ALL-B/ALL-T/AML) clusters.

We compared the performance of BMA with the spectral co-clustering [3], NMF and ONMTF [5] by using two evaluation metrics: accuracy (ACC) corresponding to the percentage of well-classified elements and the normalized mutual information (NMI) [13]. In Table 1, we observe that BMA outperforms the spectral co-clustering, NMF and ONMTF in all situations. Furthermore, BMA needs only few iterations to achieve $|\gamma^{(t+1)} - \gamma^{(t)}| \simeq 0$. The convergence behaviour is empirically illustrated on CSTR dataset (Fig. 3).

7 Conclusion

Our main contribution in this paper is the proposition of an efficient BMA co-clustering algorithm. Our contribution is firstly an extension of the works of [14,16], to co-clustering context. More precisely, our proposed approach, not only learns two similarity matrices from rows and columns of data matrix, but uses these two matrices in an iterative process that converges to a matrix $\hat{A}$ in which each row and each column is represented by its prototype. The co-clustering solution is given by the first left and right eigenvectors of $\hat{A}$ while overcoming the knowledge of the number of co-clusters.

Acknowledgments. This work has been funded by AAP Sorbonne Paris Cité.

References

1. Cheng, Y., Church, G.M.: Biclustering of expression data, pp. 93–103. AAAI (2000)
2. Cho, H., Dhillon, I., Guan, Y., Sra, S.: Minimum sum-squared residue co-clustering of gene expression data. In: Proceedings of the Fourth SIAM International Conference on Data Mining, pp. 114–125 (2004)
3. Dhillon, I.: Co-clustering documents and words using bipartite spectral graph partitioning. In: Proceedings of the International Conference ACM SIGKDD, San Francisco, USA, pp. 269–274 (2001)
4. Dhillon, I., Mallela, S., Modha, D.S.: Information-theoretic coclustering. In: Proceedings of KDD 2003, pp. 89–98 (2003)
5. Ding, C., Li, T., Peng, W., Park, H.: Orthogonal nonnegative matrix tri-factorizations for clustering. In: Proceedings of KDD 2006, Philadelphia, PA, pp. 635–640, September 2006
6. Golub, G.H., van Loan, C.F.: Matrix Computations, 3rd edn. Johns Hopkins University Press, Baltimore (1996)
7. Govaert, G., Nadif, M.: Block clustering with Bernoulli mixture models: comparison of different approaches. Comput. Stat. Data Anal. **52**, 2333–3245 (2008)
8. Govaert, G., Nadif, M.: Latent block model for contingency table. Commun. Stat. Theor. Methods **39**, 416–425 (2010)
9. Govaert, G., Nadif, M.: Co-clustering: Models, Algorithms and Applications. Wiley, New York (2013)
10. Hartigan, J.A.: Direct clustering of a data matrix. J. Am. Stat. Assoc. **67**(337), 123–129 (1972)
11. Labiod, L., Nadif, M.: Co-clustering for binary and categorical data with maximum modularity. In: ICDM 2011, pp. 1140–1145 (2011)
12. Labiod, L., Nadif, M.: Co-clustering under nonnegative matrix tri-factorization. In: Arik, S., Huang, T., Lai, W.K., Liu, Q. (eds.) ICONIP 2015. LNCS, vol. 9492, pp. 709–717. Springer, Heidelberg (2011). doi:10.1007/978-3-642-24958-7_82
13. Strehl, A., Ghosh, J.: Cluster ensembles - a knowledge reuse framework for combining multiple partitions. J. Mach. Learn. Res. **03**, 583–617 (2002)
14. Wang, F., Li, P., König, A.C., Wan, M.: Improving clustering by learning a bi-stochastic data similarity matrix. Knowl. Inf. Syst. **32**(2), 351–382 (2012)
15. Yoo, J., Choi, S.: Orthogonal nonnegative matrix tri-factorization for co-clustering: multiplicative updates on Stiefel manifolds. Inf. Process. Manag. **46**(5), 559–570 (2010)
16. Zass, R., Shashua, A.: A unifying approach to hard and probabilistic clustering. In: ICCV, pp. 294–301 (2005)

Sequential Cost-Sensitive Feature Acquisition

Gabriella Contardo[1]([✉]), Ludovic Denoyer[1], and Thierry Artières[2]

[1] Sorbonne Universités, UPMC Univ Paris 06, UMR 7606, LIP6, 75005 Paris, France
`gabriella.contardo@lip6.fr`
[2] Ecole Centrale Marseille-Laboratoire d'Informatique Fondamentale
(Aix-Marseille University), Marseille, France

Abstract. We propose a reinforcement learning based approach to tackle the cost-sensitive learning problem where each input feature has a specific cost. The acquisition process is handled through a stochastic policy which allows features to be acquired in an adaptive way. The general architecture of our approach relies on representation learning to enable performing prediction on any partially observed sample, whatever the set of its observed features are. The resulting model is an original mix of representation learning and of reinforcement learning ideas. It is learned with policy gradient techniques to minimize a budgeted inference cost. We demonstrate the effectiveness of our proposed method with several experiments on a variety of datasets for the sparse prediction problem where all features have the same cost, but also for some cost-sensitive settings.

1 Introduction

We are concerned here with budgeted learning, where we want to design algorithms that perform optimal compromises between (small) test cost and (high) accuracy. Most of today's machine learning approaches usually assume that the input (i.e. its features) is fully observable for free. However, it is often a strong assumption: indeed, each feature may have to be acquired and this acquisition usually has a *cost*, e.g. computational or financial. Hence, in many applications (e.g. personalized systems), the prediction performance may be seen as a trade-off between the said prediction accuracy (as in classical machine learning settings), and the cost of the information (i.e. features) needed to perform this prediction[1]. A natural approach to optimize such a trade-off is to rely on feature selection through e.g. L1 regularization [2] or dimensionality reduction. But it is likely that an optimal feature selection should be sample dependent. A better solution should then be adaptive, i.e. the features should be acquired depending on what has been previously gathered and observed, which asks for a sequential acquisition process. Medical diagnosis illustrates this case, where a doctor, to set a diagnosis, only asks for the results of a few exams, which depend on

---

[1] We consider here that the computation cost (time spent to compute the prediction based on the acquired features values) is negligible w.r.t to the acquisition cost, as it is usually done in cost-sensitive approaches – see Sect. 4.

H. Boström et al. (Eds.): IDA 2016, LNCS 9897, pp. 284–294, 2016.
DOI: 10.1007/978-3-319-46349-0_25

the patient and his previous results on other exams. Moreover, it may happen that the acquisition cost varies from a feature to another, as in medical diagnosis again, where some medical results are cheap to acquire (e.g. blood analysis), while other can be quite expensive (e.g. fMRI exams). In this *cost-sensitive* case, lowering the acquisition cost is not only a matter of number of features gathered.

We consider the challenging setting that may be characterized by the following properties: (i) optimality is defined as a trade-off between prediction quality and acquisition cost, (ii) each feature may have a different acquisition cost, (iii) prediction may be made from a partially observed input -i.e with only a subset of its features-, (iv) the optimal subset of features to acquire (to perform accurate prediction) depends on the input sample.

We present in this paper a stochastic sequential method that relies on both reinforcement learning through the use of policy gradient inspired techniques and representation-learning to tie the prediction and acquisition tasks together. Section 2 describes our proposal. We first introduce the generic formulation of our sequential modeling framework and explain how it may be optimized through gradient descent. We then detail how it is mixed with representation learning to enable dealing with our setting. We next report in Sect. 3 experimental results gained in various settings. Finally Sect. 4 situates our work with respect to the main approaches in the literature.

2   Cost Sensitive Classification as a Sequential Problem

We consider the classification problem of mapping an input space $\mathcal{X}$ to a set of classes $\mathbb{Y}$, where examples $x \in \mathcal{X}$ have n features (x_i denotes the i-th feature of x) (we focus on classification for clarity but our work may be applied straightforwardly to other tasks like regression or ranking). We consider that our model produces a score for each possible category (i.e. positive scores for true categories, and negative scores for wrong ones), the quality of the prediction being measured through a loss function $\Delta : \mathbb{R}^Y \times \mathbb{Y} \to \mathbb{R}^+$ (e.g. hinge loss), where we consider the prediction function to output a score for each class (with Y being the cardinality of $\mathbb{Y}$), and we assume that this loss function is differentiable almost everywhere on its first component. This corresponds to the classical context of numerical classifiers like SVM or neural networks.

We focus on predictors that iteratively acquire new features of an input x and that finally perform prediction from the observed partial view of x. To do so, we consider sequential methods that acquire features from x through a series $a = (a_t)_{t=1..T}$ of T acquisition steps (T is a hyper-parameter of the approach) encoded as binary vectors $a_t \in \{0; 1\}^n$ indicating which features are acquired at time t: $a_{t,i} = 1$ iff feature i is acquired. The final prediction is made based on the set of features that have been acquired along the acquisition process that we note $a = (a_1, ..., a_T)$. Noting $\bar{a} \in \{0; 1\}^n$ the vector whose i-th component equals $\bar{a}_i = max(a_{1,i}, ..., a_{T,i})$, i.e. it is 1 iff feature i has been acquired at any step of the process, the final prediction is noted as $d(x[\bar{a}])$ where d is the prediction function and $x[\bar{a}]$ stands for the partial view acquired on x along acquisition

---

Algorithm 1. The sequential inference algorithm

procedure INFERENCE(x, T)
    $a_0 = 0$
    for $t = 1..T$ **do**
        Sample a_t from $\pi(a_t | x[(a_1, .., a_{t-1})])$
        Acquire $x[a_t]$ where new features are such that $a_{t,i} = 1$
    end for
    return $\hat{y} = d_\theta(x[\bar{a}])$
end procedure

---

sequence a. Note that this formalism allows the model to acquire many features at each timestep – while classical existing sequential features acquisition models usually only allow to get the features one by one as explained in Sect. 4, resulting in a high complexity.

Quite generally, we consider that feature acquisition is a stochastic process that we want to learn, and that every a_t is sampled following an **acquisition policy** denoted $\pi(a_t | a_1, ..., a_{t-1}, x)$, which corresponds to the **probability** of acquiring the features specified in a_t, given all previously acquired features. This policy is jointly learned with the prediction function d. The inference algorithm goes like the one described in Algorithm 1. Many feature acquisitions approaches can be expressed within this formalism. For example, static (e.g. not adaptive) feature selection corresponds to one step models ($T = 1$), while decision trees may be thought as acquiring a new feature one at a time that deterministically depend on the values of the features that were previously observed.

We now introduce our objective function. Considering that the feature acquisition cost might not be uniform, we note $c_i \geq 0$ the acquisition cost of feature i and c the vector of all features' costs. The overall acquisition cost for classifying an input x given an acquisition sequence a is then equal to $\bar{a}^\mathsf{T}.c = \sum_{i=1}^{n} \bar{a}_i \times c_i$.

The cost-sensitive and sequential feature acquisition learning problem may then be cast as the minimization of the following loss function $\mathcal{J}$, which depends on the prediction function d and on the policy π:

$$\mathcal{J}(d, \pi) = \mathbb{E}_{(x,y) \sim p(x,y)} \left[\mathbb{E}_{a \sim \pi(a|x)} \left[\Delta(d(x[\bar{a}]), y) + \lambda \bar{a}^\mathsf{T}.c \right] \right] \tag{1}$$

where λ controls the trade-off between prediction quality and feature acquisition cost, $p(x, y)$ is the unknown underlying data distribution, and $\mathbb{E}_{a \sim \pi(a/x)}[.]$ stands for the expectation on the sequence of acquisition a given a particular input sample x and the acquisition policy induced by π.

The empirical loss $\mathcal{J}^{emp}(d, \pi)$ is defined on a training set of ℓ samples $\left\{ (x^1, y^1), ..., (x^\ell, y^\ell) \right\}$ as:

$$\mathcal{J}^{emp}(d, \pi) = \frac{1}{\ell} \sum_{k=1}^{\ell} \mathbb{E}_{a \sim \pi(a|x^k)} [\Delta(d(x^k[\bar{a}]), y^k) + \lambda \bar{a}^\mathsf{T}.c] \tag{2}$$

2.1   Policy-Gradient Based Learning

In order to simultaneously learn the policy π and the prediction function d, we propose to define these two functions as differentiable parametric functions d_θ and π_γ, which allows us to use efficient stochastic gradient descent optimization methods. The parameter sets θ and γ are learned by optimizing the empirical cost in Eq. 2 (details on π and d are given later in this Section). We explain now how optimization is performed.

Let us rewrite the empirical loss in Eq. 2 for a single training example (x, y) (to improve readability), $\mathcal{J}^{emp}(x, y, \gamma, \theta)$:

$$\mathcal{J}^{emp}(x, y, \gamma, \theta) = \mathbb{E}_{a \sim \pi_\gamma(a|x)} \left[\Delta(d_\theta(x[\bar{a}]), y) + \lambda \bar{a}^\mathsf{T}.c \right] \tag{3}$$

To overcome the non differentiability of the max operator in $\bar{a}$ we propose to upper bound $\bar{a}^\mathsf{T}.c$ with $\sum_{t=1}^{T} a_t^\mathsf{T}.c$ and to perform the gradient descent over this bound. This bound is exactly equal to $\mathcal{J}^{emp}$ when a feature can be acquired only once along an acquisition sequence a. In our implementation we chose not to impose such a constraint which yields this rather tight and easier to optimize (smooth) upper bound.[2] The upper bound on the empirical risk may be rewritten as (omitting details):

$$\mathcal{J}^{emp}(x, y, \gamma, \theta) \leq \mathbb{E}_{a \sim \pi_\gamma(a|x)} \left[\Delta(d_\theta(x[\bar{a}]), y) \right] + \lambda \mathbb{E}_{a \sim \pi_\gamma(a|x)} \left[\sum_{t=1}^{T} a_t^\mathsf{T}.c \right]$$

$$= \mathbb{E}_{a \sim \pi_\gamma(a|x)} \left[\Delta(d_\theta(x[\bar{a}]), y) \right] + \lambda \sum_{t=1}^{T} \sum_{i=1}^{n} \pi_\gamma(a_{t,i} = 1|x).c_i \tag{4}$$

where $\pi_\gamma(a_{t,i} = 1|x)$ is the probability of acquiring the ith feature at time-step t. The first term stands for the **prediction quality** while the second term is the upper bound on the cost of the **acquisition policy**. The gradient of this upper bound can be written as follows:

$$\nabla_{\gamma,\theta} \hat{\mathcal{J}}(x, y, \gamma, \theta) = \nabla_{\gamma,\theta} \mathbb{E}_{a \sim \pi_\gamma(a|x)} \Delta(d_\theta(x[\bar{a}]), y) + \lambda \nabla_{\gamma,\theta} \sum_{t=1}^{T} \sum_{i=1}^{n} \pi_\gamma(a_{t,i} = 1|x).c_i \tag{5}$$

The gradient of the prediction quality term may be computed using policy-gradient based techniques [12,20] (we do not provide details here for space constraint, the final form is detailed later in Eq. 7) and the gradient of the acquisition policy term can be evaluated as follow by using Monte-Carlo approximation over M trail histories, where a is sampled w.r.t $\pi_\gamma(a|x)$:

$$\nabla_{\gamma,\theta} \sum_{t=1}^{T} \sum_{i=1}^{n} \pi_\gamma(a_{t,i} = 1|x).c_i \approx \frac{1}{M} \sum_{m=1}^{M} \sum_{t=1}^{T} \sum_{i=1}^{n} c_i \nabla_{\gamma,\theta} \pi_\gamma(a_{t,i} = 1|a_1, ..., a_{t-1}, x) \tag{6}$$

---

[2] However note that during test-time, e.g. in our experimental results in Sect. 3, when a feature is acquired several times (i.e. at different steps), we count its cost in evaluation only once.

Algorithm 2. Inference algorithm with representation-based components

procedure INFERENCE WITH REPRESENTATION($x, (p, \theta, \beta, \gamma, T)$)

    $a_0 = 0$

    $z_1 = 0 (\in \mathbb{R}^p)$

    for $t = 1..T$ **do**

        Sample a_t from $f_\gamma(z_t)$

        Acquire $x[a_t]$ where new features are such that $a_{t,i} = 1$

        $z_{t+1} \leftarrow \Psi_\beta(z_t, x[a_t])$

    end for

    return $\hat{y} = d_\theta(z_{T+1})$

end procedure

2.2 Representing Partially Acquired Data

The last component that completes our proposal (and makes it fully learnable with gradient descent) is a mechanism allowing to iteratively build a representation of an input along the acquisition process, starting with z_1, then z_2, up to z_{T+1}. The successive representations $\{z_t\}$ of x all belong to a common representation space $\forall t, z_t \in Z = R^p$ (with $p \approx 20$ in our experiments). This representation space allows expressing any partially observed input x. The inference process - see Algorithm 2 – starts with a null representation of x at step 1, $z_1 = 0$. Then this representation is refined every iteration t according to $z_t = \Psi_\beta(z_{t-1}, x[a_{t-1}])$, i.e. an aggregation between the previous representation and the newly acquired features. The final prediction is performed from the finally obtained representation of x: $\hat{y} = d_\theta(z_{T+1})$. Doing so one may define a prediction function operating on Z, $d : Z \rightarrow \mathbb{R}^Y$ which is then callable on any partially observed input. We operate the same way for the acquisition policy and we define $\pi_\gamma(a_t | a_1, \ldots, a_{t-1}, x) = f_\gamma(z_t)$, where $f_\gamma : Z \rightarrow [0, 1]^n$.

When reintroducing these functions and the representations z_t into the loss, we get the following gradient estimator:

$$\nabla_{\gamma, \theta, \beta} \hat{\mathcal{J}}(x, y, \gamma, \theta, \beta) \approx \frac{1}{M} \sum_{m=1}^{M} \left[\Delta(d_\theta(z_{T+1}), y) \sum_{t=1}^{T} \nabla_{\gamma, \theta} \log f_\gamma(z_t) \right.$$

$$\left. + \nabla_{\gamma, \theta}(\Delta(d_\theta(z_{T+1}), y) + \lambda \sum_{t=1}^{T} \sum_{i=1}^{n} \nabla_{\gamma, \theta} f_{\gamma, i}(z_t).c_i \right]$$

$$(7)$$

with a_t sampled w.r.t. $f_\gamma(z_t)$, and $f_{\gamma, i}$ is the i-th component of the output of f_γ. Note that this gradient can be efficiently computed using back-propagation techniques as it is usually done when using recurrent neural networks for example.

Various instances of the proposed framework can be described, depending on the choices of the f_γ, Ψ_β and d_θ functions. We tested two non-linear functions as aggregation function Ψ_β, RNN cells and Gated Recurrent Units (GRUs [6]), and used linear functions for d_θ. Regarding f_γ, we propose to use a **Bernouilli-based**

sampling model (B-REAM): it samples a_t as a set of a bernoulli distribution, i.e. each component i of f_γ corresponds to the probability of sampling feature x_i. This allows to sample multiple features at each time-step, which is an interesting and original property regarding state of the art, and can be implemented using linear functions followed by a sigmoid activation function. Note that one can learn a unique function f_γ or one can learn a distinct function f_γ for every step (i.e. with its own set of parameters γ^t), which is what we did in our experiments.

With our implementation choices, the final representations, hence the final prediction, which are obtained after a sequence of T acquisition steps, are thus highly nonlinear function of the input, giving this model a deep network's like capacity.

Table 1. Accuracy at different *cost* levels i.e. the amount (%) of features used. The accuracy is obtained through a linear interpolation on accuracy/cost curves. The same subset of train/validation/test data have been used for all models for each dataset. Acquiring 25 % of the features is equivalent for these datasets to using from 4 features (on *letter*) to 41 features (on *musk*).

Corpus name	Nb. Ex	Nb. Feat	Nb. Cat	Model	Amount of features used (%)			
					90 %	75 %	50 %	25 %
Letter	6661	16	26	SVM L_1	0.483	0.330	0.236	0.142
				C4.5	**0.823**	**0.823**	**0.823**	**0.484**
				GreedyMiser	0.749	0.401	0.275	0.156
				B-REAM	0.738	0.695	0.660	0.441
Pendigits	2460	16	10	SVM L_1	0.795	0.555	0.327	0.245
				C4.5	0.944	0.944	0.944	**0.796**
				GreedyMiser	0.858	0.678	0.649	0.375
				B-REAM	**0.975**	**0.963**	**0.948**	0.782
Cardiotocography	685	21	10	SVM L_1	0.683	0.580	0.496	0.338
				C4.5	0.775	0.775	0.775	0.771
				GreedyMiser	**0.827**	**0.818**	0.751	0.480
				B-REAM	0.807	0.807	**0.800**	**0.809**
Statlog	1105	60	3	SVM L_1	0.775	0.741	0.703	0.630
				C4.5	0.823	0.823	0.823	0.823
				GreedyMiser	0.851	0.846	0.831	0.765
				B-REAM	**0.864**	**0.864**	**0.860**	**0.851**
Musk	2175	166	2	SVM L_1	0.950	0.950	0.942	0.921
				C4.5	0.942	0.942	0.942	0.942
				GreedyMiser	0.950	0.950	0.951	0.952
				B-REAM	**0.968**	**0.969**	**0.970**	**0.963**

3   Experiments

We present in this section a series of experiments on feature-selection problems and on cost-sensitive setting, conducted on a variety of datasets on the mono-label classification problem.

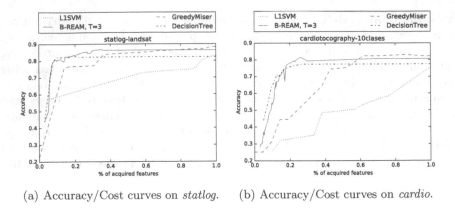

(a) Accuracy/Cost curves on *statlog*.   (b) Accuracy/Cost curves on *cardio*.

Fig. 1. Accuracy/Cost curves on two different datasets of UCI, comparing L1SVM, GreedyMiser, B-REAM with 3 steps.

Experimental protocol: Due to the bi-objective nature of the problem (maximizing accuracy while minimizing the acquisition cost), it is not possible to do cross-validation on multiple batches. We use the following experimental validation protocol, where each dataset has been split in training, validation and testing sets, each split corresponding to one third of the examples: (1) A set of models is learned **on the training set** with various hyper-parameters values. (2) We select the models that are on the **Pareto front** of the accuracy/cost points inferred on the **validation set** from the previously learned models. (3) We compute accuracy and cost for each of the "Pareto" models on the **test set**, which are the results reported here.

We have launched a variety of experiments to evaluate our stochastic bernouilli-based acquisition model **B-REAM**. We used a least-square loss function Δ. The experimental results have been obtained with the software provided at http://github.com/ludc/csream and are fully reproducible.

Our method is compared with three state-of-the-art features selection approaches: (i) **SVM** L_1 is a L_1 regularized linear SVM. (ii) **Decision Trees** can be seen as particular cases of sequential adaptive predictive models[3] (iii) **Greedy Miser** [23] is a recent cost-sensitive model that relies on several weak classifiers (Decision Trees) where the acquisition cost is integrated as a local and a global constraint[4].

Feature Selection Problem: In this setting, we consider that all the features have the same cost, i.e. $\forall i, c_i = 1$. We therefore express the cost directly as the percentage of feature gathered regarding the total number of features. It thus corresponds to a problem of adaptive sparse classification.

---

[3] These two baselines don't allow to integrate a specific cost per feature during learning.

[4] We used the MATLAB implementation provided by the authors http://www.cse.wustl.edu/~xuzx/research/code/code.html.

The results obtained on different UCI datasets are summarized in Table 1 for various percentages amount of acquisition. Conjointly, Fig. 1 presents the associated accuracy/cost curves on two of these datasets for better illustration. For example, on dataset *cardio* (Fig. 1b), the model B-REAM learned with 3 steps of acquisition obtains an accuracy of approximately 70 % for a cost of 0.2 (i.e. acquiring 20 % of the features on average), while GreedyMiser reaches 45 % accuracy for the same amount of features.

Overall, the results provided in Table 1 illustrate the competitiveness of our approaches in regard to state of the art models (GreedyMiser and other baselines). Yet it is interesting to note that naive baseline such as a Decision Tree can achieve quite good results on few datasets (e.g. *letter*), and may remain competitive nonetheless on the others. But, on average, B-REAM exhibits a high ability to adaptively select the "good" features, and to simultaneously use the gathered information for prediction.

Cost-Sensitive Setting: This section focuses on the *cost-sensitive* setting, where each feature is associated with a particular cost. We propose to study the ability of our approach to tackle such problems on two artificially generated cost-sensitive datasets (from UCI) and on two cost-sensitive datasets of the literature [15]. Figure 2 illustrates the performance on these 4 different datasets. The X-axis corresponds to the acquisition cost which is the sum of the costs of the acquired features during inference on the test set. On the 4 datasets, one can see that our B-REAM approach obtain similar results or outperforms GreedyMiser (to which we compare our work since it has been designed for cost sensitive feature acquisition as well). We can observe an interesting behaviour on the two real medical datasets: there exist cost thresholds to reach a given level of accuracy (e.g. Fig. 2d, when $cost \approx 23$, or Fig. 2c when $cost \approx 14$). This phenomenon is due to the presence of expensive features that clearly bring relevant information. A similar behaviour is observed with GreedyMiser and with B-REAM, but the latter seems more agile and able to better benefit from relevant expensive features[5]. We suppose that this is due to the use of reinforcement-learning inspired learning techniques which are able to optimize a long-term objective i.e. the cumulative some of costs over an acquisition trajectory.

4   Related Work

The feature acquisition problem has been studied by different approaches in the literature. The first propositions were static methods (*feature selection*), where

---

[5] Note that due to the small size of the real-world datasets (*hepa* and *pima*) the performance curve is not monotonous. Actually the difference between the pareto front on the validation set and the resulting performance on the test set suffers from a "high" variance. Moreover, this variance cannot be reduced by averaging over different runs because resulting accuracy/cost curves are composed of points at different cost/accuracy levels and cannot be matched easily. Yet these curves show significative trends in our opinion.

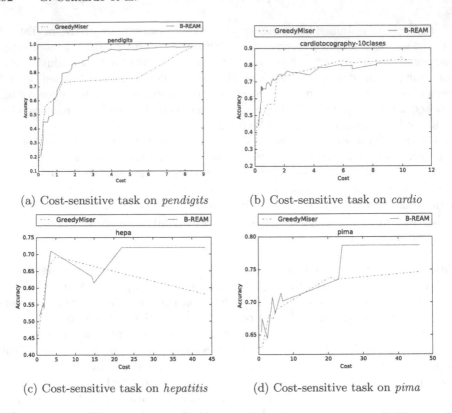

(a) Cost-sensitive task on *pendigits*   (b) Cost-sensitive task on *cardio*

(c) Cost-sensitive task on *hepatitis*   (d) Cost-sensitive task on *pima*

Fig. 2. Accuracy/Cost in the cost-sensitive setting. Top: results on two UCI datasets, in a, b, artificially made cost-sensitive by defining the cost of a feature i as $c_i = \frac{i}{n}$, where n is the total number of features. Bottom: Results on two medical datasets, with real costs as given in [15] for c, d.

there is only one step of acquisition and the subset acquired is therefore common to all inputs. [8] presents various methods in this settings such as *filter models* (e.g. variable ranking), and *wrapper approaches* like [11]. Integration of the feature selection in the learning process has been proposed for example in [2,18,19], by using resp. l_1-norm and l_0-norm in SVM. Adaptive acquisition approaches were then proposed, for example by estimating the "usefulness" (*information value*) of the features, as in [3] which present a specific data structure to do so. Using an estimation of the gain a feature would yield has also been proposed in [4] with greedy strategies to learn a naive Bayes classifier. Reinforcement learning has also been proposed in this setting, to learn a value-function of the information gain [17]. In parallel, several methods relying on decision trees have been presented as they provide efficient adaptive acquisition properties. They are for example used as weak classifiers learned with constraints on the features used in [21,23]. Cascade architecture, e.g. [16] or more recently [22], are another important part of the feature acquisition literature, and they usually enable the possibility of early-stopping in the acquisition. The objective is then to learn which features to acquire at each stage of the cascade using for example additive regression method [5]. Block acquisition has been proposed in [13] but the groups of features are pre-assigned.

Closer to our work, several methods using a Markov Decision Process formalization or reinforcement learning techniques have been presented. Partially-observable MDP with a myopic algorithm is presented in [10], while [1] propose a Markov Decision Directed Acyclic Graph to design a controller that decides between evaluating (a feature), skipping it or classifying. [14] also present a MDP-based model that choose between classifying or acquiring the "next feature" at each step. Regarding reinforcement methods, algorithms to learn acquisition policies have been proposed for example using imitation policies [9], however this requires an oracle to guide learning. [7] presents a method where the "state" of the process is represented as a vector of the acquired features (built following a pre-defined heuristic), this representation state is then used to learn and follow the acquisition policy. Visual attention models such as [12], which often rely on policy-gradient, are also close to our work, while being specific to a particular type of inputs (images). They generally follow a recurrent architecture and aim at predicting locations of a patch of pixels to acquire, instead of a subset of features. Regarding these various methods, our approach differs on several aspects. It is one of the only method, to the best of our knowledge, that relies on representation-learning and reinforcement learning and provides adaptive and batch cost-sensitive acquisition of features without suffering from the combinatorial problem, and without making assumption on the nature of the (partially observed) input.

5   Conclusion

We presented a generic framework to tackle the problem of adaptive cost-sensitive acquisition. The B-REAM model is based on both reinforcement learning and representation learning techniques, resulting in a stochastic cost-sensitive acquisition model able to acquire block of features. We also showed that the model performs well on different problem settings. This framework allows us to imagine different research directions. We are currently investigating the integration of real-world budgets like CPU consumption or energy footprint. Moreover, it would be an interesting line of future work to see if this type of approach can be learned in a unsupervised way - like auto-encoders - allowing to transfer the features acquisition policy to multiple tasks.

Acknowledgments. This article has been supported within the Labex SMART supported by French state funds managed by the ANR within the Investissements d'Avenir programme under reference ANR-11-LABX-65. Part of this work has benefited from a grant from program DGA-RAPID, project LuxidX.

References

1. Benbouzid, D., Busa-Fekete, R., Kégl, B.: Fast classification using sparse decision dags. In: ICML (2012)
2. Bi, J., Bennett, K., Embrechts, M., Breneman, C., Song, M.: Dimensionality reduction via sparse support vector machines. JMLR **3**, 1229–1243 (2003)
3. Bilgic, M., Getoor, L.: Voila: efficient feature-value acquisition for classification. In: Proceedings of AAAI, vol. 22, p. 1225 (2007)
4. Chai, X., Deng, L., Yang, Q., Ling, C.X.: Test-cost sensitive naive bayes classification. In: Data Mining, ICDM 2004 (2004)
5. Chen, M., Weinberger, K.Q., Chapelle, O., Kedem, D., Xu, Z.: Classifier cascade for minimizing feature evaluation cost. In: AISTATS, pp. 218–226 (2012)
6. Cho, K., van Merriënboer, B., Bahdanau, D., Bengio, Y.: On the properties of neural machine translation: encoder-decoder approaches. arXiv preprint (2014). arXiv:1409.1259
7. Dulac-Arnold, G., Denoyer, L., Preux, P., Gallinari, P.: Sequential approaches for learning datum-wise sparse representations. Mach. Learn. **89**(1–2), 87–122 (2012)
8. Guyon, I., Elisseeff, A.: An introduction to variable and feature selection. JMLR **3**, 1157–1182 (2003)
9. He, H., Daumé III., H., Eisner, J.: Cost-sensitive dynamic feature selection. In: ICML Workshop: Interactions Between Inference and Learning, Edinburgh (2012)
10. Ji, S., Carin, L.: Cost-sensitive feature acquisition and classification. Pattern Recogn. **40**(5), 1474–1485 (2007)
11. Kohavi, R., John, G.H.: Wrappers for feature subset selection. Artif. Intell. **97**(1), 273–324 (1997)
12. Mnih, V., Heess, N., Graves, A., et al.: Recurrent models of visual attention. In: NIPS (2014)
13. Raykar, V.C., Krishnapuram, B., Yu, S.: Designing efficient cascaded classifiers: tradeoff between accuracy and cost. In: 16th ACM SIGKDD (2010)
14. Trapeznikov, K., Saligrama, V.: Supervised sequential classification under budget constraints. In: AISTATS (2013)
15. Turney, P.D.: Cost-sensitive classification: empirical evaluation of a hybrid genetic decision tree induction algorithm. J. Artif. Intell. Res. **2**, 369–409 (1995)
16. Viola, P., Jones, M.: Robust real-time object detection. Int. J. Comput. Vis. **4**, 51–52 (2001)
17. Weiss, D.J., Taskar, B.: Learning adaptive value of information for structured prediction. In: NIPS (2013)
18. Weston, J., Elisseeff, A., Schölkopf, B., Tipping, M.: Use of the zero norm with linear models and kernel methods. JMLR **3**, 1439–1461 (2003)
19. Weston, J., Mukherjee, S., Chapelle, O., Pontil, M., Poggio, T., Vapnik, V.: Feature selection for SVMS. In: NIPS (2000)
20. Wierstra, D., Foerster, A., Peters, J., Schmidhuber, J.: Solving deep memory POMDPs with recurrent policy gradients. In: Sá, J.M., Alexandre, L.A., Duch, W., Mandic, D. (eds.) ICANN 2007. LNCS, vol. 4669, pp. 697–706. Springer, Heidelberg (2007). doi:10.1007/978-3-540-74690-4_71
21. Xu, Z., Huang, G., Weinberger, K.Q., Zheng, A.X.: Gradient boosted feature selection. In: ACM SIGKDD (2014)
22. Xu, Z., Kusner, M.J., Weinberger, K.Q., Chen, M., Chapelle, O.: Classifier cascades and trees for minimizing feature evaluation cost. JMLR **15**(1), 2113–2144 (2014)
23. Xu, Z., Weinberger, K., Chapelle, O.: The greedy miser: learning under test-time budgets. arXiv preprint (2012). arXiv:1206.6451

Explainable and Efficient Link Prediction in Real-World Network Data

Jesper E. van Engelen[1], Hanjo D. Boekhout[1], and Frank W. Takes[1,2(✉)]

[1] LIACS, Leiden University, Leiden, The Netherlands
{jvengele,hdboekho,ftakes}@liacs.nl
[2] AISSR, University of Amsterdam, Amsterdam, The Netherlands

Abstract. Data that involves some sort of relationship or interaction can be represented, modelled and analyzed using the notion of a network. To understand the dynamics of networks, the *link prediction* problem is concerned with predicting the evolution of the topology of a network over time. Previous work in this direction has largely focussed on finding an extensive set of features capable of predicting the formation of a link, often within some domain-specific context. This sometimes results in a "black box" type of approach in which it is unclear how the (often computationally expensive) features contribute to the accuracy of the final predictor. This paper counters these problems by categorising the large set of proposed link prediction features based on their topological scope, and showing that the contribution of particular categories of features can actually be explained by simple structural properties of the network. An approach called the *Efficient Feature Set* is presented that uses a limited but explainable set of computationally efficient features that within each scope captures the essential network properties. Its performance is experimentally verified using a large number of diverse real-world network datasets. The result is a generic approach suitable for consistently predicting links with high accuracy.

1 Introduction

Many real-world phenomena, structures and interactions can be described by networks. Examples include links on the web, social interactions through online media, co-authorship of scientific papers, connectedness of physical devices and the spread of diseases. The field of *network science* [15] (or social network analysis [24]) is concerned with mining patterns and structures in these networks. Dynamic networks uncover a new property of networks to study: the process of their *evolution*. Is it possible to predict which links will form between pages on the web, to estimate the probability of two researchers co-authoring a paper in the future, to predict whether a friendship between users in a social network will at some point be formed; or to say which links are missing on Wikipedia? In short: can we predict how a network will evolve in the future? In this paper, we specifically consider the *link prediction* problem: based on the topology of a network at a certain time, we want to predict exactly which links will form in the

© Springer International Publishing AG 2016
H. Boström et al. (Eds.): IDA 2016, LNCS 9897, pp. 295–307, 2016.
DOI: 10.1007/978-3-319-46349-0_26

future. We wish to do so for *real-world networks* representing actual interactions, relations or communication within a real system or environment [25]. Although the underlying data of these networks is diverse, it is well-known that they have a common structure: there is a power law degree distribution (they are scale-free), there is a substantially higher than random number of closed triangles (measured by the clustering coefficient) and the average distance between two nodes is very low, typically between four and eight. This is altogether frequently referred to as the small-world property of real-world networks [13]. To ensure the generalisability of a link prediction technique across real-world networks, we consider generic methods that can be defined for any type of real-world network and are thus only based on the *structure* (*topology*) of the network.

In previous work, a number of topological measures have been proposed for predicting whether links will form between two nodes in a network. We refer to these measures as *features*: properties of the network that can be used to predict whether a link will form. Link prediction then becomes a *supervised learning task*: develop a classifier, trained on features derived from the current network, able to predict which links will form in the future. The goal is not to compare different supervised learning algorithms. Instead, we want to provide a suitable alternative for the "black box" type of approach resulting from simply reusing the large number of (often computationally expensive) features proposed in previous work [2,7,16–19,21,23]. Therefore, we propose a categorization of existing and new link prediction features into sets of features operating at distinct topological scopes of the network, followed by an experimental assessment of their performance. As a result, we are able to choose a much smaller subset of useful features: the *Efficient Feature Set*. It provides a number of advantages in terms of explainability, efficiency, consistency and general applicability.

The remainder of this paper is structured as follows. We explain how we approach the link prediction problem in Sect. 2, followed by outlining previous work in Sect. 3. We elaborate on link prediction features in Sect. 4, after which we introduce the proposed *Efficient Feature Set* in Sect. 5. In Sect. 6 we conduct experiments on real-world dynamic network data to test our method. Conclusions and suggestions for future work are given in Sect. 7.

2   Problem Statement

In an undirected, unweighted network or graph $G = (V, E)$, we have a set of nodes V and a set of links E, where $\{u, v\} \in E$ denotes whether a link exists between nodes $u, v \in V$. In the case of a directed network, the nodes of a link are ordered pairs and $(u, v) \in E$ denotes that there is a directed link from node u to v, but not necessarily that there is a link from v to u. For brevity, we henceforth refer to the number of nodes $|V|$ as n and the number of edges $|E|$ as m. For undirected networks, m denotes the number of undirected edges. In weighted networks, each link is assigned a weight. We denote this weight as w_{uv} for a link $(u, v) \in E$. In undirected networks, one weight is assigned to each undirected link such that $w_{uv} = w_{vu}$ for all links $\{u, v\} \in E$.

Link Prediction Problem: Given a network $G_\tau = (V_\tau, E_\tau)$, denoting the network at time τ, predict the newly formed links in the evolved network $G_{\tau'} = (V_{\tau'}, E_{\tau'})$ at time $\tau' > \tau$, i.e., predict the contents of $E_{\tau'} \backslash E_\tau$.

Depending on the considered network, directed or undirected links are predicted. Note that we always consider predicting whether a link will form, and not what weight it has when it is formed. To do link prediction as defined above, we must have at least two observations of the same network at different points in time. The link prediction problem can then be seen as a supervised binary classification task, where, given a set of features x, the value of y has to be predicted. Vector x is the value of particular features we deem suitable for predicting whether a link (u, v) will be formed at time τ' (with $\tau' > \tau$) and y is a binary value indicating whether a link (u, v) has formed ($y = 1$) at τ' or not ($y = 0$). We can then use metrics common in data mining to assess the quality of some classifier and thus the accuracy of the performed link prediction task. Ultimately, we aim for a classifier which satisfies the following properties:

1. *Explainable* in its performance based on simple topological network features.
2. *Efficient* in terms of computational complexity (and thus running time).
3. *Accurate* in providing correct link predictions.
4. *Consistent* in its accuracy relative to larger feature sets.
5. *General*, yielding reliable results across different real-world networks.

The five requirements above will be used to validate the performance of different link prediction feature sets in Sect. 6.

3   Related Work

The roots of link prediction lie in the context of information retrieval [3], focussing on the prediction or retrieval of missing data elements rather than future network links. The term "link prediction" [18] was given to describe the problem of predicting the formation of links and thus the evolution of a network. Early algorithms focused on single measures, such as the work of Sarukkai using Markov chains [23]. Later, it was tackled as a supervised learning task, for example using linear regression, including the work by Popescul et al. [21] and O'Madadhain et al. [19]. The seminal paper of Liben-Nowell and Kleinman [16] introduced a large set of features to combine using supervised learning. In fact, many features used in link prediction stem from other subfields of data mining, including the popular Katz measure [12] and Adamic/Adar measure [1], both discussed later.

To exploit the large number of features nowadays available, several frameworks and algorithms have been devised, using existing classifiers from popular data mining tools [9] combined with domain-independent feature sets. Lichtenwalter et al. propose a framework in [17] and show that in supervised contexts, existing methods are less accurate than their proposed PropFlow measure. Several frameworks specifically focus on easy-to-compute features, see for example [2] or the excellent extensive overview of further related work provided in [7].

This paper attempts to distinguish itself from previous work by aiming to satisfy each (rather than a subset) of the five goals explained in Sect. 2.

4   Link Prediction Methods and Approaches

The link prediction methods proposed in literature (e.g. [7,16]) and throughout this section are formulated as topological *features* of the network. We distinguish between *node features*, *neighbourhood features*, and *path features*. Each feature assigns a score $S(u, v)$ to a candidate pair (u, v), which can be used by a classifier to determine the probability of a link forming between source u and target v.

For convenience in the following definitions, for directed networks, we define $\Gamma(v) = \{u \in V : (u, v) \in E \vee (v, u) \in E\}$ as the neighbourhood of v. We can then also define the out-neighbourhood and the in-neighbourhood of v as $\Gamma_{\text{out}}(v) = \{w \in V : (v, w) \in E\}$ and $\Gamma_{\text{in}}(v) = \{u \in V : (u, v) \in E\}$, respectively. For undirected networks, we define $\Gamma(v) = \{v \in V : \{u, v\} \in E\}$. The *degree* $d(v)$ of a node is simply the size of its neighborhood: $d(v) = |\Gamma(v)|$. The in-degree and out-degree for a directed network are then respectively $d_{\text{in}}(v) = |\Gamma_{\text{in}}(v)|$ and $d_{\text{out}}(v) = |\Gamma_{\text{out}}(v)|$. Below, to improve readability, where applicable, we choose not to formally define the trivial extension of each measure to the equivalent in- and out-measures for directed networks.

4.1   Node Features

Node features are derived from the properties of a node and its links, only considering the node currently under evaluation.

- *Degree:* The node degree feature simply uses $d(u)$ and $d(v)$, i.e., the degree of source node u and target node v.
- *Volume:* The node volume measures the total weight of all incoming or outgoing links (or both) of both the source and the target node. For source u it is defined as $\sum_{t \in \Gamma(u)} w_{ut}$. The target node volume is defined analogously.

4.2   Neighbourhood Features

Neighbourhood features also consider patterns and relations of the direct neighbours of the source and target node.

- *Total neighbours:* This measure counts the total number of distinct neighbours that exist for the candidate pair and is defined as $|\Gamma(u) \cup \Gamma(v)|$.
- *Common neighbours:* Here we compute the number of neighbours two nodes have in common. Formally, we write $|\Gamma(u) \cap \Gamma(v)|$.
- *Transitive common neighbourhood:* This is a variation on the common neighbours measure intended for use in directed networks. It determines the number of neighbours to which u has a link and that have a link to v: $|\Gamma_{\text{out}}(u) \cap \Gamma_{\text{in}}(v)|$.

- *Jaccard coefficient:* This coefficient proposed in [10] considers the number of common neighbours two nodes have, relative to the total number of distinct neighbours they have. Formally, we write $|\Gamma(u) \cap \Gamma(v)|/|\Gamma(u) \cup \Gamma(v)|$.
- *Transitive Jaccard coefficient:* To capture link direction in directed networks, we propose to combine the concept of the transitive common neighbourhood with the Jaccard coefficient: $|\Gamma_{\text{out}}(u) \cap \Gamma_{\text{in}}(v)|/|\Gamma_{\text{out}}(u) \cup \Gamma_{\text{in}}(v)|$.
- *Adamic/Adar:* This measure, introduced by Adamic and Adar in [1], considers properties shared by two nodes, favouring properties that not many other nodes have. In a network, such a property can be the set of out-neighbours of u and v. Below, the number of in-neighbours of w signals the same feature, so we sum these ratios for u and v, obtaining: $\displaystyle\sum_{w \in (\Gamma_{\text{out}}(u) \cap \Gamma_{\text{out}}(v))} \frac{1}{\log(\Gamma_{\text{in}}(w))}$.
- *Preferential attachment:* This concept is used in the well-known Barabási-Albert graph generation model [4] to model the creation of a network. It is based on the observation that nodes that already have a high degree are more likely to attract new links than nodes with a lower degree. We use this property in link prediction by computing the product of degrees of the source and target node. So for the out-degree, it is defined as $|\Gamma_{\text{out}}(u)| * |\Gamma_{\text{out}}(v)|$.
- *Opposite direction link:* We consider a feature introduced in [7] for directed networks, where the probability of a link (u, v) is assumed to be dependent on the existence of (v, u), captured in a binary feature value of 0 or 1.

4.3 Path Features

The entire topology of the network is considered in path features: not only direct neighbours, but also nodes further away are considered in the evaluation.

- *Shortest path length:* This measure indicates the length of a shortest path from the source node to the target node, i.e., the minimal number of edges that have to be traversed to reach node v starting in node u. This measure is commonly referred to as the *distance*, denoted $d(u, v)$.
- *Number of shortest paths:* This metric proposed in [17] ranks candidate links based on how many shortest paths of length $d(u, v)$ exist from u to v. Parameter ℓ_{MAX} defines the maximum distance $d(u, v)$ that is considered. We denote this measure by $\text{paths}_{u,v}^{(\ell_{\text{MAX}})}$.
- *Restricted Katz measure:* Introduced by Katz [12], it awards importance to the number of paths between two nodes as a predictor of the likelihood of a link, but exponentially decreases the importance as the path length grows, determined by the parameter $\beta < 1$; typical values of β are around 0.05 [12]. In accordance with [17] we again use ℓ_{MAX}, giving: $\displaystyle\sum_{\ell=0}^{\ell_{\text{MAX}}} \beta^{\ell} |\text{paths}_{u,v}^{(\ell)}|$.
- *PropFlow:* The last measure we utilise is PropFlow, introduced by Lichten-walter et al. in [17], relating the probability of a link forming between nodes u and v to the probability that a random walk starting in node u ends up in node v, considering all walks from u of at most length ℓ_{MAX}. In weighted networks, it assigns the probabilities for following each link proportional to

their weight. Starting at u, the score update rule for each neighbouring node t in an iteration is $S(u, t) = S(u, t) + S(u, v) * (w_{ut})/(\sum_{x \in \Gamma(u)} w_{ux})$.

5   Efficient Feature Set

In this section, we explain our approach to selecting features for link prediction, taking into account the first two quality goals posed in Sect. 2. As stated before, one of the problems with gathering a vast number of features and simply using all of them, is that it does not lead to an *explainable* method, as the classifier considers so many features that it becomes impossible to distinguish between well and poor performing features, and that it is hard if not impossible to understand exactly how different features use the topological structure of the network to predict future links. We solve the above mentioned problem by grouping our features based on their *topological scope*, explicitly distinguishing between:

- *Individual properties* (evaluating properties of the source and target node).
- *Local properties* (considering similarities and differences between the neighbourhoods of the source and target node).
- *Global properties* (considering paths between the source and target node).

These notions correspond to the different levels at which small world networks are typically studied: the micro level, meso level and macro level [24]. More importantly, they coincide with the grouping of features used in Sect. 4, namely node features, neighbourhood features and path features. We present the categorized features in Table 1. The "Compl." column contains information on the computational complexity of computing the feature for a single candidate pair, expressed as a function of the number of nodes n and links m.

The second goal that we achieve with EFS deals with *efficiency*. We hypothesise that, when constructing a set of features for training a classifier, choosing features whose category is already sufficiently represented in the feature set will only marginally increase prediction accuracy, whereas computation time may increase substantially. Based on this, we construct a subset of features in which we sufficiently represent each feature group. We propose the *Efficient Feature Set* (EFS), which is constructed by choosing a small subset of features that captures the widest variety of topological properties possible. To do so, we first split the full feature set along the first *variety* dimension: topological scope (as explained above). From each category, we then select the features that exhibit a high degree of diversity within the second dimension: the *balance* between undirectedness vs. directedness, weighted vs. unweighted and absolute vs. relative counts.

From the node features, we therefore include in EFS the degree and the in- and out-volume of both the source and the target node. For the neighbourhood features, we wish to capture both the absolute and the relative size of the joined neighbourhood. So, we include the transitive Jaccard coefficient to further support directed networks and the opposite direction link feature as a direct indicator of reciprocation. From the path features, we capture general propagation properties of the network by using the undirected shortest path length, and

Table 1. List of candidate pair features, categorized by topological scope.

Node features				Neighbourhood features			
Feature	Variant	Compl.	EFS	Feature	Var.	Compl.	EFS
Degree (source)	-	$O(1)$	✓	Total neighbours	-	$O(m/n)$	
Degree (source)	d_{in}	$O(1)$		Total neighbours	Γ_{in}	$O(m/n)$	
Degree (source)	d_{out}	$O(1)$		Total neighbours	Γ_{out}	$O(m/n)$	
Degree (target)	-	$O(1)$	✓	Common neighbours	-	$O(m/n)$	✓
Degree (target)	d_{in}	$O(1)$		Common neighbours	Γ_{in}	$O(m/n)$	
Degree (target)	d_{out}	$O(1)$		Common neighbours	Γ_{out}	$O(m/n)$	
Volume (source)	-	$O(m/n)$		Trans. comm. neigh.	-	$O(m/n)$	
Volume (source)	d_{in}	$O(m/n)$	✓	Jaccard Coeff.	-	$O(m/n)$	✓
Volume (source)	d_{out}	$O(m/n)$	✓	Jaccard Coeff.	Γ_{in}	$O(m/n)$	
Volume (target)	-	$O(m/n)$		Jaccard Coeff.	Γ_{out}	$O(m/n)$	
Volume (target)	d_{in}	$O(m/n)$	✓	Trans. Jacc. Coeff.	-	$O(m/n)$	✓
Volume (target)	d_{out}	$O(m/n)$	✓	Adamic/Adar	-	$O(m/n)$	
Path features				Pref. attachment	-	$O(1)$	
Feature	Param.	Compl.	EFS	Pref. attachment	Γ_{in}	$O(1)$	
Shortest path length	-	$O(m+n)$	✓	Pref. attachment	Γ_{out}	$O(1)$	
Num. shortest paths	$\ell_{MAX}=3$	$O(m+n)$		Opp. direction link	-	$O(1)$	✓
Restricted Katz	$\ell_{MAX}=3$, $\beta=0.05$	$O(m+n)$					
PropFlow	$\ell_{MAX}=3$	$O(m+n)$	✓				

finally we capture properties of weighted and directed graphs by using PropFlow. In Table 1, the column "EFS" summarizes the selected features. In terms of efficiency, EFS attains an immediate speed-up of two or more in each category. We performed experiments to empirically verify the contribution and performance of the three proposed feature categories. For space and readability reasons, these results are presented in Sect. 6.3, followed by the results of using only EFS in Sect. 6.4.

6 Experiments and Results

In this section, we discuss the considered network datasets in Sect. 6.1, after which we outline preprocessing steps followed by the encountered class imbalance problem. Next we explain the experimental setup in Sect. 6.2 before evaluating the results in Sects. 6.3 and 6.4.

6.1 Network Datasets

An overview of the considered networks is given in Table 2, listing the network name, source, type and network properties such as the number of nodes, the number of links and the clustering coefficient (in the "CC" column). The "Type" column indicates the directedness (D for directed, U for undirected) and weighting of the network (- for unweighted, + for weighted). Finally, the average shortest path length is listed in the "Dist" column. The diverse real-world network datasets cover a range of different networks (see the sources listed in Table 2 for details). We generated the `liacs` weighted scientific collaboration network from raw data on co-authorship of researchers involved with the computer science

Table 2. Characteristics of network datasets used for testing.

Dataset	Type	Nodes	Links	CC	Type	Dist	3Γ
digg [6]	News communication	30,398	86,404	0.01	+ D	4.68	45%
fb-links [26]	Social friendship	63,731	817,035	0.22	− U	4.31	88%
fb-wall [26]	Social communication	46,952	274,086	0.11	+ D	5.71	61%
infectious [11]	Disease spread	410	2,765	0.46	+ U	3.57	83%
liacs	Scientific collaboration	1,036	4,650	0.84	+ U	3.86	100%
lkml-reply [14]	Email communication	27,927	242,976	0.30	+ D	5.19	99%
slashdot [8]	Web communication	51,083	131,175	0.02	+ D	4.59	75%
topology [27]	Network topology	34,761	107,720	0.29	+ U	3.78	97%
ucsocial [20]	Social communication	1,899	20,296	0.11	+ D	3.07	99%
wikipedia [22]	Information network	100,312	746,114	0.21	− D	3.83	89%

institute LIACS in the period 2005–2014, a link denoting the joined publication count.

For each network, we choose τ (see Sect. 2) such that 95% of the links were formed before time τ. This is because we are predicting individual links as opposed to macroscopic evolution, and thus require a relatively developed state of the network for training.

The number of candidate pairs for which we could give a prediction is very large: with n nodes and m directed links at time τ, potentially $(n \cdot (n-1)) - m$ links could be formed at time τ'. The number of links that are actually formed between τ and τ' is in practice only a tiny fraction. Calculating features for all possible node pairs would be infeasible in terms of computation time. To address this, we look at column "3Γ" of Table 2, showing the percentage of nodes formed between nodes at distance 3 or less. In our networks, on average 84% of all newly formed links were at undirected distance 3 of one another, so we limit our predictions to candidate pairs at that distance. This results in the omittance of a substantial portion of the node pairs where no links are formed while retaining a large majority of all positive instances. As the considered networks are all small-world networks in which the average pairwise distance is very low, we further reduce the class imbalance by randomly removing negative instances as extensively discussed and suggested in [7].

From preliminary experiments we found that shifting the ratio between the number of negative and positive class instances did not significantly influence classification performance. From this range of acceptable balances, we have chosen a class balance of nine negative instances for each positive instance.

6.2   Experimental Setup

To determine the accuracy, we want to capture the relation between true positive and false positive rates, so we use the well-known ROC-curves, plotting these two rates against each other. The area under the ROC-curve (AUROC) can be used to assess the quality of the predictor. As explained before, our work focuses on the comparison between different feature sets to obtain a robust classifier, and not on the comparison of supervised learning algorithms. Therefore we use random forests, which are repeatedly and consistently identified as well-performing general purpose supervised learning algorithms, see for example the discussion in [5]. We use the implementation of [9] with an ensemble of 50 random decision trees. Along similar lines, we abstract away from specific hardware and software, comparing the efficiency of different feature sets based on the computational complexity as listed in Table 1.

6.3   Results — Topological Scopes

The ROC-curves of the predictions of the three feature sets discussed in Sect. 5 are depicted in Fig. 1, and the corresponding AUROC values are listed in Table 3. The variance in the performance of the single feature class predictors underlines the degree to which the performance of these feature sets alone is significant, justifying the construction of EFS in Sect. 5. For instance, in Fig. 1 we see that for the wikipedia dataset, the set of node features outperforms all other individual feature sets, whereas it is outperformed by all other feature sets in the liacs dataset. Trivially, we find that the full feature set is the best predictor in terms of accuracy and consistency, performing well across all datasets. We note that in one case (ucsocial), the set of node features outperformed the set of all features by 0.007, likely due to the randomness in the random forest classifier (an optimization step beyond the interest and scope of this paper). Below, we discuss the main results for each of the feature sets, relating their performance at the particular topological scope to the considered network datasets.

Node Features. The node features appear to perform better than might be expected from such local metrics. An interesting observation is the exceptionally good performance on the networks modelling online conversations (lkml-reply, ucsocial, slashdot and digg). This might be because users who are active in replying to messages and receiving replies to their messages are likely to remain active in the future. As expected, the performance of node features appears negatively correlated with the mean distance in the graph: as this distance grows, the node features classifier performance decreases.

Neighbourhood Features. In directed networks, we observe that a high rate of reciprocity gives good performance, which may be because in many real-world networks, non-reciprocated links tend to be reciprocated at some point in the future, captured by the *opposite direction link* feature. Indeed, in the fb-wall, ucsocial and lkml-reply networks, each having a reciprocity rate of around 65 %, we observe that the neighbourhood features generate an AUROC of 99.2 % of the AUROC of the full feature set against an overall average 97.2 % for the neighbourhood features in directed networks. The digg network, having a reciprocity rate of just 2 % yields the second to worst performance. The low performance of neighborhood features on the infectious network can be explained by its degree distribution which as opposed to all other networks, does not follow a power law, but is instead distributed around the average (see Fig. 2).

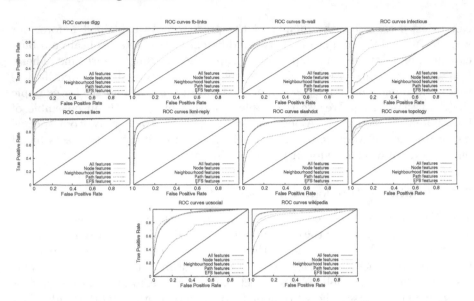

Fig. 1. ROC curves for the network datasets in Table 2, for each feature set.

Table 3. AUROC for each network and each set of features.

Features	digg	fb-links	fb-wall	infectious	liacs	lkml	slash-dot	topology	ucsocial	wikipedia
Full	0.830	0.933	0.887	0.967	0.997	0.975	0.928	0.967	0.913	0.970
Node	**0.827**	0.700	0.710	**0.955**	0.969	0.971	**0.922**	0.949	0.911	**0.941**
Neighbourhood	0.761	**0.911**	**0.866**	0.794	**0.986**	**0.974**	0.920	**0.961**	**0.920**	0.926
Path	0.632	0.897	0.819	0.579	0.979	0.925	0.777	0.940	0.673	0.827
EFS	0.825	0.930	0.876	0.958	0.995	0.973	0.921	0.965	0.910	0.967
EFS Performance	**99.4 %**	**99.6 %**	**98.8 %**	**99.1 %**	**99.8 %**	**99.8 %**	**99.2 %**	**99.8 %**	**99.7 %**	**99.7 %**

Path Features. By themselves, the path features are likely too global to sufficiently capture local network patterns. Nevertheless, they do add value to the full feature set as well as EFS. Most notably, the performance of path features increases as the mean path length grows, highlighting the importance of this feature set in less dense networks.

6.4 Results — EFS

We evaluate our Efficient Feature Set using the five criteria for a link prediction technique outlined in Sect. 2. In Sect. 5, we already elaborated on the explainability and efficiency of EFS. Here we continue by substantiating the claim that EFS also meets the accuracy, consistency and generality criteria.

Accuracy. The results in Table 3 show that EFS is able to accurately predict links, achieving results very similar to the accuracy achieved using the full feature set. The bottom row of Table 3 shows how on average EFS achieves an AUROC of a remarkable 99.5 % of the AUROC value of the full feature set.

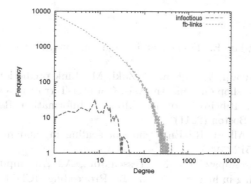

Fig. 2. Degree distribution of datasets `infectious` and `fb-links`.

Consistency. Not only did the EFS-based classifier yield very high AUROC values relative to the full feature set, it did so with a high degree of consistency: the lowest EFS AUROC measured was 98.8 % and the highest AUROC measured was 99.8 % of the full feature set AUROC value.

Generality. We have tested EFS on a broad range of diverse networks, as can be seen in Table 2. The networks vary in number of nodes and links, clustering coefficient, diameter, directedness, weightedness and many other properties. The Efficient Feature Set performs well regardless of these differences in network properties. Even the absence of a power law in the degree distribution, such as in the `infectious` network, does not influence the prediction ability of the Efficient Feature Set. It is indeed a generic way of predicting links in real-world networks.

7 Conclusion

The proposed Efficient Feature Set (EFS) is a relatively small set of structural network features, categorized based on the topological scope of the network at which they each uniquely capture dynamics. Together, the feature categories can be used in a supervised learning framework to predict future links in real-world networks in an efficient and explainable way. Experiments show that the approach reaches over 99 % of the accuracy of a much larger and more complex set of features. EFS is explainable as the contribution of its feature categories can be linked to the topological properties of the considered networks. Furthermore, the method shows consistent performance with respect to larger feature sets, independent of the network size, hinting towards high scalability. EFS works well independent of network density and type, demonstrating that it is a generic approach to predict links in real-world network data that is evolving over time.

In future work, we want to investigate the order in which links appear and how suitable EFS is for predicting the timestamp of a link. Furthermore, we will look at whether the same approach could be applied to the removal of links, allowing not only the expansion but also the contraction of networks to be predicted.

References

1. Adamic, L.A., Adar, E.: Friends and neighbors on the web. Soc. Netw. **25**(3), 211–230 (2003)
2. Al Hasan, M., Chaoji, V., Salem, S., Zaki, M.: Link prediction using supervised learning. In: Workshop on Link Analysis, Counter-Terrorism and Security (2006)
3. Baeza-Yates, R., Ribeiro-Neto, B.: Modern Information Retrieval, 2nd edn. Addison-Wesley, Boston (2011)
4. Barabási, A.-L., Albert, R.: Emergence of scaling in random networks. Science **286**(5439), 509–512 (1999)
5. Caruana, R., Karampatziakis, N., Yessenalina, A.: An empirical evaluation of supervised learning in high dimensions. In: Proceedings ICDM, pp. 96–103 (2008)
6. Choudhury, M.D., Sundaram, H., John, A., Seligmann, D.D.: Social synchrony: predicting mimicry of user actions. In: Proceedings ICCSE, pp. 151–158 (2009)
7. Fire, M., Tenenboim-Chekina, L., Puzis, R., Lesser, O., Rokach, L., Elovici, Y.: Computationally efficient link prediction in a variety of social networks. ACM Trans. Intell. Syst. Technol. (TIST) **5**(1), 10 (2013)
8. Gómez, V., Kaltenbrunner, A., López, V.: Statistical analysis of social network discussion threads in Slashdot. In: Proceedings WWW, pp. 645–654 (2008)
9. Hall, M., Frank, E., Holmes, G., Pfahringer, B., Reutemann, P., Witten, I.H.: The WEKA data mining software. SIGKDD Explor. Newslett. **11**(1), 10–18 (2009)
10. Huang, Z., Li, X., Chen, H.: Link prediction approach to collaborative filtering. In: Proceedings DLT, pp. 141–142 (2005)
11. Isella, L., Stehlé, J., Barrat, A., Cattuto, C., Pinton, J.-F., den Broeck, W.V.: What's in a crowd? Analysis of face-to-face behavioral networks. J. Theor. Biol. **271**(1), 166–180 (2011)
12. Katz, L.: A new status index derived from sociometric analysis. Psychometrika **18**(1), 39–43 (1953)
13. Kleinberg, J.: The small-world phenomenon: an algorithmic perspective. In: Proceedings STOC, pp. 163–170 (2000)
14. KONECT. Linux mailing list replies network (2015). http://konect.uni-koblenz.de
15. Lewis, T.G.: Network Science: Theory and Applications. Wiley, New York (2011)
16. Liben-Nowell, D., Kleinberg, J.: The link prediction problem for social networks. In: Proceedings CIKM, pp. 556–559 (2003)
17. Lichtenwalter, R.N., Lussier, J.T., Chawla, N.V.: New perspectives and methods in link prediction. In: Proceedings KDD, pp. 243–252 (2010)
18. Lü, L., Zhou, T.: Link prediction in complex networks: a survey. Physica A Stat. Mech. Appl. **390**(6), 1150–1170 (2011)
19. O'Madadhain, J., Hutchins, J., Smyth, P.: Prediction and ranking algorithms for event-based network data. SIGKDD Explor. Newslett. **7**(2), 23–30 (2005)
20. Opsahl, T., Panzarasa, P.: Clustering in weighted networks. Soc. Netw. **31**(2), 155–163 (2009)
21. Popescul, A., Ungar, L.H.: Statistical relational learning for link prediction. In: IJCAI Workshop on Learning Statistical Models from Relational Data (2003)
22. Preusse, J., Kunegis, J., Thimm, M., Gottron, T., Staab, S.: Structural dynamics of knowledge networks. In: Proceedings ICWSM (2013)
23. Sarukkai, R.R.: Link prediction and path analysis using Markov chains. Comput. Netw. **33**(1), 377–386 (2000)
24. Scott, J.: Social Network Analysis. Sage, London (2012)

25. Takes, F.W.: Algorithms for analyzing and mining real-world graphs. Ph.D. thesis, Leiden University (2014)
26. Viswanath, B., Mislove, A., Cha, M., Gummadi, K.P.: On the evolution of user interaction in Facebook. In: Proceedings WOSN, pp. 37–42 (2009)
27. Zhang, B., Liu, R., Massey, D., Zhang, L.: Collecting the internet AS-level topology. SIGCOMM Comput. Commun. Rev. **35**(1), 53–61 (2005)

DGRMiner: Anomaly Detection
and Explanation in Dynamic Graphs

Karel Vaculík[(✉)] and Luboš Popelínský

KD Lab, FI MU Brno, Brno, Czech Republic
{xvaculi4,popel}@fi.muni.cz

Abstract. Ubiquitous network data has given rise to diverse graph mining and analytical methods. One of the graph mining domains is anomaly detection in dynamic graphs, which can be employed for fraud detection, network intrusion detection, suspicious behaviour identification, etc.
Most existing methods search for anomalies rather on the global level of
the graphs. In this work, we propose a new anomaly detection and explanation algorithm for dynamic graphs. The algorithm searches for anomaly patterns in the form of predictive rules that enable us to examine the
evolution of dynamic graphs on the level of subgraphs. Specifically, these
patterns are able to capture addition and deletion of vertices and edges,
and relabeling of vertices and edges. In addition, the algorithm outputs
normal patterns that serve as an explanation for the anomaly patterns.
The algorithm has been evaluated on two real-world datasets.

Keywords: Graph mining · Data mining · Dynamic graphs · Rule mining · Anomaly detection · Outlier detection · Anomaly explanation

1 Introduction

A large amount of real-world graphs is dynamic by nature, i.e. they evolve
through time. These graphs can exhibit anomalous behaviour on various levels:
from single vertices and edges through subgraphs to whole graphs. Examples
include macious network attacks, frauds in trading networks, opinion spam, and
many others. Anomalies, however, do not have to be necessarily negative. For
example, novel patterns of behaviour in communication and interaction networks
can be considered as an improvement over other patterns.

Most of the existing approaches for anomaly detection in dynamic graphs
search for anomalous vertices, edges, or graph snapshots [1]. When searching
for anomalies on local level of the graph, single vertices or edges without the
structural context may not provide a satisfactory explanation. Methods based on
tensor decomposition [5,7,10] are able to find groups of anomalous vertices and
edges, but it is hard to capture the evolution on the local level in detail. There
are only few methods for subgraph patterns and they typically impose various
restrictions on the form of the patterns. For example, the method presented in
[4] assumes that the vertices are immutable.

© Springer International Publishing AG 2016
H. Boström et al. (Eds.): IDA 2016, LNCS 9897, pp. 308–319, 2016.
DOI: 10.1007/978-3-319-46349-0_27

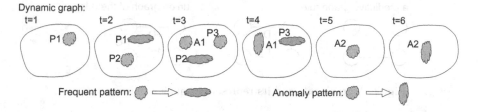

Fig. 1. An illustration of a dynamic graph, a frequent pattern with occurrences P1, P2, P3, and an anomaly pattern with occurrences A1, A2. (Color figure online)

In this work, we propose a new method for anomaly detection in labeled dynamic graphs that is able to capture the evolution on the subgraph level. This method was built on our previous work, DGRMiner algorithm [12] for frequent pattern mining in dynamic graphs. Specifically, the patterns are in the form of predictive rules expressing how a subgraph can be changed into another subgraph by adding new vertices and edges, deleting specific vertices and edges, or relabeling vertices and edges. DGRMiner is able to mine patterns from a single dynamic graph and also from a set of dynamic graphs. Figure 1 illustrates the idea of anomaly detection by exploiting these frequent predictive rules. In the first phase, we found a frequent pattern depicting a transformation of the *green* subgraph into the *blue* one. This pattern has three occurrences in the input dynamic graph: P1, P2 and P3. However, not all occurrences of the *green* subgraph, i.e. the antecedent, are transformed into the *blue* subgraph. There are situations in which it is transformed into the *orange* subgraph (A1 and A2). We mark such a transformation as an anomaly pattern and use the original frequent one as an explanation of it.

The remainder of this paper is organised as follows. Section 2 briefly describes the frequent pattern mining method used in DGRMiner. The new method for anomaly detection and explanation is proposed in Sect. 3. Section 4 presents an experimental evaluation. Properties of the new method are discussed in Sect. 5. Finally, related work and conclusion can be found in Sects. 6 and 7, respectively.

2   DGRMiner: Frequent Pattern Mining

In our previous work [12], we presented DGRMiner algorithm for frequent pattern mining. Given a dynamic graph or a set of graphs, a minimum support value σ_{min}, and a minimum confidence value $conf_{min}$, the task is to find all patterns for which $\sigma \geq \sigma_{min}$ and $conf \geq conf_{min}$. Patterns are predictive graph rules in this case. An example of such a rule is depicted in Fig. 2. It represents a change of one graph into another. In this case, the edge with label B was relabeled to E, the edge with label C was deleted, and a new edge with label D was added. The same type of changes is allowed for vertices. *Support* of a rule is computed as the number of snapshot transitions where the rule occurred. *Confidence* is computed as the fraction of the rule support and the support of the antecedent.

a predictive graph rule: union graph of the rule:

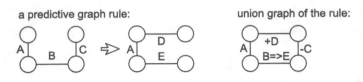

Fig. 2. A graph rule and its representation as a union graph.

2.1 Representation of Dynamic Graphs

DGRMiner is based on gSpan [13], an algorithm for frequent pattern mining in a set of undirected graphs. The main idea of DGRMiner is to transform the input dynamic graph into a new representation that can be considered as a set of static graphs. DGRMiner transforms each consecutive pair of snapshots into a single static graph by using the following procedure. It takes these two graphs and computes a union graph from them. In order to be able to transform the union graph back, the differences between those two graphs are encoded. Let us assume that a vertex with label A appears only in the first graph, i.e. this vertex represents an addition change. Such a vertex gets label $+A$ in the union graph. On the other hand, if the vertex appears only in the first graph, then it gets label $-A$. If the vertex label is changed from A to B, its union-graph label will be $A \Rightarrow B$. Unchanged vertices keep their labels, e.g. A remains A. The same encoding schema is used for edges. Frequent subgraphs found in this new set of graphs can be then extracted into predictive rules. Figure 2 shows the union-graph representation of the rule shown on the left.

2.2 Frequent Pattern Mining

DGRMiner starts by searching for single-vertex patterns. The algorithm then continues with single-edge patterns and extends these patterns recursively by one edge at a time. Support and confidence are computed for each such pattern and patterns with high enough values are outputted. If the support is too low or the algorithm already processed an isomorphic pattern before, it backtracks.

A pseudocode of DGRMiner is given in Algorithm 1. Lines with ▷ *anomaly* comment mark the extension for anomaly detection and they are described in the next section. The algorithm first transforms the input dynamic graph or a set of graphs into the union-graph representation.

Before recursive mining of larger patterns, DGRMiner outputs frequent single-vertex patterns with high enough confidence. Then for each frequent edge, it recursively calls DGR_Subgraph_Mining subprocedure, which searches for patterns growing from that edge. By using the *min* function, DGR_Subgraph_Mining first checks, whether such a pattern is processed for the first time. If it is, the algorithm continues by enumerating the pattern in relevant graphs given by $\mathbb{D}$ and searching for its children candidates. Then the algorithm extracts the antecedent from the pattern and computes the occurrences of the antecedent. These occurrences are used for confidence computation. The current pattern is saved if its confidence is high enough. At the end,

---

Algorithm 1. DGRMiner($\mathbb{DG}$)

1: convert the input dynamic graph(s) $\mathbb{DG}$ into the union-graph representation $\mathbb{D}$;
2: optional: apply a time abstraction method on union graphs;
3: output frequent *change* vertices with high enough confidence;
4: Find anomaly patterns with regard to outputted patterns   ▷ *anomaly*
5: $\mathbb{S}^1 \leftarrow$ all frequent initial edges in $\mathbb{D}$ sorted in DFS lexicographic order;
6: **for** $i \leftarrow 1$ **to** $|\mathbb{S}^1|$ **do**
7:     $p \leftarrow i$-th edge from $\mathbb{S}^1$
8:     $p.D \leftarrow$ graphs which contain p;
9:     $p.A \leftarrow$ graphs which contain antecedent of p;
10:     $\mathbb{E}_{start} \leftarrow$ first i edges from $\mathbb{S}^1$
11:     DGR_Subgraph_Mining($p,p.D,p.A,\mathbb{E}_{start}$);

Algorithm 2. DGR_Subgraph_Mining($s,\mathbb{D},\mathbb{A},\mathbb{E}_{start}$)

1: **if** $s \neq min(s, \mathbb{E}_{start})$ **then**
2:     **return**;
3: enumerate s in each graph in $\mathbb{D}$ and count its children;
4: remove children of s which are infrequent;
5: enumerate antecedent of s in graphs given by $\mathbb{A}$,
        enumerate anomaly patterns from antecedent occurrences;   ▷ *anomaly*
6: set s.A by graphs which contain antecedent of s;
7: $conf \leftarrow$ confidence of s;
8: **if** $conf \geq conf_{min}$ **then**
9:     output s;
10:     **for each** anomaly pattern a **do**   ▷ *anomaly*
11:         **if** score(a)$\geq$ min_score **then**   ▷ *anomaly*
12:             output a ▷ *anomaly*
13: sort remaining children in DFS lexicographic order;
14: **for each** child c **do**
15:     DGR_Subgraph_Mining($c,c.D,s.A,\mathbb{E}_{start}$);

DGR_Subgraph_Mining is recursively called for each frequent child of the pattern. More details on DGRMiner can be found in [12].

3   DGRMiner: Anomaly Detection and Explanation

This section describes a new module of DGRMiner for anomaly detection and explanation. Having found the frequent patterns, we are interested in patterns deviating from the frequent ones, i.e. patterns with the same antecedent but a different consequent. The frequent patterns are then used as an explanation of the anomaly patterns. For each frequent pattern being processed, we store its occurrences in the dynamic graph. Specifically, we store sets of occupied vertex and edge ids for each snapshot. When searching for anomaly patterns, we do not use vertices and edges of these occurrences. This means that occurrences of anomaly patterns and the explanatory frequent patterns are completely disjoint. This ensures that the anomaly patterns are independent of the frequent ones.

In order to decide which deviating patterns are truly anomalies, we use *outlierness* score defined as the opposite value of the confidence, i.e. $out = 1 - conf$. Given a minimum outlierness score out_{min}, we output patterns for which $out \geq out_{min}$. It is necessary to check the outlierness of all potential anomaly patterns because not all *complementary* patterns of the frequent ones have low enough confidence. For example, the frequent pattern in Fig. 1 has $conf = 3/4$, but the anomaly pattern has $conf = 2/4$. Such an anomaly pattern would not be outputted in the case of $conf_{min} = out_{min} = 3/4$.

The confidence of the pattern is computed from its support. In the following subsection, we describe how to discover the single-vertex anomalies and how to compute their support. After this simple scenario, we focus on more complex anomaly patterns, whose discovery is a more involved process.

3.1   Single-Vertex Anomalies

DGRMiner looks for single-vertex patterns that are *complementary* to the frequent ones. Single-vertex frequent patterns take one of the following forms: $-A$, $A \Rightarrow B$, $+B$. The antecedent of $-A$ is A, which is also the antecedent of patterns A (no change) and $A \Rightarrow C$ for some C. Thus, anomaly patterns, complementary to this frequent pattern, are of the form A or $A \Rightarrow C$. As for the frequent pattern $A \Rightarrow B$, the possible anomaly patterns can be A (no change), $-A$, and $A \Rightarrow C$ for some $C \neq B$. It is trival to enumerate such patterns in the input dynamic graph and compute their support because our union-graph representation allows us to obtain the antecedent labels of the vertices.

A different approach is required for frequent patterns of the $+B$ form. The antecedent of these patterns is an empty graph, whose support is the number of snapshots. There is only one anomaly pattern complementary to $+B$ and we mark it by $!B$. The meaning of $!B$ is that a vertex with label B should have been added. We use it because we need to explicitly express that such an addition did not happen. Thus, support of $!B$ is computed from graph transitions where $+B$ did not occur.

3.2   Enumeration of Anomaly Patterns in General

Enumeration of anomaly patterns with regard to larger frequent patterns follows the enumeration of antecedents, as is indicated at line 5 of Algorithm 2. Each such frequent pattern can contain multiple changes. First, suppose there are some vertices and/or edges that do not represent *addition* changes. Such elements are included in the antecedent and their occurrences can be located in the input dynamic graph. As in the case of single-vertex patterns, we can simply extract unambiguously the consequent part for each such occurrence.

The situation gets more complicated when there are also elements representing *addition* changes in the frequent pattern. Again, each such element is either found in the dynamic graph or we explicitly say that it is missing. In order to capture the parts that are different from the frequent pattern, we search for all maximal common subgraphs of the frequent pattern and the dynamic graph,

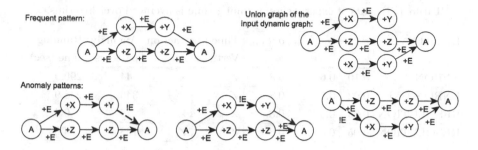

Fig. 3. An example of anomaly pattern enumeration.

given the fact that the antecedent part is already mapped to the dynamic graph. Let us illustrate this process on an example depicted in Fig. 3. Suppose that we are searching for anomalies with regard to the frequent pattern shown in the figure and we have already found the antecedent in a union graph, also shown in the figure. The antecedent consists of two vertices with label A. It is clear that the frequent pattern does not occur in this union graph. By taking the maximal common subgraphs of those two graphs, we get three anomaly patterns that differ in edge that is missing. The missing edges are depicted by dashed lines with label $!E$.

If we find an occurrence of an anomaly pattern, we use that union graph for support computation. As the frequent patterns and anomaly patterns have to be completely disjoint, there is only one scenario that has to be treated in a different way. If we are searching for anomalies with regard to a frequent pattern whose all elements denote *addition* changes, i.e. the antecedent is an empty graph, care must be taken when the support is computed. Specifically, if a union graph contains the whole *addition* frequent pattern but not its part, then this union graph cannot contain the anomaly pattern with only explicit *non-additions*. An example with such a scenario is depicted in Fig. 4. The first input union graph is an empty graph and the only anomaly pattern with regard to the frequent pattern is the one with only explicit *non-additions*. The second input union graph contains an occurrence of the frequent pattern and it does not contain

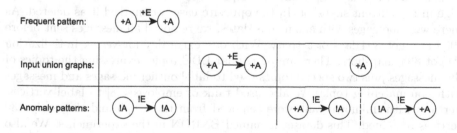

Fig. 4. An example of anomaly pattern enumeration with regard to a frequent pattern with an empty antecedent.

Table 1. Results of experiments. Running time is averaged over five runs.

Dataset	σ_{min}	$conf_{min}$	out_{min}	Time abstraction		Anomaly patterns	Running time (sec)
				Vertices	All		
ENRON	0.10	0.6	0.8	✓	×	44	290.9
ENRON UNI	0.02	0.6	0.8	✓	×	346	248.9
RESOLUTION	0.05	0.6	0.8	×	×	76	0.4
RESOLUTION	0.05	0.6	0.8	✓	✓	198	4.0

any part of the pattern besides. In this case, the only possible anomaly pattern could be with only explicit *non-additions*, but this is the special case described earlier and we do not allow such a pattern in such situations. The third input union graph contains a single vertex which is not a part of a frequent pattern occurrence and it can be used as a part of two different anomaly patterns.

4   Experiments

In this section we present results of experiments on two real-world datasets. Unfortunately, there are not many attributed dynamic graph datasets available that could be used for the experiments. The experiments were conducted by a C++ implementation on a PC equipped with CPU Intel i5-4570, 3.2 GHz, 16 GB of main memory, and running 64-bit version of Windows 10. For all experiments, we set $conf_{min} = 0.6$ and $out_{mint} = 0.8$.

Enron. Email correspondence of Enron employees [3] is used as the first dataset for our experiments. Specifically, we used a preprocessed version [8] with information about time, sender, receiver, and LDC topics of messages. Vertices of the dynamic graph represent employees. For each day, we created a snapshot and added directed edges representing the email correspondence of that day. At most one edge was added for each topic and each sender-receiver pair. The topics were used as edge labels. If an edge with a specific topic appeared in the current snapshot and it was not present in the previous one, we marked it as an added edge in the current snapshot. In the opposite case, we marked it as deleted. As there were messages with anomalous dates, we removed all messages sent before 1998 and got 894 days of activity. With one extra day for vertex initialization we got 895 snapshots. There are 32 regular LDC topics expressing the topics of the messages plus two special topics used to label outlier messages and messages with non-matching topic. We also used rank of employees [8] to label vertices. Vertices with unknown rank were removed from the graph and thus only 130 vertices remained. This dataset is named ENRON in the experiments. We also prepared a different dataset, ENRON UNI, where employee IDs were used as the vertex labels. This ensures that all patterns apply to specific employees and not to arbitrary employees of the given ranks.

As for the ENRON dataset, we set minimum support to 0.1 and performed the time abstraction method on vertices. The time abstraction of vertices allows us to ignore time connected to vertices. We have found 44 anomaly patterns, see Table 1. Two anomaly patterns are depicted in Fig. 5a. In the first example, we can see the following scenario. If an ordinary employee (Emp) sends an *outlier* email to a vice president (VP) and this vice president sends an *outlier* email to another VP, then the employee typically sends such an email to the first VP again the next day. However, it may occasionally happen that the VP sends the email instead of the employee as is depicted by the anomaly pattern. In the second example, we can see that if a VP sends such three emails, they do not send the emails again the next day. If they do, it is a rare case.

In the case of the second dataset, ENRON UNI, we set minimum support to only 0.02 because patterns connected to specific groups of people are less frequent. 346 anomaly patterns were found and two examples are shown in Fig. 5b. The difference from the previous cases is that the patterns are related to specific people. For example, the first anomaly pattern applies exactly to people with IDs 58, 63, and 146.

Resolution Proofs in Propositional Logic. We used a set of graphs representing resolution proofs in propositional logic from [11] as the next dataset. Each vertex has a set of propositional-logic literals assigned as the label. The edges are directed and have the same label in these graphs. The dynamic graphs capture the process of proof construction by students and they are evolving by vertex and edge addition or deletion, and by change of vertex labels. Time of these events was transformed into a discrete sequence. Because there were 19 different assignments in total, only dynamic graphs of the same assignment share the set of vertex labels. In order to find frequent and anomaly patterns, we restricted the dataset to only one assignment. Specifically, we took the assignment with the greatest number of solutions. This set of graphs contained 103 dynamic graphs with 2911 snapshots in total. The initial snapshot of each dynamic graph in an empty graph. This dataset is named RESOLUTION in the experiments.

For our experiments on this dataset, we set minimum support to 0.05. First, we did not use time abstraction at all and left the timestamps of vertices and edges as they were. This setting yielded 76 anomaly patterns, out of which 75 were single-vertex patterns. The remaining one described a situation in which an edge should had been added but it was not. Therefore, we repeated the experiment with time abstraction performed both on vertices and edges and got 198 anomaly patterns. Two examples are shown in Fig. 5c. The first one captures the case where students did not continue with an addition of an edge between $\{\neg b\}$ and $\{d\}$ but replaced label $\{a, b, d\}$ with $\{b, a, d\}$, which is completely unnecessary. The second example shows an odd situation in which students did not add the final edge pointing to the empty clause (depicted by a square) and they deleted the existing edge instead. Although the frequent pattern in this second example has $conf = 1.0$, we discovered this anomaly pattern because both of them can be found in certain solutions.

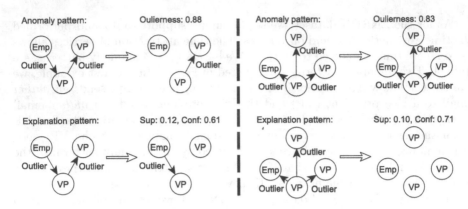

(a) Examples of two anomaly patterns in ENRON dataset.

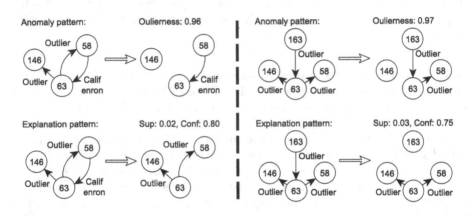

(b) Examples of two anomaly patterns in ENRON UNI dataset.

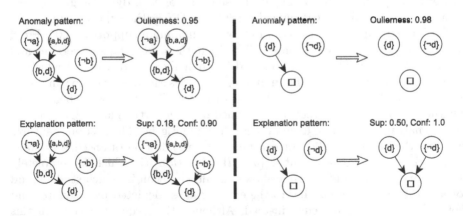

(c) Examples of two anomaly patterns in RESOLUTION dataset.

Fig. 5. Examples of anomaly patterns from experiments.

5 Discussion

Even though the enumeration of a pattern takes exponential running time in the worst case, DGRMiner can be efficient on real-world dynamic graphs as is shown in Table 1. This is mainly caused by diverse vertex and edge labels that significantly reduce the search space when subgraphs are being enumerated. On the contrary, small dense subgraphs with homogeneous labels would require the exponential running time for subgraph enumeration.

The computation of DGRMiner is driven by frequent pattern mining and anomaly patterns are mined with regard to these frequent patterns. Since multiple frequent patterns can share the same antecedent, it is possible to discover one anomaly pattern several times. Such an anomaly pattern will then be explained by multiple frequent patterns.

6 Related Work

Anomaly detection methods for dynamic graphs can be divided into several groups. One large group is comprised of methods that extract global characteristics, such as diameter, of the graphs as they evolve. The progress of these characteristics can be analysed as time series and anomalous graph snapshots or transitions between snapshots can be found. These methods typically exploit the structure of the graphs, not the labels or different attributes. Overview of these methods can be found in survey [1].

Another large group of methods is based on tensor decomposition. In this case, tensor is a generalization of the adjacency matrix. Additional dimensions store the information about time and vertex/edge attributes of the dynamic graph. Examples of algorithms from this group are STA [10], TensorSplat [5], ParCube [7], and MalSpot [6], among others. The idea of the decomposition methods is to compute factors of the given tensor and then examine the deviations from the common patterns described by main components. These methods allow us to observe anomalous patterns from the global viewpoint of various dimensions. This point of view is much coarser than the one given by DGRMiner. It does not allow us to observe the evolution on the local level in such detail.

There are several works that focus on subgraph structures and thus are more similar to our work. The idea of these methods is to monitor graph communities or clusters instead of the whole graph. One such method is Com2 [2] that uses tensor decomposition and minimum length description principle for community detection in dynamic graphs. In comparison with the already mentioned decomposition methods, Com2 allows us to work on the level of communities. Thus, we can, for example, observe the sudden changes of interactions inside a community. However, the information obtained is again coarser that the one given by DGRMiner's anomaly patterns.

A different approach is presented in [4]. Specifically, a Bayesian method is designed for detection of anomalous vertices. For each pair of vertices, the method models the communication between the vertices as a counting process. If the relationship has changed at some point in time and this change is statistically significant, the vertex pair is said to be anomalous. The method is able to model edges with labels too. The vertices are assumed to be immutable and, in general, the form of patterns is more restricted that the one given by DGRMiner.

7   Conclusion

In this paper, we developed a new algorithm for anomaly detection and explanation in dynamic graphs. This algorithm extends our previous work, DGRMiner, an algorithm for frequent pattern mining. Our method is able to capture anomaly patterns on the subgraph level in the form of predictive rules. These rules are able to capture various changes, such as addition and deletion of vertices and edges, and relabeling of vertices and edges. For each anomaly pattern, one or more frequent patterns are outputted as an explanation.

Acknowledgments. We would like to thank the IDA reviewers for valuable comments and suggestions. We would also like to thank Jan Ramon for helpful discussion on anomaly detection in graphs. This work has been partially supported by Faculty of Informatics, Masaryk University, Brno.

References

1. Akoglu, L., Tong, H., Koutra., D.: Graph-based anomaly detection and description: a survey. DAMI **28**(4) (2014)
2. Araujo, M., Papadimitriou, S., Günnemann, S., Faloutsos, C., Basu, P., Swami, A., Papalexakis, E.E., Koutra, D.: Com2: fast automatic discovery of temporal ('Comet') communities. In: Tseng, V.S., Ho, T.B., Zhou, Z.-H., Chen, A.L.P., Kao, H.-Y. (eds.) PAKDD 2014. LNCS (LNAI), pp. 271–283. Springer, Heidelberg (2014). doi:10.1007/978-3-319-06605-9_23
3. Cohen, W.W.: Enron Email Dataset. Web, Accessed 3 May 2016. www.cs.cmu.edu/~./enron/
4. Heard, N.A., et al.: Bayesian anomaly detection methods for social networks. Ann. Appl. Stat. **4**, 645–662 (2010)
5. Koutra, D., Papalexakis, E., Faloutsos, C.: Tensorsplat: spotting latent anomalies in time. In: 16th Panhellenic Conference on Informatics (PCI) (2012)
6. Mao, H.-H., Wu, C.-J., Papalexakis, E.E., Faloutsos, C., Lee, K.-C., Kao, T.-C.: MalSpot: multi2 malicious network behavior patterns analysis. In: Bailey, J., Khan, L., Washio, T., Dobbie, G., Huang, J.Z., Wang, R. (eds.) PAKDD 2016. LNCS (LNAI), vol. 9651, pp. 1–14. Springer, Heidelberg (2014). doi:10.1007/978-3-319-06608-0_1
7. Papalexakis, E.E., Faloutsos, C., Sidiropoulos, N.D.: ParCube: sparse parallelizable tensor decompositions. In: Flach, P.A., Bie, T., Cristianini, N. (eds.) ECML PKDD 2012. LNCS (LNAI), pp. 521–536. Springer, Heidelberg (2012). doi:10.1007/978-3-642-33460-3_39
8. Priebe, C.E., et al.: Scan Statistics on Enron Graphs. Web, Accessed 3 2016. http://www.cis.jhu.edu/~parky/Enron

9. Rayana, S., Akoglu, L.: Less is more: building selective anomaly ensembles (with application to event detection in temporal graphs). In: SIAM SDM, Vancouver, BC, Canada (2015)
10. Sun, J., Tao, D., Faloutsos, C.: Beyond streams and graphs: dynamic tensor analysis. In: SIGKDD, Philadelphia, PA, pp. 374–383 (2006)
11. Vaculík, K., Nezvalová, L., Popelínský, L.: Educational data mining for analysis of students' solutions. In: Agre, G., Hitzler, P., Krisnadhi, A.A., Kuznetsov, S.O. (eds.) AIMSA 2014. LNCS (LNAI), pp. 150–161. Springer, Heidelberg (2014). doi:10.1007/978-3-319-10554-3_14
12. Vaculík, K.: A versatile algorithm for predictive graph rule mining. In: Proceedings ITAT 2015: Information Technologies - Applications and Theory, pp. 51–58. Prague (2015). CEUR-WS.org
13. Yan, X., Han, J.: gSpan: Graph-based substructure pattern mining. In: IEEE ICDM 2002. Washington, DC, USA (2002)

Similarity Based Hierarchical Clustering with an Application to Text Collections

Julien Ah-Pine$^{(\boxtimes)}$ and Xinyu Wang

University of Lyon, Eric Lab,
5, Avenue Pierre Mendès France, 69676 Bron Cedex, France
{Julien.Ah-Pine,Xinyu.Wang}@univ-lyon2.fr

Abstract. Lance-Williams formula is a framework that unifies seven schemes of agglomerative hierarchical clustering. In this paper, we establish a new expression of this formula using cosine similarities instead of distances. We state conditions under which the new formula is equivalent to the original one. The interest of our approach is twofold. Firstly, we can naturally extend agglomerative hierarchical clustering techniques to kernel functions. Secondly, reasoning in terms of similarities allows us to design thresholding strategies on proximity values. Thereby, we propose to sparsify the similarity matrix in the goal of making these clustering techniques more efficient. We apply our approach to text clustering tasks. Our results show that sparsifying the inner product matrix considerably decreases memory usage and shortens running time while assuring the clustering quality.

Keywords: Agglomerative hierarchical clustering · Lance-Williams formula · Scalable hierarchical clustering · Kernel machines · Text clustering

1 Introduction

Hierarchical clustering is an important member in the clustering family. As it is able to reveal internal connections of clusters, it is more informative than its counterpart, flat clustering. Due to this advantage, it is widely applied in different domains like in documents organization where it makes it possible to highlight the relationships between topics.

There are two types of hierarchical clustering: agglomerative and divisive. Given a dataset $\mathcal{D}$ of N instances, agglomerative hierarchical clustering (AHC) recursively merges two clusters at each step, until that all instances are grouped into one cluster. Whereas, divisive hierarchical clustering (DHC) functions in the opposite way. DHC is computationally demanding, as there are $2^{N-1} - 1$ possible subdivisions into two clusters when splitting a dataset of N. Comparatively, AHC is more efficient, and thus more widely studied and applied. The result of AHC is usually represented by a dendrogram, a binary tree composed of $2N - 1$ nodes, to which a real value called height is assigned. The conventional procedure of AHC, also called the stored dissimilarities approach, takes a

© Springer International Publishing AG 2016
H. Boström et al. (Eds.): IDA 2016, LNCS 9897, pp. 320–331, 2016.
DOI: 10.1007/978-3-319-46349-0_28

pairwise dissimilarity matrix D of size N as input, initializes a binary tree with N leaves (singletons) with null height values, and iteratively adds new nodes (merged clusters) by fusing a pair of clusters (C_i, C_j) determined as follows:

$$(C_i, C_j) = \arg\min_{(C_k, C_l)} D(C_k, C_l) \tag{1}$$

AHC can be computationally costly. For the usual AHC procedure described above, the time complexity is $O(N^3)$. Other approaches, such as NN-chain based methods [1,7], have time complexity $O(N^2)$. But the drawback is that, NN-chain based methods are constrained by reducibility property, thus they cannot work with median and centroid methods. Another approach, called SparseHC [9] structures clusters with an adjacency hash map. According to its experiment results, SparseHC does decrease the memory growth, but its time complexity is improved for single link only. Besides this approach is not generic, as it is only applicable for single link, complete link and average link. Another method, CURE [4], reduces data by random sampling and partitioning. Though it decreases time complexity to $O(N_{sample}^2 log N_{sample})$, its results are indeterministic due to the random procedures. And for BIRCH [11], its time complexity is $O(N)$, but extra structure like clustering features (CF) tree has to be employed in order to store compact summaries of the original data.

In this paper, we propose an AHC approach which is generic for all usual methods. It allows a better scalability compared to the conventional AHC algorithm both in memory and processing time, and its results are deterministic unlike some aforementioned techniques. Our method is based on a new expression of the Lance-Williams (LW) formula, in which we replace dissimilarities with inner product based similarities. This change provides us with two important advantages: (1) it allows us to easily extend AHC to kernel functions; (2) it enables us to design a suitable thresholding strategy so that we can obtain a sparsified similarity matrix, resulting in a more scalable AHC procedure.

In order to illustrate the properties and benefits of our approach, we applied it to text clustering tasks. Our experiments show that results obtained by our framework are identical to the results obtained by the usual AHC methods, demonstrating their equivalence. Moreover, our experimental results show that on a largely sparsified similarity matrix, our approach is still able to cluster correctly, but with higher speed and less memory usage.

The rest of paper is structured as follows. Section 2 introduces fundamental materials on AHC. Section 3 details our approach and the mathematical proof of correctness. Experimental verification on results quality and performance improvement on real-world text clustering tasks can be found in Sect. 4. Section 5 concludes our paper with a discussion and presents future work.

2   Conventional AHC Methods and the LW Formula

There are many AHC techniques and reviews of these algorithms can be found in [8,10]. Conventional AHC methods can be classified into graph and geometric methods. Single link, complete link, average link and Mcquitty are graph methods, in which dissimilarity of two clusters is determined by the dissimilarities of instances from these clusters. Due to this property, these techniques can use graph representations that rely on a pairwise dissimilarity matrix. Whereas, for geometric methods, composed of centroid, median and Ward methods, instances are assumed to be represented in an Euclidean space, clusters are represented by prototypes and Euclidean distances between these representative vectors are used as dissimilarities.

Proposed by Lance G.N and Williams W.T in 1967 [5], the LW formula is a convenient formulation, which unifies the graph and the geometric methods mentioned above. It is used as a generic equation for updating the dissimilarity matrix D at each iteration once the newly formed cluster $C_{(ij)} = C_i \cup C_j$ given by (1) has been added to the dendrogram. According to this approach the dissimilarity between $C_{(ij)}$ and another cluster C_k is given by:

$$D(C_{(ij)}, C_k) = \alpha_i D(C_i, C_k) + \alpha_j D(C_j, C_k) + \beta D(C_i, C_j) \\ + \gamma |D(C_i, C_k) - D(C_j, C_k)| \tag{2}$$

Depending on the choice of a certain clustering method, values of parameters α_i, α_j, β and γ change accordingly. Table 1 displays parameter values in (2) that correspond to seven particular methods.

3   Our Approach

In this section, we introduce our method from three aspects: (1) we propose to renew the original LW formula using cosine similarities instead of squared Euclidean distances; (2) we extend this expression to kernel functions; (3) we introduce a simple sparsification strategy which is applied to the similarity matrix in the goal of reducing memory use and running time.

3.1   An Equivalent LW Formula Using Cosine Similarities

We suppose that the N instances of dataset $\mathcal{D}$ are represented by vectors in an Euclidean space $\mathcal{I}$ of dimension p. Cosine of the angle between two vectors is

Table 1. Lance-Williams formula: methods and parameter values.

Methods	α_i	α_j	β	γ
Single	$1/2$	$1/2$	0	$-1/2$
Complete	$1/2$	$1/2$	0	$1/2$
Average	$\frac{|C_i|}{|C_i|+|C_j|}$	$\frac{|C_j|}{|C_i|+|C_j|}$	0	0
Mcquitty	$1/2$	$1/2$	0	0
Centroid	$\frac{|C_i|}{|C_i|+|C_j|}$	$\frac{|C_j|}{|C_i|+|C_j|}$	$-\frac{|C_i||C_j|}{(|C_i|+|C_j|)^2}$	0
Median	$1/2$	$1/2$	$-1/4$	0
Ward	$\frac{|C_i|+|C_k|}{|C_i|+|C_j|+|C_k|}$	$\frac{|C_j|+|C_k|}{|C_i|+|C_j|+|C_k|}$	$-\frac{|C_k|}{|C_i|+|C_j|+|C_k|}$	0

considered as their similarity. Input S is a pairwise similarity matrix of size N. For two data points $x, y \in \mathcal{D}$, with $\langle .,. \rangle$ denoting inner product, their similarity is defined as follows:

$$S(x,y) = \langle \frac{x}{\|x\|}, \frac{y}{\|y\|} \rangle \tag{3}$$

Note that this implies that $S(x,x) = 1$ for all data point $x \in \mathcal{D}$. It is an important condition in our context to establish the new expression. We associate S with a dissimilarity matrix D whose general term is the squared Euclidean distance between normalized vectors:

$$\begin{aligned} D(x,y) &= \|\frac{x}{\|x\|} - \frac{y}{\|y\|}\|^2 \\ &= S(x,x) + S(y,y) - 2S(x,y) \tag{4} \\ &= 2(1 - S(x,y)) \tag{5} \end{aligned}$$

With the above assumptions, we provide an expression of the LW formula using S instead of D. In fact, our approach amounts to work with $-\frac{1}{2}D(C_k, C_l)$ instead of $D(C_k, C_l)$. In order to guarantee the correctness of our reasoning, we proceed by induction. In the first iteration of the AHC algorithm, there are N leaves $\{C_k\}$, each is one data point. In this case, the relationship below is straightforward:

$$\begin{aligned} \arg\min_{(C_k,C_l)} D(C_k, C_l) &= \arg\max_{(C_k,C_l)} -\frac{1}{2}D(C_k, C_l) \\ &= \arg\max_{(C_k,C_l)} S(C_k, C_l) - \frac{1}{2}(S(C_k, C_k) + S(C_l, C_l)) \tag{6} \end{aligned}$$

Next, for the subsequent iterations, we show that the LW formula can be recast as follows:

$$-\frac{1}{2}D(C_{(ij)}, C_k) = S(C_{(ij)}, C_k) - \frac{1}{2}(S(C_{(ij)}, C_{(ij)}) + (\alpha_i + \alpha_j)S(C_k, C_k)) \tag{7}$$

where:

$$S(C_{(ij)}, C_k) = \alpha_i S(C_i, C_k) + \alpha_j S(C_j, C_k) + \beta S(C_i, C_j) \tag{8}$$
$$- \gamma |S(C_i, C_i)/2 - S(C_i, C_k) - S(C_j, C_j)/2 + S(C_j, C_k)|$$
$$S(C_{(ij)}, C_{(ij)}) = (\alpha_i + \beta)S(C_i, C_i) + (\alpha_j + \beta)S(C_j, C_j) \tag{9}$$

Equation (7) together with the recurrence formulas (8) and (9) are respectively the counterparts of (1) and (2) that establish our method. With the condition that $S(C_k, C_k) = 1$ for all N singletons $\{C_k\}$, we show below that our formulation is equivalent to the usual LW formula for each clustering scheme listed in Table 1:

1. For single link and complete link, (8) reduces to $S(C_{(ij)}, C_k) = \alpha_i S(C_i, C_k) + \alpha_j S(C_j, C_k) + \beta S(C_i, C_j) - \gamma|S(C_i, C_k) - S(C_j, C_k)|$, and (9) reduces to

$S(C_{(ij)}, C_{(ij)}) = 1$, as $\alpha_j + \alpha_j + 2\beta = 1$. Since in (7) $\alpha_i + \alpha_j = 1$, then (6) with the reduced updating rules of (8) and (9) are globally equivalent to the conventional procedure. Note that for other remaining methods $\gamma = 0$, so that (8) boils down to $S(C_{(ij)}, C_k) = \alpha_i S(C_i, C_k) + \alpha_j S(C_j, C_k) + \beta S(C_i, C_j)$.

2. If we assume again that $S(C_k, C_k) = 1$ for all N singletons $\{C_k\}$ and if we replace $S(C_{(ij)}, C_{(ij)})$ with (9) as well, then the second term in the right-hand side of (7) becomes $-(\alpha_i + \alpha_j + \beta)$. By this, we can divide the remaining clustering methods into two groups: (a) average link, Mcquitty and Ward which have $-(\alpha_i + \alpha_j + \beta) = -1$, and (b) centroid and median which satisfy $-(\alpha_i + \alpha_j + \beta) > -1$:

   (a) Regarding average link, Mcquitty and Ward, it is not difficult to prove by induction that the second term in the right-hand side of (7) always equals to -1. Consequently, for these cases, (6) with the updating rules $S(C_{(ij)}, C_k) = \alpha_i S(C_i, C_k) + \alpha_j S(C_j, C_k) + \beta S(C_i, C_j)$ and $S(C_{(ij)}, C_{(ij)}) = 1$ are globally equivalent to the general procedure.

   (b) Concerning the centroid and median methods, since $\alpha_i + \alpha_j = 1$ in (7), the coefficient assigned to $S(C_k, C_k)$ vanishes. However, $\alpha_i + \alpha_j + 2\beta \neq 1$ in (9) hence $S(C_{(ij)}, C_{(ij)}) \neq 1$. Therefore, it is important to apply the weighting system determined in (9) for the global equivalence of centroid and median to hold.

We can wrap up all particular cases discussed above through the following general procedure which defines our AHC framework. At each iteration, we solve:

$$(C_i, C_j) = \arg \max_{(C_k, C_l)} S(C_k, C_l) - \frac{1}{2}(S(C_k, C_k) + S(C_l, C_l)) \tag{10}$$

After having merged (C_i, C_j) into $C_{(ij)}$, the similarity matrix S is updated by applying the two following equations:

$$S(C_{(ij)}, C_k) = \alpha_i S(C_i, C_k) + \alpha_j S(C_j, C_k) + \beta S(C_i, C_j) \tag{11}$$
$$\quad - \gamma |S(C_i, C_k) - S(C_j, C_k)|$$
$$S(C_{(ij)}, C_{(ij)}) = \delta_i S(C_i, C_i) + \delta_j S(C_j, C_j) \tag{12}$$

Table 2 lists parameter values of each method in our framework. Note that in this table, the newly introduced parameters δ_i and δ_j sum to one except for centroid and median methods. In fact, for the other methods, we could have taken any values providing that $\delta_i + \delta_j = 1$.

3.2   Extending to Kernel Functions

Our approach allows us to naturally extend AHC methods to kernel functions (see for example [2]) since most of the latter mappings are defined with respect to inner products. Consequently, in our method, broader similarity measures can be easily employed and non linearly separable cases can be addressed effectively.

Thereby, let K denote a pairwise inner product matrix (or Gram matrix) of size N whose general term for two data points $x, y \in \mathcal{D}$ is $K(x, y) = \langle \phi(x), \phi(y) \rangle$

Table 2. The cosine similarity based formula: methods and parameter values

Methods	α_i	α_j	β	γ	δ_i	δ_j
Single	1/2	1/2	0	$-1/2$	1/2	1/2
Complete	1/2	1/2	0	1/2	1/2	1/2
Average	$\frac{|C_i|}{|C_i|+|C_j|}$	$\frac{|C_j|}{|C_i|+|C_j|}$	0	0	1/2	1/2
Mcquitty	1/2	1/2	0	0	1/2	1/2
Centroid	$\frac{|C_i|}{|C_i|+|C_j|}$	$\frac{|C_j|}{|C_i|+|C_j|}$	$-\frac{|C_i||C_j|}{(|C_i|+|C_j|)^2}$	0	$\frac{|C_i|^2}{(|C_i|+|C_j|)^2}$	$\frac{|C_j|^2}{(|C_i|+|C_j|)^2}$
Median	1/2	1/2	$-1/4$	0	1/4	1/4
Ward	$\frac{|C_i|+|C_k|}{|C_i|+|C_j|+|C_k|}$	$\frac{|C_j|+|C_k|}{|C_i|+|C_j|+|C_k|}$	$-\frac{|C_k|}{|C_i|+|C_j|+|C_k|}$	0	1/2	1/2

where $\phi : \mathcal{I} \to \mathcal{F}$ is a mapping from $\mathcal{I}$ to $\mathcal{F}$ and the latter notation designates a feature space of dimension $q > p$ (q is possibly infinite).

The S matrix in our approach should contain cosine measures, and more importantly, its diagonal entries should be constant. Gaussian and Laplacian kernels satisfy this condition naturally, but for other kernels, they have to be normalized. To generalize all the cases, we obtain a cosine similarity matrix by applying for all $x, y \in \mathcal{D}$: $S(x,y) = K(x,y)/\sqrt{K(x,x)K(y,y)}$.

3.3 Sparsification of the Cosine Similarity Matrix

In general terms, S could contain negative values. In that case, let $m < 0$ be the minimal value in S and $|m|$ its absolute value. It is always possible to transform S in order to have non negative values using the following rescaling operator, $\forall x, y \in \mathcal{D}$:

$$S(x,y) \leftarrow \frac{S(x,y) + |m|}{1 + |m|} \tag{13}$$

Since this mapping is monotonically increasing, the resulting S remains an inner product matrix and, in addition, it has ones on its diagonal.

Assuming that S is non negative, we propose to apply a simple thresholding operator which depends on a parameter $\tau \in [0,1]$: any similarity value below τ[1] is considered irrelevant and it is replaced with 0[2], $\forall x, y \in \mathcal{D}$:

$$S(x,y) \leftarrow S(x,y)I_{(S(x,y)\geq\tau)} \tag{14}$$

where $I_{(S(x,y)\geq\tau)} = 1$ if $S(x,y) \geq \tau$ and $I_{(S(x,y)\geq\tau)} = 0$ otherwise.

The resulting S matrix is sparser than the original one and thus requires less memory.

Next, we propose to restrict the search for pairs of clusters to merge in (10) to the following subset: $\mathbb{S} = \{(C_k, C_l) : S(C_k, C_l) > 0\}$. This allows the running

---

[1] Note that if $\tau = 0$ then S is not sparsified.

[2] It is interesting to mention that such a thresholding operator cannot be applied to a dissimilarity matrix D, because the larger values are the less relevant ones in that case and replacing them with 0 is not sound.

time to be diminished as well, since the bottleneck procedure in the general AHC algorithm is precisely the search for the optimal proximity value, which has $O(N^2)$ time complexity. Accordingly, we propose to replace (10) with:

$$(C_i, C_j) = \arg \max_{(C_k, C_l) \in \mathbb{S}} S(C_k, C_l) - \frac{1}{2}(S(C_k, C_k) + S(C_l, C_l)) \qquad (15)$$

As we shall see in the next section, not only this approach dramatically reduces the processing time but it also allows obtaining better clustering results.

4   Experiments

The goals of our experiments are to demonstrate that: (1) our framework based on Eqs. (10), (11) and (12) is equivalent to the usual AHC procedure (1) based on the LW formula (2) under the assumptions exposed previously; (2) sparsifying the cosine similarity matrix with (14) and applying our AHC given by (11), (12) and (15) considerably decreases memory use and running time while having the capacity to provide better clustering results.

To this end, we experimented on text clustering tasks. Indeed, hierarchical clustering is particularly interesting in this case, since it allows expressing the relationships between different topics in a collection and at different granularity levels. Moreover, cosine similarities are classic proximity functions used for documents. In addition, hierarchical document organization based on the conventional AHC procedure faces the problem of scalability since text collections are usually very large. Our experiments seek to demonstrate new perspectives to overcome these limits.

It is important to note that our purpose is not to compare the different AHC methods between each other, but rather to exemplify the properties of our framework compared to the usual AHC procedure using the LW formula. As a consequence, the results obtained by the latter conventional approach are our baselines.

4.1   Datasets, Preprocessing and Evaluation Measures

We used three well-known corpora employed in text clustering benchmarks: Reuters-21578[3] (Reuters), Smart [3] and 20Newsgroups[4] (20ng) [6]. Their descriptive statistics are given in Table 3.

We used the bag-of-words approach where each document is represented by a vector in the space spanned by a set of terms. As for preprocessing we applied a rough feature selection by removing terms that appear in less than 0.2 % and more than 95 % documents of the collection. No stemming, lemmatization nor stop word removal were applied. Then, the tfidf weighting strategy was performed.

---

[3] Distribution 1.0, the ApteMod version.
[4] We used the same dataset as in http://qwone.com/~jason/20Newsgroups/.

Table 3. Descriptions of datasets.

Dataset	Nb of classes	Nb of documents	Nb of features
Reuters	10	2446	2547
Smart	3	3893	3025
20ng	15	4483	4455

The adjusted Rand index (ARI) and (the absolute value of) the cophenetic correlation (CC) between dendrograms are used to compare the clustering outputs. CC is employed to evaluate how far our dendrogram is from the one produced by the conventional AHC procedure. In this case, higher is better and a maximum value one means that the dendrograms are equivalent and thus represent the same hierarchy. ARI is an external assessment criterion that evaluates the quality of the clustering output in regard to a given ground-truth. It requires to flatten the dendrogram with the correct number of clusters, then the obtained partition and the ground-truth are compared to each other. Greater ARI values imply better clustering outputs. The maximum value one is observed when the ground-truth is perfectly recovered.

4.2 Experiments Settings and Results

Given a term-document matrix, two types of matrices are generated: the cosine similarity matrix S and the corresponding distance matrix D as defined by (5). Note that since the term-document matrix consists of non negative values then S takes values in $[0, 1]$ therefore no rescaling operator is needed.

Given a clustering method, the S matrix is taken as the input to our framework, while the related dense D matrix is input to the conventional AHC algorithm. Consequently, two dendrograms are returned and we compute the CC in order to assess the similarity between the two outputs. Two cases are of interest: (1) when $\tau = 0$ which means no sparsification and the dense S is used; and (2) when $\tau > 0$ and being increased which leads to sparser and sparser S matrices.

In addition to 0, we chose other threshold values τ as the 10th, 25th, 50th, 75th and 90th percentiles of distribution of values in S. Let k denote the rank of a percentile so that $k \in \{0, 10, 25, 50, 75, 90\}$ with the convention that the 0th percentile is 0. Accordingly, when k grows the kth percentile τ is greater and greater and the S matrix becomes sparser and sparser.

We experimented with two types of kernel: linear and Gaussian. The linear kernel is simply the inner dot product in $\mathcal{I}$ between normalized vectors as defined in (3). The Gaussian kernel between two points $x, y \in \mathcal{D}$ is given by $K(x, y) = \exp(-\gamma\|x - y\|^2)$. It corresponds to a cosine measure in $\mathcal{F}$. In our experiment we set γ to $1/p$ by default[5].

---

[5] Note that this default setting is used in popular SVM packages. Furthermore, in this case γ is very low and the Gaussian kernel provides values close to one and close to each other between pairs of points.

In Fig. 1, we show the results obtained for all seven methods on Reuters, Smart and 20ng datasets respectively. We report the curves of several measurements (y-axis) when S is progressively sparsified as the percentile rank (x-axis) increases. In addition to CC and ARI graphs (dotted lines with circle and triangle symbols respectively), the percentage of the memory cost of a sparse S with respect to the dense S, and the proportion of the running time when using a sparse S as compared to the dense S, are plotted as well (solid lines with plus symbols and dashed lines with cross symbols respectively). Therefore, the memory and processing time costs related to the full S (corresponding to the 0th percentile where $\tau = 0$) serve as baselines (with y-axis value of 100 %). In these cases, the lower the percentages the bigger the gains.

Equivalence Between Our Method and the LW Formula. In Fig. 1, for all datasets and both kernels, the CC values are all equal to one when $\tau = 0$ (0th percentile shown at the origin). This empirically demonstrates that our approach is equivalent to the AHC algorithm using the LW formula as claimed previously.

Next, as the percentile rank increases, the CC values generally decrease illustrating the fact that the dendrograms move away from the LW formula based results. However, when using the linear kernel, the CC values generally remain high even when the majority of the similarity values are removed. Concerning the Gaussian kernel, the CC values drop rapidly after having thresholded 10 % of the lowest similarities but they start increasing again after this fall.

The single link method however, presents a peculiar behavior: for all collections and both kernels, it always recover the result given by the usual AHC procedure despite the fact that 90 % of the S matrix is sparsified. In other words, our framework is able to obtain the same dendrogram provided by the original LW formula but with 90 % of memory usage and running time saved.

Impact of the Sparsification of S on Scalability. Let $M \leq N^2$ be the number of non zero cells in S. The storage cost of our approach is $O(M)$. The time complexity[6] is $O(NM)$ which indicates a linearly relationship with respect to the storage complexity.

In Fig. 1, the solid lines with plus symbols give the percentage of size of the sparse S with respect to the dense S. As expected, this quantity linearly decreases as the percentile rank k grows.

Next, the dashed lines with cross signs show the proportion of the processing time observed with a sparse S with respect to the running time noted with the dense S. We observe linear curves as well which depicts the linear relationship between the memory and time complexities as mentioned above.

The sparsification of the S matrix enables decreasing the storage complexity and the running time. Besides, it also has an impact on the clustering quality. Previously, we have noticed that CC values were decreasing as S were sparser and sparser. In the sequel, we examine some cases in which our framework wins on both sides: scalability and quality.

---

[6] Similarly to the general AHC algorithm based on a dissimilarity matrix for which $M = N(N - 1)$.

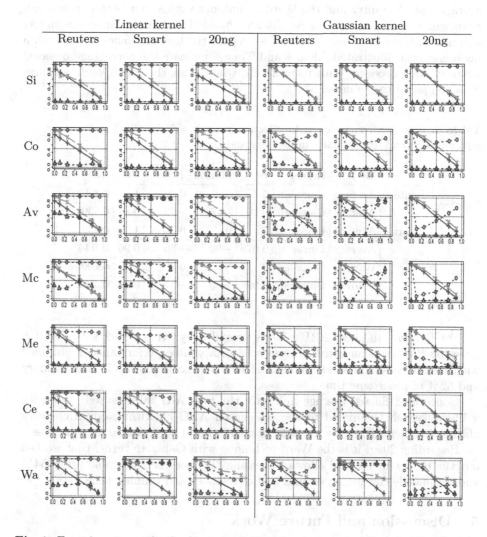

Fig. 1. Experiments results for linear and Gaussian kernels in left and right blocks. Rows correspond to AHC methods (using their abbreviations) and columns to collections. In each graph: each point corresponds to one of the measurements listed afterwards with respect to an S matrix; the x-axis correspond to percentile ranks (divided by 100) which define the threshold values τ (not shown); solid lines with plus signs represent the relative memory use, dashed lines with cross signs show the relative running time, dotted lines with circle symbols indicates the absolute value of cophenetic coefficient (CC), dotted lines with triangle symbols give the ARI values.

Impact of Sparsification of S on Clustering Quality. We focus on the quality of clustering outputs by analyzing the ARI values. We observe that average link, Mcquitty and the Ward techniques worked out better in general. Surprisingly, many of the best results are obtained with a very sparse S matrix and not with the full one. In Table 4 we report the best outcomes where such a phenomenon is illustrated. Mem% and Time% indicate the percentage of saved memory and processing time respectively, when using the corresponding sparse S as compared to the dense S.

Table 4. Best ARI results for each collection when $\tau = 0$ (baseline) and when $\tau > 0$ (sparsified S) and relative gains in memory and time.

	Method	Kernel	τ	Mem%	Time%	CC	ARI
Reuters	Average	Gaussian	0	0	0	1	**0.543**
	Average	Gaussian	0.99	−75	−62	0.81	0.539
Smart	Average	Linear	0	0	0	1	0.939
	Average	Linear	0.078	−90	−85	0.96	**0.944**
20ng	Ward	Gaussian	0	0	0	1	0.100
	Ward	Gaussian	0.99	−50	−47	0.26	**0.154**

For Reuters, the best ARI value is provided by average link with a full S given by the Gaussian kernel. However, a comparable performance is obtained with the same method and kernel but with a sparse S that saves 75 % of memory and 62 % of processing time.

Concerning Smart, average link gave the best ARI value as well, but with a linear kernel. Compared to the LW based AHC algorithm, our framework obtained higher ARI and with 90 % of memory and 85 % of running time less.

Regarding 20ng, it is the Ward technique with Gaussian kernel that worked out the best. Our method allows increasing the baseline ARI value up to 54 % and meanwhile consuming around half of memory and running time.

5   Discussion and Future Work

We have introduced an equivalent formulation of the LW formula based on cosine similarities instead of squared Euclidean distances. Our AHC procedure that relies on this formulation and a sparsified cosine similarity matrix, not only has better scalability properties but is also able to give better clustering results.

We believe that two reasons account for this phenomenon. Firstly, sparsifying the S matrix reduces the noise by removing the lowest similarity values, therefore leading to better clustering performances. Secondly, when two clusters (C_i, C_j) are merged together, their respective neighborhoods (clusters having a non null similarity value with C_i and C_j respectively) are fused as well, so that $C_{(ij)}$ has a larger neighborhood than both C_i and C_j. Furthermore, the updating rule (11) allows reinforcing the similarity value of $C_{(ij)}$ with C_k if the latter cluster belongs to both initial neighborhoods. In fact, our approach can be viewed as a sort of "transitive closure" starting with reliable seeds (the pairs with highest similarity values) and propagating similarities through "trusted" neighborhoods.

However, the main drawback of our method is that, either sparsifying S does not improve the ARI value at all (see complete link applied to Reuters with Gaussian kernel in Fig. 1 for instance), or the improvements are not regular and setting the threshold value τ becomes difficult. More theoretical investigations should be undertaken in these respects to have a better understanding of the properties of our framework.

Another line of research that we intend to pursue is to implement our approach in the manner of distributed computing to take better advantage of its scalability.

Acknowledgment. This work was supported by the french national project Request PIA/FSN.

References

1. Bruynooghe, M.: Classification ascendante hiérarchique des grands ensembles de données: un algorithme rapide fondé sur la construction des voisinages réductibles. Cahiers de L'analyse des Données **3**(1), 7–33 (1978)
2. Cristianini, N., Shawe-Taylor, J.: An Introduction to Support Vector Machines and other Kernel-Based Learning Methods. Cambridge University Press, Cambridge (2000)
3. Dhillon, I.S.: Co-clustering documents and words using Bipartite co-clustering documents and words using Bipartite spectral graph partitioning. In: Proceedings of 7th ACM SIGKDD Conference on Knowledge Discovery and Data Mining, pp. 269–274 (2001)
4. Guha, S., Rastogi, R., Shim, K.: Cure: an efficient clustering algorithm for large databases. In: ACM SIGMOD Record, vol. 27, pp. 73–84. ACM (1998)
5. Lance, G.N., Williams, W.T.: A general theory of classificatory sorting strategies ii. clustering systems. Comput. J. **10**(3), 271–277 (1967)
6. Lang, K.: NewsWeeder: learning to filter netnews. In: Proceedings of the Twelfth International Conference on Machine Learning, pp. 331–339 (1995)
7. Murtagh, F.: A survey of recent advances in hierarchical clustering algorithms. Comput. J. **26**(4), 354–359 (1983)
8. Murtagh, F., Contreras, P.: Algorithms for hierarchical clustering: an overview. Wiley Interdisc. Rev. Data Min. Knowl. Disc. **2**(1), 86–97 (2012)
9. Nguyen, T.D., Schmidt, B., Kwoh, C.K.: Sparsehc: a memory-efficient online hierarchical clustering algorithm. Procedia Comput. Sci. **29**, 8–19 (2014)
10. Xu, R., Wunsch, D., et al.: Survey of clustering algorithms. IEEE Trans. Neural Netw. **16**(3), 645–678 (2005)
11. Zhang, T., Ramakrishnan, R., Livny, M.: Birch: an efficient data clustering method for very large databases. In: ACM Sigmod Record, vol. 25, pp. 103–114. ACM (1996)

Determining Data Relevance Using Semantic Types and Graphical Interpretation Cues

Eduardo Haruo Kamioka[✉], André Freitas, Frederico Caroli,
and Siegfried Handschuh

Universität Passau, Passau, Germany
{Haruo-Kamioka,Andre.Freitas,Frederico.Caroli,
Siegfried.Handschuh}@Uni-Passau.de

Abstract. The increasing volume of data generated and the shortage of professionals trained to extract value from it, raises a question of how to automate data analysis processes. This work investigates how to increase the automation in the data interpretation process by proposing a relevance classification heuristic model, which can be used to express which views over the data are potentially meaningful and relevant. The relevance classification model uses the combination of *semantic types* derived from the data attributes and *visual human interpretation cues* as input features. The evaluation shows the impact of these features in improving the prediction of data relevance, where the best classification model achieves a F1 score of 0.906.

1   Introduction

The growing availability of data brings the demand for methods to support the automation of the data interpretation process, by automatically exploring the search spaces of possible interpretations associated with the available data. However, methods to support the automation of large-scale exploratory data analysis are still limited.

The materialisation of the vision of an *automated data analyst* requires a *heuristic model* which can optimise the exploration of the potential interpretation space of the data, detecting which data views and patterns are meaningful and potentially relevant for data consumers.

This work aims at addressing this problem by proposing a *relevance classification* approach based on the composition of *semantic types* and *visual data interpretation cues*. The main goal of the model is to provide a heuristic model which can be used for pruning the search space associated with the interpretation and identification of patterns of interest in the data.

The heuristic model is built upon the assignment of *semantic types* to data attributes which, in combination with *visual interpretation cues*, define a data relevance classification model. Both semantic types and visual interpretation cues are input as features in order to build the final *data interpretation relevance classifier*.

© Springer International Publishing AG 2016
H. Boström et al. (Eds.): IDA 2016, LNCS 9897, pp. 332–342, 2016.
DOI: 10.1007/978-3-319-46349-0_29

The proposed model lies on the intuition that the semantic types associated with attributes can be used to infer their compatibility to form a meaningful *data view*. Additionally, coarse-grained visual interpretation cues over the final visualisation output (mediated by a specific visualisation type) are used as evidence to detect salient potential patterns of interest within the data. We assess the human interpretation process by systematically and manually classifying meaningful and relevant data views for different domains.

The contributions of this work are: *(i)* the definition of a data interpretation relevance model based on the combination of semantic types and visual interpretation cues; *(ii)* an evaluation of the proposed model and of the impact of semantic and visual cues and *(iii)* the determination of the best classification model through a systematic analysis of different classifiers.

2   Related Work

This work concentrates on the area of automated and intelligent data analysis. In Grosse *et al.* (2012); Duvenaud *et al.* (2013); Lloyd *et al.* (2014) the concept of an *automatic statistician* is introduced. The automatic statistician framework introduces a process to explore the compositionality of a large space of models structures to find the applicable model to predict, classify or extrapolate based on new unseen data. Our approach differs as we explore the compositionality of data views and visual patterns to classify data relevance. Another proposed model is AIDE, which provides a semi-automated process, which relies on planning data analyses steps by a determined combination of data type and user interaction (St. Amant and Cohen (1998); St. Amant and Cohen (1997)). AIDE limits its application as a fully autonomous system, requiring corrections executed by the user without training the system to correct itself automatically. Our approach focuses on an automatic classification approach for the selection of relevant data views.

Regarding exploratory data analysis, two works are considered. The first work focuses on automated knowledge discovery workflow composition through ontology-based planning (Záková *et al.* (2011)). It differs from our approach in the semantic representation model where the extraction of semantic features from WordNet hypernyms and distributional word vectors target a more generic semantic representation solution (open vocabulary). The proposed model in this work builds upon Bremm *et al.* (2011) which focused on assisted data descriptors selector based on visual comparative data analysis. It aims at facilitating the user's access to the data analysis process. This data is used to link the description of features combinations and resulting functions with clear meaning by a human data analyst, selecting the views and output interpretability. This approach differs from our work as we explore the combination of semantic types and high-level visual interpretation cues to classify data views representing relevant meaning.

3   Relevance Classification Model

3.1   Proposed Approach

The proposed relevance classification model consists of four main steps:

- Automatic pair-wise selection of attribute combination into a *data view*.
- Extraction of *descriptive statistical features* and *semantic type features*.
- Extraction of *visual interpretation cues*.
- Classification of the *relevance* of the data view.

Figure 1 presents an overview of the relevance classification model.

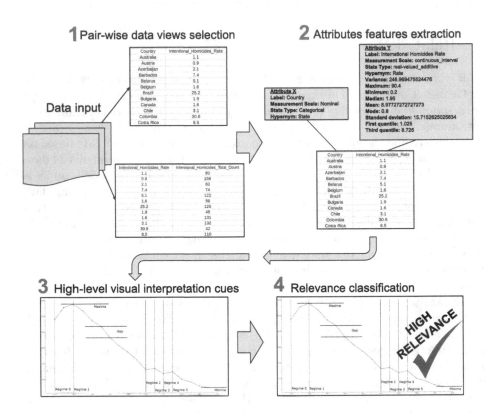

Fig. 1. Overview of our proposed approach.

We say that a data view is *relevant* when a visual interpretation provides a clear trend or pattern which is easily recognisable by a human, with or without previous knowledge about the data being analysed from two or more attributes in a dataset.

To classify the relevance when analysing plots of pairs of attributes (data views) we defined target classes based in the human process of data analysis.

During the data analysis the analyst explore visualisations in order to understand the data. To achieve a clear meaning the analyst should take decisions, such as the application of operations (e.g. group by, sort by), changing the visualization plot type, and including more data attributes or dimensions. Thus, the classes defined are the representation of decisions required by the data analyst at each step of the exploration process. These classes are:

- Class 1 - Clear meaning - Generic (Very intuitive - you do not need to know the dataset/context to understand)
- Class 2 - Clear meaning - Dataset Context (you should know the dataset to understand)
- Class 3 - Data non-relevant for data analysis (or for the analysis in question)
- Class 4 - Label not equal to data semantics (Inconsistent data)
- Class 5 - Change visualisation (plot type or axis - makes sense, but if change it's better)
- Class 6 - Add operations (ex: group by, sort by, etc.)
- Class 7 - Additional data attributes needed for the interpretation
- Class 8 - Add operations and/or more data attributes and/or visualisation.

Figure 2 shows an example of a plot that requires additional attributes to present a clear meaning. Figure 3 shows an example of a plot requiring an additional attribute, and/or an additional operation, and/or a change in the visualisation plot to present a clear meaning. Figure 4 shows an example of clear meaning.

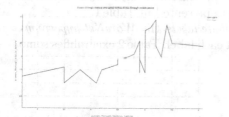 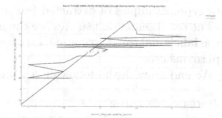

Fig. 2. Mobile devices and desktop/laptop devices by country sorted.

Fig. 3. Mobile devices and desktop/laptop devices by country without a sort operation.

In our approach we assume that the datasets have no missing values. Any missing values from the collected data is striped out before we start processing them.

3.2 Semantic Features

To represent the semantic type of an attribute, this work considers two approaches. The first one uses WordNet[1] hypernyms. WordNet is a lexical thesaurus for the English language, where words are organized into a lexical semantic network. WordNet also specifies the sets of hypernyms associated to a word (its taxonomical structure), where Y is a hypernym of X if every X is a (kind of) Y. We use these hypernyms as the semantic feature of words.

The other approach is the use of distributional vectors, extracted using the Word2vec framework (Mikolov *et al.* (2013)). These vectors encode co-occurrence statistics of words, relying on the linguistic notion that the context of a word defines the semantics of it (Harris (1954)). Distributional vectors are normally used as semantic representation of words.

For our work, we always assume that an attribute label has a descriptive meaning. For example, a label will never be 'X1' or 'Y'. This assumption is necessary if we want to automatically assign semantic features to a label.

3.3 Attribute Feature Extraction

In order to simulate the data analysis steps executed by humans, we developed a feature extraction process. The process explores the compositionality of statistical data types, the semantic representation of attributes, associated data operations and basic plotting resources.

For the extraction of *descriptive statistical features*, we consider the univariate analysis for description of the distribution, central tendency and the dispersion for each data attribute, also classifying the measurement scale and statistical data type. Examples of extracted features are presented in Table 1.

The set of semantic features are the *attribute labels*, the *WordNet hypernyms* of the labels and the *distributional vectors* of each label. Table 2 exemplifies some hypernyms used.

We end up with the following features:

- Statistical features:
  - Mean, median, first quartile, third quartile, mode (for categorical data), standard deviation, variance;
  - Measurement scale (nominal, ordinal, continuous interval, continuous ratio);
  - Statistical data type (categorical, ordinal, real, binary, multiclass, count);
- Semantic features:
  - Data labels;
  - WordNet hypernyms;
  - Distributional vector representations of data attributes labels;

The process of assigning hypernyms to the attributes consists in the identification of the head word of the phrase associated with the attribute label

---
[1] http://wordnet.princeton.edu.

Table 1. Examples of extracted descriptive statistical features.

Attribute label	Measurement scale	Statistical Data Type	Median	Mean	First quart	Third quart	Mode	Std.Deviation	Variance
Access Through Mobile Device	continuous interval	real-valued additive	2	2.125	1.575	2.525	1.9	0.689	0.475
SepalLength	continuous interval	count	5.8	5.843	5.1	6.4	5	0.828	0.685
Intentional Homicides Total Count	ordinal	count	88	89.53	48	130	28	47.18	2226.037
Intentional Homicides Year	ordinal	ordinal-integer number	2012	2011.7	2012	2012	2012	0.569	0.324
Country	nominal	categorical	-	-	-	-	-	-	-
having IP Address	nominal	binary	1	0.314	-1	1	1	0.949	0.901
URL Length	nominal	multiclass	-1	-0.633	-1	-1	-1	0.766	0.586
Price	continuous ratio	real-valued multiplicative	6.985	7.042	6.508	7.441	6.204	0.634	0.402
alcohol-use	continuous ratio	count	2	2.176	0.5	77.5	49.3	26.878	722.473

Table 2. Examples of extracted semantic features.

Label	WordNet hypernym
Access Through Mobile Device	activity
Country	area, social unit
Intentional Homicides Rate	rate
Intentional Homicides Total Count	count
Intentional Region	area
Intentional Homicides Year	time period
Homicides Total	total sum
Gun Homicides Sources and Notes	source, note
GDP Rank	rank
GDP Int.dolar	monetary unit

(when the label contains multiple words). A *word sense disambiguation* process selects the associated sense of the word considering the other words within the phrase as its context. Afterwards, the associated hypernym is assigned. The level of taxonomic abstraction is assigned to two taxonomic hops.

3.4   Visual Interpretation Cues

Another fundamental component of the proposed model consists in simulating the human visual interpretation process when analysing a data view (the combination of pairs of attributes).

The visual attention mechanism associated with the process of human data interpretation focuses on targeting the detection of outliers, coarse-grained variation regimes, clusters, periodicity, among others. These are examples of high-level visual features which provide an entry point to the interpretation of the data.

For the purpose of this approach, we identified a set of ten *high-level visual interpretation cues* described below. These cues are then used as features for our classifiers.

- Whether the function is a pair of numerical data or a pair containing at least one categorical data;
- Gaps;

- Quantity of existent gaps;
- Measure of numerical correlation;
- Whether the correlation is positive or negative;
- Whether the function is linear or nonlinear;
- Derivative regimes;
- Quantity of derivative regimes;
- Maxima/minima;
- Periodicity.

We define a gap as any considerable difference of value magnitude between consecutive data points. We call a "considerable difference" any value greater than the arithmetic mean of all the differences from consecutive or non consecutive data points.

Figure 4 show some high-level visual cues in the context of a data view.

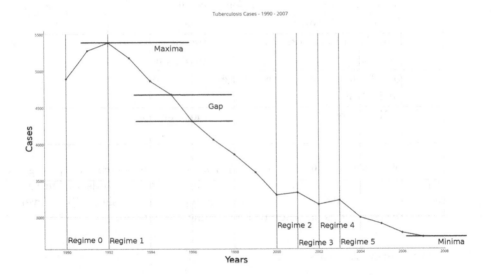

Fig. 4. Example of a two-dimensional numerical function describing the number of tuberculosis cases since 1990 up to 2007 with examples of visual interpretation cues.

4 Evaluation

Twenty-three machine learning models were trained to address the relevance classification problem, considering as an outcome one of the eight classes previously presented. We applied machine learning models based on *(i)* linear, *(ii)* non-linear, *(iii)* non-linear with decision trees, *(iv)* non-linear with boosting and *(v)* neural networks approaches. The application of more than one machine learning model is intended to assess the behaviour of our dataset and evaluate the impact of semantic and visual features in the predictive modelling.

4.1   Creation of the Relevance Gold-Standard

Our gold-standard dataset consists of 20 open datasets commonly used for data analysis and machine learning tasks[2], available at the UCI repository[3], Plotly[4], EDX Analytics Edge[5], and William B. King R Tutorials[6]. All attribute pairs from the collected datasets are then plotted and classified as one of the eight classes already presented previously, resulting in 2989 attribute pairs.

Two types of visualization plots were applied, the *bar plot* and *scatterplot with lines*, following the rules: *(i) bar plot* is applied when at least one data type is qualitative and *(ii) scatterplot with lines* is applied when the X axis is a quantitative data type and Y axis is a quantitative data type. Based on the plotting, a human classified the relevance class of each attribute pair.

4.2   Experiments

For the training of the machine learning models, we represented the extracted features in five scenarios:

- using only statistical features;
- using visual interpretation cues;
- using WordNet hypernym and visual interpretation cues;
- using distributional vectors of the attribute labels and visual interpretation cues;
- using distributional vectors of random words and visual interpretation cues.

We use the last two scenarios to validate our assumption that semantic vectors of the labels would improve the predictability accuracy of classification. The variation of the semantic types is used to evaluate the impact of different semantic features representations.

In all of the experiments we used 80 % of our dataset instances in the training phase, using the remaining 20 % to validate the resulting models.

4.3   Results

The best classification results are achieved with the feature combination of hypernyms as semantic types and visual interpretation cues (Random Forest achieves a 11.87 % improvement in F1 score over the best result of the scenario without these features). Distributional semantic vectors also impact in the classification of the results (5.1 % improvement in F1 score over the best result of the scenario without distributional semantic vectors). The best F1 score (0.906) was achieved by using a Boosted C5.0 classifier, using distributional vectors of attributes labels

---

[2] http://github.com/ekamioka/unipassau-ada.
[3] http://archive.ics.uci.edu/ml/.
[4] http://plot.ly/.
[5] https://www.edx.org/course/analytics-edge-mitx-15-071x-2#!.
[6] http://ww2.coastal.edu/kingw/statistics/R-tutorials/multregr.html.

Table 3. Results of the relevance classification. Best AUC and F1 score for each classifier is highlighted. HypVis - With hypernyms and visual interpretation cues, Vis - Just with visual interpretation cues, WO - No semantic features and no visual interpretation cue, VecVis - With distributional representations of the labels and visual interpretation cues, VecrVis - With distributional representations of random words and visual interpretation cues. The last line shows the percentual improvement for each feature set.

Algorithm	HypVis		Vis		WO		VecVis		VecrVis	
	AUC	F1 Score	AUC	F1 Score	AUC	F1 Score	AUC	F1 Score	AUC	F1 Score
Latent Discriminant Analysis	**0.742**	0.658	0.664	0.578	0.469	0.323	0.735	**0.667**	0.659	0.531
Linear Support Vector Classification	**0.455**	0.430	0.398	0.517	0.399	0.502	0.384	0.536	0.283	**0.554**
Stochastic Gradient Descent	**0.522**	0.035	0.478	0.055	0.466	0.065	0.518	0.014	0.394	0.017
SGDClassifier with kernel approximation	**0.496**	0.380	0.470	0.385	0.423	**0.460**	0.492	0.373	0.485	0.344
R Kernel Support Vector Machine	**0.868**	**0.754**	0.687	0.622	0.811	0.656	0.807	0.623	-	-
Naive Bayes	0.647	**0.617**	0.671	0.543	0.632	0.545	**0.680**	0.530	0.644	0.500
R k-Nearest Neighbors 3 (k=5)	0.798	0.698	0.723	0.603	0.687	0.426	0.782	0.610	**0.866**	**0.767**
Sklearn k-Nearest Neighbors (k-10)	0.320	**0.680**	0.314	0.665	**0.330**	0.645	0.320	**0.680**	0.320	**0.680**
R Classification and Regression Trees(CART)	0.793	0.620	0.784	0.629	0.713	0.571	0.651	0.538	**0.797**	**0.652**
Sklearn DecisionTreeClassifier	0.272	**0.763**	0.276	0.741	0.301	0.751	0.273	0.762	0.330	0.688
ExtraTreesClassifier	0.282	0.752	0.306	0.699	0.322	0.651	0.268	**0.762**	**0.326**	0.534
C4.5 Weka	0.633	0.074	0.658	0.074	0.729	0.667	**0.873**	**0.825**	0.679	0.082
PART Weka	0.688	0.540	0.892	**0.738**	0.804	0.657	**0.905**	0.702	0.510	0.346
R Random Forest	**0.969**	**0.895**	0.882	0.765	0.915	0.779	0.919	0.794	0.781	0.667
Sklearn RandomForestClassifier	0.263	**0.792**	0.282	0.762	0.287	0.757	0.263	0.775	**0.349**	0.614
R Gradient Boosted Machine	**0.646**	0.375	0.543	**0.757**	0.551	0.425	0.577	0.170	0.625	0.318
R Boosted C5.0	0.904	0.767	0.932	0.862	0.899	0.800	0.925	**0.906**	**0.951**	0.875
R eXtreme Gradient Boosting	0.916	0.759	0.914	0.783	0.878	0.772	0.925	**0.851**	**0.926**	0.818
Sklearn AdaBoostClassifier with: 10 max_depth	0.278	0.780	0.275	0.778	0.288	0.777	0.261	**0.791**	**0.340**	0.667
R Simple Neural Networks - nnet	0.559	**0.757**	**0.759**	0.729	0.716	0.518	-	-	-	-
H20 Deep Learning	0.958	**0.867**	0.944	0.741	**0.968**	0.731	0.875	0.642	-	-
Multi Layer Perceptron - tanh	**0.538**	0.358	0.536	0.076	0.457	0.431	0.341	**0.490**	0.508	0.237
Multi Layer Perceptron - relu	0.452	0.428	0.362	**0.485**	0.522	0.020	0.483	0.394	**0.500**	0.376
Wins percentage	**34.78%**	**34.78%**	4.35%	13.04%	13.04%	8.69%	13.04%	**34.78%**	**34.78%**	17.39%

and visual interpretation cues. The full comparative analysis of different classification methods and features are fully presented in Table 3.

Considering the imbalanced problem in the classification dataset, we noted a better classification performance in nonlinear models and ensemble-based models, which implements resampling techniques and combinations, thus rebalancing the classes at learning time (Chawla (2005)).

Other classifiers that do not implement some type of rebalancing have a hard time classifying some instances. For instance, the samples labeled as *Class 7*, a class that has only 7 occurrences in our dataset, are rarely classified correctly. On the other hand, our most common class (*Class 2*), represents 52.12 % of our dataset.

To further interpret our classifier results we hand-picked some examples. Those examples are presented in Figs. 5, 6, 7 and 8.

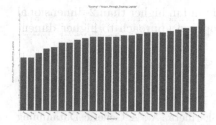

Fig. 5. Country by Access_Through_Desktop_Laptop. Correct classification of Class 1 (Clear meaning and very intuitive).

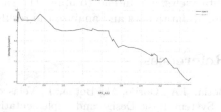

Fig. 6. CPI_all by Unemployment in a dataset about Elantra Sales. Correct classification of Class 2 (Clear meaning dependent of dataset context).

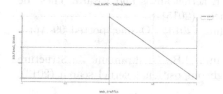

Fig. 7. Website_traffic by SSLfinal_state in a dataset about detection of phishing attack in webpages. Correct classification of Class 8 (Requires additional operations and/or data attributes/dimensions and/or different plot type to depict a clear meaning).

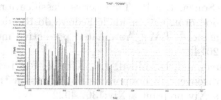

Fig. 8. TAX by TOWN in dataset about Boston price location. Class 5 (Change visualisation) misclassified as Class 2 (Clear meaning dependent of dataset context).

5   Conclusion and Future Work

This work proposes a classification model for data relevance using the combination of *semantic types* and *high-level visual interpretation cues*. After performing a systematic comparative analysis of different classifiers, the proposed model achieves a 0.906 F1 score using a Boosted C5.0 classifier, using distributional vectors of attributes labels and visual interpretation cues. The relevance classification model can be used to classify relevance of new data views. Additionally, the evaluation shows that semantic features and visual interpretation cues have a clear impact on classification performance.

Our approach currently does not cover use cases where missing values are present. The treating of missing values is crucial for real-world applications. This limitation should be addressed in future work.

Also, in practice, relevant patterns can be found in higher than 2-dimensional data views. We intend to apply the same proposed approach to higher dimensional data views and analyze the results.

References

Botia, J.A., Garijo, M., Bot'ia, J., Velasco, J., Skarmeta, A.: A Generic Datamining System. Basic Design and Implementation Guidelines (1998)

Bremm, S., von Landesberger, T., Bernard, J., Schreck, T.: Assisted descriptor selection based on visual comparative data analysis. In: Computer Graphics Forum, vol. 30, pp. 891–900. Wiley Online Library (2011)

Chawla, N.V.: Data mining for imbalanced datasets: an overview. In: Maimon, O., Rokach, L. (eds.) Data Mining and Knowledge Discovery Handbook, pp. 853–867. Springer, New york (2005)

de Souza, D.F.P.: Time-series classification with kernelcanvas and wisard. Thèse de doctorat, Universidade Federal do Rio de Janeiro (2015)

Dinsmore, T.W.: Automated predictive modelling (2014). [Online; posted 09-April-2014]

Duvenaud, D., Lloyd, J.R., Grosse, R., Tenenbaum, J.B., Ghahramani, Z.: Structure discovery in nonparametric regression throughcompositional kernel search (2013). arXiv preprint arXiv:1302.4922

Grosse, R., Salakhutdinov, R.R., Freeman, W.T., Tenenbaum, J.B.: Exploiting compositionality to explore a large space of model structures (2012). arXiv preprint arXiv:1210.4856

Harris, Z.S.: Distributional structure. Word 10(2–3), 146–162 (1954)

Lloyd, J.R., Duvenaud, D., Grosse, R., Tenenbaum, J.B., Ghahra-mani, Z.: Automatic construction and natural-language description of nonparametric regression models (2014). arXiv preprint arXiv:1402.4304

Lubinsky, D., Pregibon, D.: Data analysis as search. J. Econometrics 38(1–2), 247–268 (1988)

Manyika, J., Chui, M., Brown, B., Bughin, J., Dobbs, R., Roxburgh, C., Byers, A.H.: Big data: the next frontier for innovation, competition, and productivity (2011)

Mikolov, T., Chen, K., Corrado, G., Dean, J.: Efficient estimation of word representations in vector space (2013). arXiv preprint arXiv:1301.3781

Spott, M., Nauck, D.: Towards the automation of intelligent data analysis. Appl. Soft Comput. 6(4), 348–356 (2006)

St. Amant, R., Cohen, P.R.: Interaction with a mixed-initiative system for exploratory data analysis. In: Proceedings of the 2nd International Conference on Intelligent User Interfaces, pp. 15–22. ACM (1997)

St. Amant, R., Cohen, P.R.: Intelligent support for exploratory data analysis. J. Comput. Graph. Stat. 7(4), 545–558 (1998)

Záková, M., Křemen, P., Železný, F., Lavrač, N.: Automating knowledge discovery workflow composition through ontology-based planning. Autom. Sci. Eng., IEEE Trans. 8(2), 253–264 (2011)

A First Step Toward Quantifying the Climate's Information Production over the Last 68,000 Years

Joshua Garland[1,3](✉), Tyler R. Jones[2], Elizabeth Bradley[1,3], Ryan G. James[4], and James W.C. White[2]

[1] Department of Computer Science, University of Colorado, Boulder, CO, USA
garland.joshua@gmail.com
[2] University of Colorado, INSTAAR, Boulder, CO, USA
[3] Santa Fe Institute, Santa Fe, NM, USA
[4] Department of Physics, University of California, Davis, CA, USA

Abstract. Paleoclimate records are extremely rich sources of information about the past history of the Earth system. We take an information-theoretic approach to analyzing data from the WAIS Divide ice core, the longest continuous and highest-resolution water isotope record yet recovered from Antarctica. We use weighted permutation entropy to calculate the Shannon entropy rate from these isotope measurements, which are proxies for a number of different climate variables, including the temperature at the time of deposition of the corresponding layer of the core. We find that the rate of information production in these measurements reveals issues with analysis instruments, even when those issues leave no visible traces in the raw data. These entropy calculations also allow us to identify a number of intervals in the data that may be of direct relevance to paleoclimate interpretation, and to form new conjectures about what is happening in those intervals—including periods of abrupt climate change.

1 Introduction

The Earth system contains a vast archive of geochemical information that can be utilized to understand past climate change. Using continually improving analytical techniques, records of change have emerged from corals, marine and lake sediments, tree rings, cave formations, pollen distribution, and the ice sheets. These heterogeneous data sets paint an intricate history of climate change on Earth, often being linked in time by common features, but also containing distinct information about local, regional, and global processes.

To our knowledge, no one has applied information-theoretic techniques to these data—an approach that holds promise for improving climatic interpretations. Knowledge as to where information is created in the climate system, and how it propagates through that system, could reveal and elucidate triggers, amplifiers, sources of persistence, and globalizers of climate change [2,23].

© Springer International Publishing AG 2016
H. Boström et al. (Eds.): IDA 2016, LNCS 9897, pp. 343–355, 2016.
DOI: 10.1007/978-3-319-46349-0_30

For example, ice cores provide high-resolution proxies for hydrologic cycle variability, greenhouse gases, temperature, and dust distribution, among others. The spatiotemporal information captured in these records of climate change could reveal intricacies about the Earth climate system.

This paper is about one piece of that question: what the Shannon entropy rate of the water isotope signals in a specific Antarctic ice core tells us—about that data, about the past conditions at the core site, and about the overall climate. In an ice core, layers capture information about the local conditions at the time of deposition. A depth-wise series of measurements of some chemical or physical property of the ice, then, is effectively a time-series trace of those conditions. Water isotopes are a particularly useful property to study because they are good proxies for temperature and atmospheric circulation that result from variability in the hydrologic cycle. The time scale is unknown, though, and understanding the specific form of the relationship between the measured quantity and different aspects of the climate system requires forensic reasoning. These issues are discussed further in Sect. 2.

The Shannon entropy rate is a potentially useful way to carry out forensic reasoning about the climate system. It measures the average rate at which new information—unrelated to anything in the past—is produced by the system that generated the time series. If that rate is very low, the current observation contains a lot of information about the past and the signal is perfectly predictable. If that rate is very high, all of the information in the observation is completely new: i.e., the past tells you nothing about the future. Calculated over time-series data from ice cores, this quantity—described in Sect. 3—allows one to explore temporal correlations in the climate, which are critically important in understanding the underlying spatiotemporal mechanisms of this complex dynamical system. The results of these calculations, described in Sect. 4, are quite promising; they not only corroborate known facts, but also suggest new and sometimes surprising geoscience, and pave the way towards more-advanced interhemispheric entropy comparisons that could elucidate some of the deeper questions posed above about the larger climate system.

2   Paleoclimate: Dynamics and Data

At long time scales, the climate alternates between glacial and interglacial periods. A few of these cycles are shown in Fig. 1: the warm Holocene in which we live, which began ≈ 12,000 years before present (12 ka), then the last glacial period from 110–12 ka and the Eemian interglacial period from 135–110 ka. (NB: time runs backwards in most paleoclimate data analysis: the plots in this paper start at the current era and move into the past, from left to right.) There is finer-grained structure in the record as well: the Younger Dryas "cold snap" between 12.8–11.5 ka, for instance, which interrupted the slow temperature rise into the Holocene [11]. There are also meaningful differences between records in different parts of the world.

Greenland cores like the NGRIP one in Fig. 1, for example, preserve strong signatures of Dansgaard-Oeschger (DO) events [9], where the temperature rises

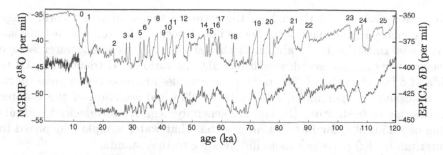

Fig. 1. Climate records from Greenland (NGRIP; top, in blue) and Antarctica (EPICA; bottom, in red). The horizontal axis is time before present in thousands of years. The quantities $\delta^{18}O$ and δD, as explained at more length in the text, are temperature indicators. The numbers identify Dansgaard-Oeschger events. (Color figure online)

9–16 °C over a span of decades or even years, then slowly falls over the course of $\approx$ centuries, and finally decays rapidly back to the baseline. These events, shown with superimposed numbers in Fig. 1, are thought to involve large-scale redistribution of oceanic heat. The trigger is rapid warming in the north Atlantic, perhaps because of injection of fresh water from ice-sheet melting; this disturbs the deep "conveyor belt" currents, causing Antarctica to cool some 200 years later, which is evident in ice cores from that region in the form of an AIM event (Antarctic isotope maxima) [22]. There are many other meaningful features in these data, as well; see [9,16] for good reviews.

Modern ice cores, from which data sets like the one in Fig. 1 are derived, cover timespans of up to 800,000 years. These can reach over 3 Km in length and are typically analyzed on a scale of cm—and, for some properties, mm. Each sample may involve dozens of measurements: different kinds of ions and isotopes, dust levels, conductivity, and so on. Some of the more useful of these are the stable and radiogenic isotopes, the amount and type of dust (which are correlated to the energy and humidity of the atmosphere), and the conductivity. The dynamic ranges of these measurements can be huge: sulfate levels go up by a factor of 1000 when a volcano erupts, for instance. Noise levels vary greatly across the different measurements, but those levels are not well established—and indeed are the subject of some important arguments about how to distinguish signal from noise. And of course the analysis equipment affects the data, sometimes without leaving any visually obvious trace in that data. That issue will return later in this paper.

The study reported here involves data from the 3405 m long West Antarctic Ice Sheet Divide core (WDC), which was gathered and analyzed by a team involving authors Jones and White [21,22]. This core, which covers a period of roughly 68 ka, is the highest-resolution and longest continuously measured record of its kind ever recovered from Antarctica. The high accumulation rate at the WAIS Divide—about 23 cm/yr in recent times—results in annual isotopic signals that persist for the last $\approx$ 16 thousand years, as well as signals at three years and greater that persist throughout the entire 68 thousand year record. These high-frequency signals have never before been interpreted across the last glacial-interglacial transition in Antarctica. In this paper, we focus on the water isotope measurements in this record: specifically δD, the ratio of ^{2}H (deuterium, D) to ^{1}H, and δ^{18}O, the ratio of ^{18}O to ^{16}O. Their values are reported in mille (parts per thousand, or "per mil"), relative to a calibrated standard of the isotopic composition of fresh water [1], and are generally negative for glacier ice. A δD value of -250 mille, for instance, means that that water sample is depleted in deuterium by 250 parts per thousand, relative to that standard.

Both δD and δ^{18}O are good proxies for temperature at the time of deposition of the associated section of the core [10], but the underlying mechanisms are not straightforward. The initial values of these ratios are known from ocean chemistry. The heavier isotopes (deuterium, ^{18}O) precipitate out preferentially, at known rates, as the water is carried to the polar regions in the form of vapor and the air mass cools. Other factors also affect that air mass along the way, however, so δD and δ^{18}O are not simple functions of temperature. And some of those effects—as well as some of the post-depositional processes that affect these two quantities once they are embedded in an ice core—are different for the two isotopes because of their differing molecular weights [8]. One method for understanding these molecular differences, for example, is to study the secondary measure $dxs = \delta$D $- 8 \times \delta^{18}$O, which is considered to be an effective proxy for kinetic effects during evaporation at the moisture source (largely a function of sea surface temperature), or a reflection of changing moisture sources over time.

All of those measurements are on a depth scale; to do any kind of time-series analysis, one must convert them to an age scale. This requires an "age model" for the core: a mapping of depth to age. Constructing this mapping requires a subtle, complicated combination of data analysis and scientific reasoning. Layers can be counted, for instance, but only to a maximum of 40–50 ka because the upper layers compress the ice underneath, thinning the layers to the point that they are unrecognizable. The measurements in the core play a key role in age-model construction: the astronomically based "Milankovitch" theory of ice ages predicts how δ^{18}O should vary through time, for instance. But ocean δ^{18}O also depends on the total volume of land ice on Earth[1], so this quantity is also a useful climate proxy. And near the base of the ice sheet, the ice often melts and/or deforms, making dating—or any kind of data analysis—very difficult. For the WAIS Divide core, the construction of the age model required several person-years of effort. The top 31.2 ka of the core was dated by four different individuals and one HMM-based software tool [24]; this procedure entailed visual identification of annual fluctuations in several different chemical traces along thousands of meters of core, followed by cross-corroboration between different proxies and different daters [20,21]. From 31.2–67.8 ka, the age scale was based on stratigraphic matching to "gold standard" Greenland ice cores and cross referenced using uranium/thorium ratios from cores drilled from cave features [5]. This represents the state of the art for data analysis in this field.

---

[1] Since ice sheets preferentially collect ^{16}O, while oceans preferentially collect ^{18}O.

3   Calculating the Rate of Information Production

The rate at which new information appears in a time series has been shown to be an effective method for signaling regime shifts: e.g., epileptic seizure detection in EEG signals [6], bifurcations in the transient logistic map [6], and recognizing voiced sounds in a noisy speech signal [3]. Estimating that quantity from an arbitrary, real-valued time series can be a real challenge, however. Most approaches to this problem use the Shannon entropy rate [15,19] and thus require categorical data: $x_i \in S$ for some finite or countably infinite *alphabet* S. This is an issue in the analysis of the type of high-resolution data produced by an ice-core lab because symbolization introduces bias and is fragile in the face of noise [4,13].

Permutation entropy (PE) [3] is an elegant solution to this problem. It symbolizes the time series in a manner that follows the intrinsic behavior of the system under examination. This method is quite robust in the face of noise and does not require any knowledge of the underlying mechanisms of the system. Rather than calculating statistics on sequences of values, as is done when computing the Shannon entropy in the standard way, permutation entropy looks at the statistics of the *orderings* of sequences of values using ordinal analysis. Ordinal analysis of a time series is the process of mapping successive elements of a time series to value-ordered permutations of the same size. For example, if $(x_1, x_2, x_3) = (7, 2, 5)$ then its *ordinal pattern*, $\phi(x_1, x_2, x_3)$, is 231 since $x_2 \leq x_3 \leq x_1$. The ordinal pattern of the permutation $(x_1, x_2, x_3) = (7, 5, 2)$ is 321.

Given a time series $\{x_i\}_{i=1,...,N}$, there is a set S_ℓ of all $\ell!$ permutations π of order ℓ. For each $\pi \in S_\ell$, one defines the relative frequency of that permutation occurring in $\{x_i\}_{i=1,...,N}$:

$$p(\pi) = \frac{|\{i|i \leq N - \ell, \phi(x_{i+1}, \ldots, x_{i+\ell}) = \pi\}|}{N - \ell + 1} \tag{1}$$

where $p(\pi)$ quantifies the probability of an ordinal and $|\cdot|$ is set cardinality. The permutation entropy of order $\ell \geq 2$ is:

$$PE(\ell) = -\sum_{\pi \in S_\ell} p(\pi) \log_2 p(\pi) \tag{2}$$

Since $0 \leq PE(\ell) \leq \log_2(\ell!)$ [3], it is common in the literature to normalize permutation entropy as follows: $\frac{PE(\ell)}{\log_2(\ell!)}$. With this convention, "low" PE is close to 0 and "high" PE is close to 1.

PE runs into trouble if the observational noise is larger than the trends in the data, but smaller than its larger-scale features. *Weighted permutation entropy* (WPE) [12] addresses this issue by taking the weight of a permutation into account:

$$w(x_{i+1}^\ell) = \frac{1}{\ell} \sum_{j=i}^{i+\ell} \left(x_j - \bar{x}_{i+1}^\ell\right)^2 \tag{3}$$

where x_{i+1}^ℓ is a sequence of values $x_{i+1}, \ldots, x_{i+\ell}$, and $\bar{x}_{i+1}^\ell$ is the arithmetic mean of those values. The weighted probability of a permutation is defined as:

$$p_w(\pi) = \frac{\displaystyle\sum_{i \leq N-\ell} w(x_{i+1}^\ell) \cdot \delta(\phi(x_{i+1}^\ell), \pi)}{\displaystyle\sum_{i \leq N-\ell} w(x_{i+1}^\ell)} \tag{4}$$

where $\delta(x, y)$ is 1 if $x = y$ and 0 otherwise. Effectively, this weighted probability emphasizes permutations that are involved in "large" features and de-emphasizes permutations that are small in amplitude, relative to the features of the time series. The standard form of weighted permutation entropy is:

$$\mathrm{WPE}(\ell) = -\sum_{\pi \in \mathcal{S}_\ell} p_w(\pi) \log_2 p_w(\pi), \tag{5}$$

which can also be normalized by dividing by $\log(\ell!)$, to make $0 \leq \mathrm{WPE}(\ell) \leq 1$.

In practice, calculating permutation entropy and weighted permutation entropy involves choosing a good value for the word length ℓ. The primary consideration in that choice is that the value be large enough to allow the discovery of forbidden ordinals, yet small enough that reasonable statistics over the ordinals can be gathered. If an average of 100 counts per ordinal is considered to be sufficient, for instance, then $\ell = \mathrm{argmax}_{\hat{\ell}}\{N \gtrsim 100\hat{\ell}!\}$. In the literature, $3 \leq \ell \leq 6$ is a standard choice—generally without any formal justification. In theory, the permutation entropy should reach an asymptote with increasing ℓ, but that can require an arbitrarily long time series. In practice, the right thing to do is to calculate the *persistent* permutation entropy by increasing ℓ until the result converges, but data length issues can intrude before that convergence is reached. We used that approach to choose $\ell = 4$ for the calculations in this paper. This value represents a good balance between accurate ordinal statistics and finite-data effects.

WPE is a powerful technique, but it is not without issues. The choice of the ℓ value is one; another is the notion of significance. As is the case with many nonlinear measures on data, it is quite difficult to define what qualifies as a significant change in a WPE plot. One way to tell if a particular feature (e.g., jump, spike, valley) is important is by first understanding the time scales of the system and comparing them to the size of the window of data over which the WPE calculation is performed. This can help establish whether a change on that time scale makes sense. It is also important to remember that for a particular system, small-scale fluctuations over short time intervals may indeed signal some small event. To distinguish between signal and noise, it is important to vary the window size (as the data allows) and see if the result persists. We take that approach in the calculations reported in the following section.

4   Results

Ice cores are sampled at evenly spaced intervals in depth, but these measurements are spaced nonlinearly (and unevenly) in time because of the progressive downcore thinning of the ice and differing annual accumulation rates of snow. To create evenly spaced time-series data for δD, $\delta^{18}O$, and dxs, we first used the age model described at the end of Sect. 2 to convert depths to ages, and then re-mapped the data to a constant temporal spacing of 1/20th of a year using linear interpolation. The effective resolution of the data is 0.005 m. In the upper portions of the ice core, annual layer thicknesses are about 20 cm, so there are roughly 40 data points per year. At greater depths in the core, an annual layer may only be 4 cm thick, yielding eight data points per year. The accuracy involved in interpolating these unevenly spaced data to a uniform spacing of 1/20th year varies over the depth of the core; this matter, and its potential effects on the results, are discussed further at the end of this section. The specific age scale spacing of 1/20th per year was chosen because it preserves the structure and amplitude of the data—that is, there are no instances of significantly reduced amplitude in the signal, or losses in spectral power.

We then used the normalized version of Eq. (5) with $\ell = 4$ to calculate WPE in 500-year long windows across each of those time-series traces. Each point in the resulting calculation captures the rate of information production over the previous 500 years. The sliding-window nature of this calculation is intended to bring out the fine-grained details of the information mechanics of the system. Since WPE's statistics are built up over the full span of the data that is passed to it, performing that calculation over a longer segment of the climate data—one that spanned different regimes—would intermingle the mechanics of those different regimes. In the case of the questions that we were asking about the WAIS Divide core, the minimum scale of the interesting events was 100–500 years. We ran all of the calculations reported here for that range of window sizes and observed no change in the results. (All of these window sizes, incidentally, satisfied the theoretical data-length requirements for successful WPE calculations.)

Figure 2 shows the δD data from the WDC, along with the WPE of that trace, calculated as described above. The δD WPE is ≈ 0.2 from 10–60 ka, indicating

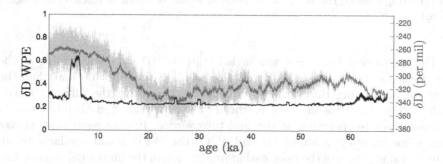

Fig. 2. The deuterium/hydrogen ratio (δD) measured from the WAIS Divide Core. The original data is shown in grey, the smoothed data (500-year moving average) in red, and the weighted permutation entropy (WPE) calculated from the original data in black. (Color figure online)

that δD values depend strongly on their previous values during this period—i.e., that very little new information is produced by the system at each time step. A very interesting feature here is the large jump in WPE between 5–8 ka. As it turns out, an older instrument was used to analyze the ice in this region. The WPE results clearly show that that instrument introduced noise into the data: i.e., every measurement contains completely new information, unrelated to the previous ones. As can be seen from examination of the red and grey traces in the figure, that noise was not visually apparent in the δD data itself, so the instrument issue was not detected immediately by the laboratory team. The fact that WPE brings out the disparity between the two instruments so clearly is a major advantage. (Indeed, that revelation has caused author White's team to re-examine the data in the depth ranges where the blips occur in the WPE results, near 17, 26, and 30 ka.) Another interesting feature of Fig. 2 is the rise in δD WPE from 62–68 ka. This may be due to geothermal heat at the base of the ice sheet, which causes water isotopes to diffuse in that region, thereby injecting new information into the oldest section of the time series. This matter is discussed at more length below.

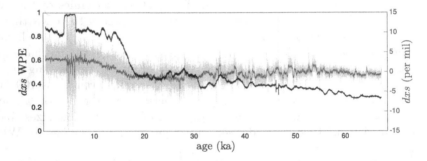

Fig. 3. The secondary measure dxs calculated from the δD and δ^{18}O measurements in the WAIS Divide Core. Original and smoothed values of dxs are shown in green and grey respectively; the WPE of the original signal is shown in black. (Color figure online)

The instrument issue that was invisible in the δD data **was** visible in the dxs trace, as shown in Fig. 3. Recall that $dxs = \delta$D $- 8\delta^{18}$O, and that the intent of performing this weighted sum is to deconvolve the differential scientific effects that are at work in these two quantities and thereby zero in on sea-surface temperature at the moisture source. It may seem that as dxs is an affine transformation of δD and δ^{18}O, the WPE would similarly be an affine transformation of the information production of the individual signals, i.e., $\mathrm{WPE}(\ell; dxs) = a\mathrm{WPE}(\ell; \delta$D$) + b\mathrm{WPE}(\ell; \delta^{18}O)$. However, *weighted* permutation entropy is not preserved by affine transformations—and this is useful. Among other things, it allows us to leverage the information production of dxs to gain insight into the second-order dynamics that the calculations of dxs is intended to get at: specifically, the interplay between kinetic fractionation effects and moisture source conditions.

Careful comparison of the three WPE traces in Fig. 4 illustrates the utility of that reasoning. Recall that both δD and δ^{18}O of a section of the core are, to first approximation, proxies for temperature at the time of the deposition of that material, but that thermodynamics and the difference in atomic weight causes them to behave slightly differently. (The dxs calculation, again, is intended to get at these second-order effects.) As expected, the δD and δ^{18}O WPE traces are largely similar, except for a small but fairly consistent vertical offset from 0–17 ka. This offset indicates that the information production of δ^{18}O increased towards the end of the last glacial period, suggesting that, from an information-theoretic perspective, the second-order thermodynamic effects have been playing a greater role in the climate dynamics since 17,000 years ago. On the right-hand side of the figure, however, the similarity fails. δ^{18}O WPE rises slowly from 50–62 ka, at which point δD WPE rises somewhat as well. This could be a thermal diffusion effect due to geothermal heat at the bedrock interface, which will inject noise into the data—and differentially affect the isotopes due to their different molecular masses. It could also be a scaling issue; signal at those depths is attenuated by the effects of time and depth, and also smoothed by the finite resolution of the analysis equipment.

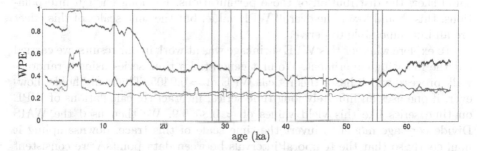

Fig. 4. WPE of δD (red), δ^{18}O (blue), and dxs (green); Dansgaard-Oeschger events are marked with black vertical lines. (Color figure online)

The Dansgaard-Oeschger events described in Sect. 2 are climatologically important and scientifically interesting, but their mechanics and dynamics are not completely understood. During the early stages of the collaboration that produced this paper, the geoscientists on the team conjectured that these events would inject new information into the time series. As is clear from Fig. 4, however, that is not the case. That is, while DO events may reflect changes in the dynamics of the climate (cf., recent work on "critical slowing down" [7,14,17]), they are not associated with changes in the *information production* of that system. Rather, they appear to be just part of the normal operating procedure of the climate system. We are currently looking at shorter windows to see whether WPE reveals "triggers" or other early-warning signals for these important events.

The results shown in Fig. 4 catalyzed a number of other new hypotheses about climate science. The WAIS Divide ice core derives moisture mainly from a Pacific Ocean source. During the beginning of the deglaciation at $\approx$ 19 ka, the dxs WPE and $\delta^{18}O$ WPE begin to increase, in line with accumulation[2]. At that time, changing climatic conditions in West Antarctica may have produced storm tracks that delivered precipitation from more-diverse locations. For example, increased sea ice extent during the glacial period may have limited storm tracks to those originating primarily in the central Pacific, but upon sea ice decline during deglaciation, more local storm tracks, and possibly storm tracks from the west Pacific and even the Indian Ocean, could have contributed precipitation to the ice core site. Simply put, we suggest that more accumulation means more storms originating from more locations. Of course, from a fractionation standpoint, the increasingly variable location of moisture formation would have to introduce differing kinetic effects that injected more information into the system; we would not expect this as the physics of evaporation should be the same.

It is worth thinking about whether the preprocessing step outlined in the first paragraph of this section—which is the standard approach in this field if one wants an ice-core data set with even temporal sampling—could have disturbed the information mechanics of the data. The ramps introduced by linear interpolation introduce repeating, predictable patterns in the π of Sect. 3, which could skew the distribution of those permutations. For long enough interpolations, this should lower the overall WPE value, but the time scales of this effect are all but impossible to derive.

To explore whether this WPE shrinkage was at work in our results, we carried out the following experiment. We first generated a time series using a random-walk process, which has a theoretical WPE of $\approx$ 0.9405. That only holds, however, if one uses an infinitely long time series; in practice, calculations of WPE on time series like this yield values of $\approx$ 0.85–0.9. We then used the WAIS Divide core age model to invert the time scale of this trace, downsampling it nonlinearly so that the temporal intervals between data points were consistent with an 0.005 m spacing. This is an effective ansatz for a data set from that core: closely spaced points near the beginning of the trace, where the core is less dense and a year's worth of material is thicker, and spreading apart roughly exponentially later in the time series, which corresponds to the highly compressed ice deep in the core. We then subjected that trace to the same preprocessing steps outlined at the beginning of this section and finally computed its WPE. The results showed a correct baseline of 0.82–0.87 early in the time series, where the interpolation interval is small, followed by a slow decrease starting around 12 ka.

That decrease suggests that one should be careful comparing WPE values of a single trace across wide temporal ranges—especially when one is working deep

---

[2] The accumulation data from the WDC has not yet been released publicly, so we cannot include a plot of it here, but there are some *extremely* interesting correspondences that we hope to be able to include in a few months, when we are allowed to share those data.

in the core. However, the claims offered in this paper concern (a) features high up in the core (viz., the instrument issue) (b) narrow features deeper down, and (c) broad time-scale comparisons of traces produced with identical interpolation processes (Fig. 4). Moreover, we are not completely convinced of the accuracy of our synthetic experiment; note, for example, that the δ^{18}O and δD WPE traces in Fig. 4 actually *rise* at lower depths—the region where interpolation effects should, theoretically, cause a dropoff. It may be the case that the geothermal effects inject much more information into the system than the results presented here suggest, or it may be that we do not completely understand the interpolation effects. We are currently exploring this from several angles: by simply downsampling the WDC data, rather than interpolating it, and by comparing WPE of different WDC traces that were sampled at different intervals. There are other issues as well. Gases diffuse through the material in the core: more readily at the top, where the material is less dense, and more slowly lower down, where the snow has been compressed into solid ice. Since diffusion effectively introduces white noise, it should raise WPE. However, this effect would be quite difficult to deconvolve from the data in an ice core, since a given segment of that core has undergone a continuum of diffusion processes at different temporal and spatial scales during its history.

5   Conclusion and Future Work

Paleoclimatic analyses require complicated forensic reasoning to determine the timing and phasing of past events. In ice-core science, a particularly meaningful paleoclimatic indicator is found in the measurements of water isotopes; these are useful for understanding temperature, atmospheric circulation, and oceanic conditions. Interpreting records of the history of these isotopes requires knowledge of the thermodynamics of the climate system, along with a precise age scale. The former is governed by physics; the latter requires intelligent data analysis performed by multiple individuals and software tools.

In this paper, we used data from the WAIS Divide ice core to analyze information production over the last 68,000 years. Through permutation entropy techniques (WPE), we found that information production in δD, δ^{18}O, and dxs is consistent with thermodynamic expectations. The WPE of δ^{18}O and dxs share common features, likely because these parameters are more sensitive to kinetic fractionation effects in ocean moisture source regions. Conversely, the δD is likely more responsive to temperature-related equilibrium effects. We identify a number of intervals in the data that may be of direct relevance to paleoclimate interpretation. In the deepest sections of the core (> 60 ka), divergence of WPE for δD and δ^{18}O may help identify time periods when geothermal heat flux causes differential solid diffusion of water molecules. Throughout the last glacial period (> 12 ka), rather constant WPE values suggest no information production during large-scale abrupt warming events in Greenland that also appear as isotope maxima in Antarctica: this is an unexpected result. Across the glacial-interglacial transition ($\approx 19 - 11$ ka) and into the Holocene (< 11 ka), increases in WPE for all variables may signal more variability in the hydrologic cycle—for example, receding sea ice may establish more-diverse moisture sources across latitudes and ocean basins. In the last 2000 years, an increase in WPE (the small rise that is visible on the left-hand edge of Fig. 4, in all three traces) may signal anthropogenic effects on the climate system. These comparisons deserve more rigorous treatment, and will be included in future work.

The overall results of this study constitute only one kind of information calculation on a few isotopes in a single ice core; there are additional options that could broaden the scope in useful and meaningful ways. For example, comparison of WPE from the WAIS Divide with that of other Antarctic ice cores may elucidate the thermodynamics of varying moisture sources in the Indian, Pacific, Atlantic, and Southern Oceans. It may even provide information about atmospheric processes that affect the isotopic signal. Comparison of WPE of varying WAIS Divide proxies, such as water isotopes and accumulation, may also hold important clues to information production in the climate system. Comparison of Antarctic and Greenland cores could provide interhemispheric viewpoints that can inform us about abrupt climate change events (e.g., trigger mechanisms). Other information-theoretic measures, such as mutual information or transfer entropy [18], may be even more useful than WPE in these kinds of comparisons. None of these sorts of calculations have previously been done on paleoclimatic data sets—let alone multiple ones—and the initial findings presented here hold promise for improving climatic interpretations.

References

1. http://tinyurl.com/hz8xelf
2. Alley, R., et al.: Abrupt climate change. Science **299**, 2005–2010 (2003)
3. Bandt, C., Pompe, B.: Permutation entropy: a natural complexity measure for time series. Phys. Rev. Lett. **88**(17), 174102 (2002)
4. Bollt, E., Stanford, T., Lai, Y.C., Życzkowski, K.: What symbolic dynamics do we get with a misplaced partition?: on the validity of threshold crossings analysis of chaotic time-series. Physica D **154**(3), 259–286 (2001)
5. Buizert, C.: The WAIS divide deep ice core WD2014 chronology–Part 1: methane synchronization (6831 ka BP) and the gas age-ice age difference. Clim. Past **11**, 153–173 (2015)
6. Cao, Y., Tung, W., Gao, J., Protopopescu, V., Hively, L.: Detecting dynamical changes in time series using the permutation entropy. Phys. Rev. E **70**(4), 046217 (2004)
7. Dakos, V., et al.: Slowing down as an early warning signal for abrupt climate change. Proc. Nat. Acad. Sci. **105**(38), 14308–14312 (2008)
8. Dansgaard, W.: Stable isotopes in precipitation. Tellus **16**, 436–447 (1964)
9. Dansgaard, W., et al.: Evidence for general instability of past climate from a 250-kyr ice-core record. Nature **364**, 218–220 (1993)
10. Dansgaard, W., Johnsen, S., Clausen, H., Gundestrup, N.: Stable isotope glaciology. Medelelsser om Gronland **197**, 1–53 (1973)
11. Dansgaard, W., White, J., Johnsen, S.: The abrupt termination of the younger Dryas climate event. Nature **339**, 532–534 (1989)

12. Fadlallah, B., Chen, B., Keil, A., Príncipe, J.: Weighted-permutation entropy: a complexity measure for time series incorporating amplitude information. Phys. Rev. E **87**(2), 022911 (2013)
13. Kraskov, A., Stögbauer, H., Grassberger, P.: Estimating mutual information. Phys. Rev. E **69**(6), 066138 (2004)
14. Lenton, T., et al.: Early warning of climate tipping points from critical slowing down: comparing methods to improve robustness. Philos. Trans. R. Soc. Lond. A: Math. Phy. Eng. Sci. **370**(1962), 1185–1204 (2012)
15. Mantegna, R., Buldyrev, S., Goldberger, A., Havlin, S., Peng, C., Simons, M., Stanley, H.: Linguistic features of noncoding DNA sequences. Phys. Rev. Lett. **73**(23), 3169–3172 (1994)
16. Petit, J., et al.: Climate and atmospheric history of the past 420,000 years from the Vostok ice core, Antarctica. Nature **399**, 429–436 (1999)
17. Scheffer, M., et al.: Early-warning signals for critical transitions. Nature **461**, 53–59 (2009)
18. Schreiber, T.: Measuring information transfer. Phys. Rev. Lett. **85**(2), 461–464 (2000)
19. Shannon, C.: Prediction and entropy of printed English. Bell Syst. Tech. J. **30**(1), 50–64 (1951)
20. Sigl, M., et al.: The WAIS Divide deep ice core WD2014 chronology–part 2: Annual-layer counting (031 ka BP). Clim. Past **12**, 769–786 (2016)
21. WAIS Divide Project Members: Onset of deglacial warming in west antarctica driven by local orbital forcing. Nature **500**, 440–444 (2013)
22. WAIS Divide Project Members: Precise interpolar phasing of abrupt climate change during the last ice age. Nature **520**, 661–665 (2015)
23. White, J., et al.: Abrupt impacts of climate change: Anticipating surprises. In: EGU General Assembly Conference Abstracts, vol. 16, p. 17028 (2014)
24. Winstrup, M., et al.: An automated approach for annual layer counting in ice cores. Clim. Past **8**, 1881–1805 (2012)

HAUCA Curves for the Evaluation of Biomarker Pilot Studies with Small Sample Sizes and Large Numbers of Features

Frank Klawonn[1,2(✉)], Junxi Wang[1], Ina Koch[3], Jörg Eberhard[4], and Mohamed Omar[5]

[1] Biostatistics, Helmholtz Centre for Infection Research,
Inhoffenstr. 7, 38124 Braunschweig, Germany
frank.klawonn@helmholtz-hzi.de
[2] Department of Computer Science, Ostfalia University of Applied Sciences,
Salzdahlumer Str. 46/48, 38302 Wolfenbüettel, Germany
[3] Institute for Molecular Bioinformatics, Johann Wolfgang Goethe-University,
Robert-Mayer-Str. 11-15, 60325 Frankfurt, Germany
[4] Department of Prosthetic Dentistry and Biomedical Materials Science,
Hannover Medical School, Carl-Neuberg-Str. 1, Hannover, Germany
[5] Trauma Department, Hannover Medical School,
Carl-Neuberg-Str. 1, Hannover, Germany

Abstract. Biomarker studies often try to identify a combination of measured attributes to support the diagnosis of a specific disease. Measured values are commonly gained from high-throughput technologies like next generation sequencing leading to an abundance of biomarker candidates compared to the often very small sample size. Here we use an example with more than 50,000 biomarker candidates that we want to evaluate based on a sample of only 24 patients. This seems to be an impossible task and finding purely random-based correlations is guaranteed. Although we cannot identify specific biomarkers in such small pilot studies with purely statistical methods, one can still derive whether there are more biomarkers showing a high correlation with the disease under consideration than one would expect in a setting where correlations are purely random. We propose a method based on area under the ROC curve (AUC) values that indicates how much correlations of the biomarkers with the disease of interest exceed pure random effects. We also provide estimations of sample sizes for follow-up studies to actually identify concrete biomarkers and build classifiers for the disease. We also describe how our method can be extended to other performance measures than AUC.

1 Introduction

A biomarker is a measurable value that is an indicator for a biological state. In recent years, the search for biomarkers for diseases has gained high interest in medicine. A well-known biomarker is the so-called prostate-specific antigen (PSA) which was or is sometimes still used as a biomarker for prostate cancer

H. Boström et al. (Eds.): IDA 2016, LNCS 9897, pp. 356–367, 2016.
DOI: 10.1007/978-3-319-46349-0_31

although its reliability and usefulness is sometimes doubted [1]. Along with the advancement of high-throughput technologies like microarrays, next generation sequencing and mass spectrometry, that allow to measure the whole or large parts of the genome, transcriptome, proteome or metabolome, came a strong hope to find a single biomarker for each disease or state of a disease to be diagnosed with very high certainty. However, this dream did not come true and it seems to be unrealistic from today's point of view. Biological systems are probably too complex for simple single-cause single-effect associations. Nevertheless, there are biomarker candidates that show a high correlation with specific diseases but are not reliable enough to function as predictors for the presence of a specific disease alone. Therefore, instead of relying on a single biomarker, the idea is to combine biomarkers that are not good enough for the diagnosis of a specific alone but can jointly provide a diagnosis with high certainty. An example of such a combination is the Enhanced Liver Fibrosis (ELF) score [2] that uses a linear combination of (log-)values of three single biomarkers to predict fibrosis stages in chronic liver disease patients. The EFL score was derived from a quite limited number of standard blood values of altogether 479 patients. No high-throughput technology was involved.

The use of high-throughput technologies for finding reliable combinations poses new challenges. First of all, in contrast to standard blood values, patient data based on high-throughput technologies are not commonly available in hospitals. This means, they have to be generated separately. Secondly, although the prices for generating data from high-throughput technologies are constantly decreasing, it is still quite expensive and also time-consuming to carry out these experiments. This implies high costs for such data. The advantage and the curse at the same time is that such experiments easily yield thousands or even far beyond 10,000 possible candidates for biomarkers. The sample size in expensive pilot studies is usually very limited, sometimes less than 20. From a machine learning or classification point of view one then tries to derive a classifier from a data set with more than 1000 attributes (biomarker candidates) and perhaps only around 20 or 30 instances. Finding random associations and overfitting is therefore hard to avoid.

Pilot studies with a small sample size and a large number of biomarker candidates are – as the name already points out – not intended to finally mark down a biomarker combination for clinical use but to check whether there are potential biomarkers that make a more expensive follow-up study with a larger sample size worthwhile. In this paper, we propose methods to assess biomarker pilot studies with respect to their potential to yield promising results when they are extended by follow-up studies. Section 2 gives a formal definition of our problem and provides an illustrating example. Section 3 describes how a pilot study can be analysed with respect to the potential to find reliable biomarker combinations in a follow-up study. A rough estimation of the required sample size is provided in Sect. 4. The ideas of Sect. 3 based on simple area under the ROC curve (AUC) values are extended to other measures for classifiers in Sect. 5.

The final conclusions address the problem of possible high correlations between biomarker candidates.

2   Problem Formalisation and an Example

From a formal point of view, we face the following problem. We have n instances – usually patients – from which we have measured m attributes (biomarker candidates). The patients are assigned to c different classes, i.e. different diagnoses or different states of a certain disease. The number of classes c is usually small, in many cases even $c = 2$ where we only want to distinguish patients suffering from a certain disease from patients who do not have this disease. Typically, we have $m \ggg n$. Our ultimate goal is to find a classifier that can predict the class (diagnosis) based on the values of the m attributes. For reasons of simplicity, we assume we do not have to deal with missing values[1]. Due to the fact that we have to face $m \ggg n$, we cannot directly build a standard classifier based on the given data set. A feature selection technique is required to reduce the number of attributes drastically. As a possible way to evaluate the predictive power of a classifier, we could apply cross-validation and because of the small sample size we would prefer to use leave-one-out cross-validation (the jackknife method). It must be emphasised that when we want to evaluate a classifier, we must not separate feature selection from the classifier. It was already noted in [3,4] that first applying feature selection on the whole data set and then evaluate a classifier using only the selected features based on cross-validation can lead to a strong model selection bias, since the actual model consists of the classifier *and* the (pre-)selected features. To illustrate this problem, we have carried out the following simulation. We have generated $m = 1000$ random attributes following a standard normal distribution for different sample sizes n. The we have randomly assigned the n instances to two classes, $n/2$ instances to each class, i.e. if we consider the two classes as healthy vs. sick, we have a prevalence of 50 %. This means that correlations between the 1000 attributes and the classes are purely random. We have then carried out the following two experimental settings.

(a) We have first selected the best 20 attributes from the whole data set and then trained classifiers and evaluated them based on the leave-one-out method (LOO). This is how it should not be done!
(b) Within the leave-one-out method, i.e. when the test sample had already been removed from the training data set, we have selected the best 20 attributes and trained the classifiers without the sample that was left out for testing.

The selection of the "best" attributes was based on a very simple strategy. We chose 20 attributes with the highest area under the ROC curve (AUC) values. As classifiers we used support vector machines (SVM), random forests (RF)

---

[1] This is a more or less realistic assumption for microarray and next generation sequencing data but not for data from mass spectrometry.

Table 1. Percentage of correctly classified instances in a completely random data set with 1000 features when feature selection is applied to the whole data set before (before LOO) and within leave-one-out cross-validation (during LOO). The sample size with two classes is given by n. Prevalence is 50 %. Classifiers are support vector machines (SVM), random forests (RF) and linear discriminant analysis (LDA).

n	before LOO			during LOO		
	SVM	LDA	RF	SVM	LDA	RF
20	100	80	95	65	65	55
30	97	70	97	43	43	67
40	95	83	85	53	55	45
50	84	80	82	26	20	22
100	82	80	76	47	46	48
150	79	79	79	60	61	60
200	72	70	70	44	48	44

and linear discriminant analysis (LDA). Table 1 shows the percentage of correctly classified instances for the leave-one-out evaluation. Since the data set is completely random with a prevalence of 50 %, we would expect to classify about half of the instances correctly. One can see easily see that this is not true for the (inappropriate) method explained in (a) where the feature selection is carried out on the whole data set before leave-one-out cross-validation. Even for a sample size of 200, around 70 % of the instances are still correctly classified, a value that is never achieved for any sample size n and any classifier with the correct method (b). It is noteworthy that for $n = 50$ the correct method by chance performs even far worse than random guessing.

As an example for illustration purposes of our approach we use a data set from $n = 24$ patients[2] who had undergone a surgery for a hip prosthesis which later on caused problems. The final goal is to classify whether the problems are caused by a low-grade periprosthetic hip infection or by aseptic hip prosthesis failure, i.e. to see whether the problems come from an infection or not. A microarray kit was used to obtain $m = 50,416$ biomarker candidates based on measured genes and RNA values [5]. When we apply the above mentioned method (b) to this data set, we obtain rates of correctly classified instances of around 50 % which corresponds to random guessing. Even changing the number of selected biomarker candidates – for instance choosing only the top 10 or 4 instead of 20 – in the leave-one-out cross-validation loop does not lead to an improvement. Should we draw the conclusion that this pilot study has failed and it is not worthwhile to consider a follow-up study? An answer to this question will be provided in the following section.

---

[2] The data set is currently submitted to a medical journal.

Table 2. Top 10 AUC values and their p-values for the hip prosthesis infection data set.

Biomarker	AUC	p-value (raw)	p-value (corrected)
1	0.951388889	0.0000333	1
2	0.944444444	0.0000496	1
3	0.944444444	0.0000496	1
4	0.930555556	0.0001028	1
5	0.930555556	0.0001028	1
6	0.930555556	0.0001028	1
7	0.930555556	0.0001028	1
8	0.930555556	0.0001028	1
9	0.923611111	0.0001442	1
10	0.923611111	0.0001442	1

3   HAUCA Curves

One might argue that our feature selection method using only the AUC values is too simple. Indeed, there are more sophisticated techniques, especially those that do not simply rank the single features but look directly for combinations of features. However, it is extremely difficult to choose among feature subsets from more than 50,000 features. We do not want to dive into advanced feature selection methods here. In any case, it would be unrealistic to build a classifier based on a data set with 24 instances (patients).

Although the concept of AUC is sometimes criticised [6], especially because it does not take the prevalence into account, and it is restricted to two-class problems, it is still a meaningful approach [7] and we will take a closer look at the AUC values in our data set. The second column of Table 2 shows the 10 highest AUC values for our example data set – computed on the whole data set.

AUC values over 0.9 are definitely interesting although such a value might not be sufficient for a medical test. Nevertheless, a combination of biomarkers with such high AUC values might lead to a classifier with sufficient predictive power. However, we have seen in the previous section that this does not really apply in the case of our data set. The classifiers we had mentioned using the biomarker candidates with the highest AUC values – based on the feature selection and cross-validation strategy (b) – were not better than random guessing. So can we conclude that the high AUC values occur just by chance?

Fortunately, there is a method to compute the probability that an attribute with n values randomly assigned to two classes with a given prevalence exceeds a given AUC value [8]. The computation is mainly based on the same statistic that is used for the Wilcoxon-Mann-Whitney-U test. We just need to be able to compute the quantile of the statistic of the Mann-Whitney-U test. For a sample of size n with an absolute prevalence of n_+ of the disease in the sample, the probability that a single biomarker candidate with randomly assigned values exceeds an AUC value of a is

$$F_U((n - n_+) \cdot n_+ \cdot (1 - a), n - n_+, n_+) \tag{1}$$

where F_U is the cumulative distribution function of the Wilcoxon-Mann-Whitney-U statistic [8]. Actually this probability should be multiplied by two (and cut off at 1 in case it exceeds 1) because an AUC value close to 0 is also of high interest because it indicates a high, but negative correlation between the value of the biomarker candidate and the disease.

The corresponding probabilities for the top 10 AUC values in hip prosthesis infection data set are given in the third column "p-value (raw)" of Table 2. These probabilities can be interpreted as p-values for the hypothesis test with the null hypothesis that the classes are randomly assigned to the instances or that all biomarker candidates have random values that are not correlated with infection. Although the raw p-values in the third column of Table 2 seem to be small enough to reject the null hypothesis, we must take into account that we have applied multiple testing, i.e. we have applied the test to all biomarker candidates, so that the test was repeated $m = 50,416$. Therefore, a correction for multiple testing is needed. No matter which correction method we choose – here we have applied Bonferroni-Holm correction [9] – all p-values are changed to 1, losing their significance as can be seen from the last column of Table 2. So does this support what we have already observed when we constructed the classifiers, i.e. that the data set does not seem to indicate any non-random correlation between the biomarker candidates and infection?

The answer is no. There is another way of looking at the AUC values. Because of the large number of biomarker candidates, we would expect high AUC values just by chance. But how many high AUC values could we expect if the data were completely random? According to Eq. (1) we can compute the probability that an attribute with n values randomly assigned to two classes with a given prevalence exceeds a given AUC value, we can also calculate the number of expected biomarker candidates exceeding a given AUC value in a random data set. It is simply the probability for the AUC times the number of biomarker candidates m, here $m = 50,416$.

The bottom black curve in Fig. 1 shows on the y-axis how many biomarker candidates one could expect to exceed a given AUC value marked on the x-axis in a random data set that contains the same number of biomarker candidates as our hip prosthesis infection data set. The top blue curve in the figure shows how many biomarker candidates exceed the corresponding AUC threshold in our real data set on infection after hip prosthesis surgery. One can clearly see that there are many more biomarker candidates in the AUC range from 0.85 to over 0.9 than one would expect in a random data set.

Of course, the expected number of high AUC values in a random data set is just a bottom line for comparison with the AUC values found in the real data set. We also provide a 95 % upper confidence band for the number of high AUC values in a random data set. This confidence band is indicated by the middle red

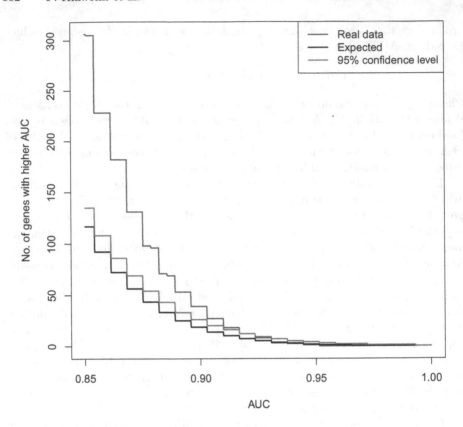

Fig. 1. HAUCA curves for the hip prosthesis infection data set showing on the y-axis how many biomarker candidates exceed a given AUC threshold indicated on the x-axis. Top blue curve: real data. Bottom black curve: Expected number in a random data set. Middle red curve: 95 % upper confidence band for a random data set. (Color figure online)

line in Fig. 1. Again, one can observe that the number of biomarker candidates in the range between 0.85 and 0.9 exceeds even this 95 % upper confidence band.

The computation of the confidence band is based on the following considerations. For any given AUC value a we know the probability p_a that a biomarker candidate with random values would obtain an AUC value larger than a. If we have m independent random biomarker candidates the probability that exactly k random biomarker candidates have a higher AUC value than a follows a binomial distribution $B(m, p_a)$, so that we simply have to compute the 95 % quantile of this binomial distribution to obtain the value of the 95 % upper confidence band at the AUC value a. Of course, one could choose other values than 95 % for the confidence band and just replace the values in the graph by the corresponding quantile of the binomial distribution.

We call the curves shown in Fig. 1 high AUC abundance (HAUCA) curves. The HAUCA curves in Fig. 1 clearly indicate that there is more than just a random correlation between the biomarker candidates and infection with the hip prosthesis infection data set. Taking a closer look at the AUC value of 0.85 in the HAUCA curves, we can see that the real data set contains over 300 biomarker candidates with an AUC value higher than 0.85, whereas one would expect in a random data set clearly less than 150. Even the 95 % upper confidence bound at an AUC value of 0.85 does not reach the value 150. This means that the high correlation with infection of about half of the biomarker candidates in the real data set with AUC value greater than 0.85 cannot be explained by pure random effects. We cannot identify which biomarker candidates are the right ones. But there should be some valid biomarkers that once – once they are identified in a follow-up study – can be used to build a classifier.

So from this pilot study on infection after hip prosthesis surgery we cannot confirm any specific biomarker candidates. But we can nevertheless say that there must be very good candidates and it is worthwhile to extend the pilot study to a larger sample. Of course, one could also look at the functional annotations of the genes (biomarker candidates) with high AUC values and see which ones are associated with infection processes to make a pre-selection of promising biomarker candidates for an extended study. But this is out of the scope of our purely statistics oriented discussion here.

Figure 2 shows another example of HAUCA curves for data from a biomarker study published in [10][3]. The data set contains information about the microbiome in the mouth of $n = 19$ patients of which 9 suffered from periodontitis. The microbiome was characterised by the abundance of $m = 242$ operational taxonomic units (OTUs). Here again the HAUCA curves clearly indicate that there is more than just random correlation between OTUs and periodontitis. In this case, the classifiers based on leave-one-out cross-validation and feature selection within the cross-validation loop could even provide for 14 out of 19 patients the correct diagnosis which is far away from being of clinical use but also indicates a correlation between OTUs and periodontitis.

4   Sample Size Estimation for Follow-Up Studies

In the previous section, we have seen that based on very small pilot studies we can at least find good indicators for a connection between the set of biomarker candidates and the disease under consideration. But our approach is neither capable to identify specific biomarkers nor to construct a reliable classifier that supports the diagnosis which is no surprise given the small sample size compared to the number of biomarker candidates. The question that arises here is how to determine the sample size of a follow-up study that should either confirm the validity of biomarkers with high AUC values or even construct a classifier based on a combination of good biomarker candidates.

---

[3] The HAUCA curves were neither available nor discussed in the paper [10].

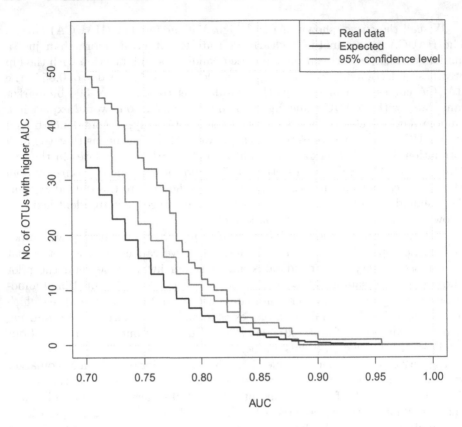

Fig. 2. HAUCA curves for the OTU data set.

It is quite simple to specify the sample size for confirming the validity of a high AUC value for a biomarker. For instance, the top biomarker candidate in Table 2 has an AUC value of 0.95 whose low raw p-value turns into a non-significant p-value after correction for multiple testing. Assuming that the AUC value of this biomarker would remain at 0.95 in a follow-up study with a larger sample size, we can compute the resulting p-value in this study after correction for multiple testing using Eq. (1) (multiplied by two to account for biomarkers that show a negative correlation with infection). We simply need to increase n and n_+ proportionally – assuming the prevalence in the follow-up study remains the same – until the resulting probability is small enough, so that it remains significant after correction for multiple testing, in our example after multiplication with $m = 50,416$. Already at $n = 30$, the p-value after correction for multiple testing drops below 5%. In the same way, the two last candidates in Table 2 with an AUC value of 0.92 would need a sample size of at least $n = 36$ to be confirmed given the AUC and the prevalence remain the same for the larger sample size in the follow-up study.

Apart from confirming high AUC values of single biomarker candidates we are also interested in how many of such potentially valid biomarkers we need to combine for a classifier to obtain sufficiently reliable predictions. This, of course, highly depends on the correlation between the good biomarker candidates. In the worst case, their correlation equals 1 and their combination does not lead to any improvement compared to the single biomarkers. In order to get a rough idea of how well a classifier based on a combination of good biomarkers could perform, one could exploit the ideas from [11] although the underlying assumptions are quite restrictive. There it is assumed that the values for the biomarkers follow normal distributions and the normal distributions for the two classes differ. The paper [11] provides a method how to compute the AUC value of linear discriminant analysis based on the AUC values of the single biomarkers and their correlations within the two classes. Of course, the estimation of the correlation based on the pilot study is not very reliable. But the proposed procedure of estimating the AUC values of the biomarker combination still provides a rough judgment how well the biomarker combination could perform for later prediction.

5   Alternatives to AUC

As mentioned already in the beginning of Sect. 3, AUC values are neither the only nor the best performance measure of scores used for classification. In principle, one could replace AUC by other performance measures, for instance entropy, accuracy or the area above cost curves [12]. These performance measures also have the advantage that they are not restricted to binary classification problems but are also applicable in the context of multiclass classification problems. Of course, for multiclass classification problems one could also use extensions of AUC to more than two classes as described in [13–17]. However, for all these measures it is no longer obvious how the corresponding p-values can be computed that are needed for the equivalent to the HAUCA curves.

A possible solution is an estimation of these p-values based on Monte-Carlo or permutation tests. For a Monte-Carlo test, one would generate a large number of biomarkers with random values and compute the values of the corresponding performance measure. Then the p-value of a biomarker candidate in the real data set is the proportion of random biomarkers with a better value for the performance measure than the considered biomarker in the real data set. For a permutation test, one would randomly permute the classes while fixing the values of the biomarker candidates to compute values of the performance measures for random biomarkers. The p-value of a biomarker candidate in the real data set is then computed in the same way as for the Monte-Carlo test. Estimating the p-values based on Monte-Carlo or permutation tests requires a large number of simulations implying high computational costs. For instance, in the example of the hip prosthesis infection data set the best AUC value has a p-value of approximately $3 \cdot 10^{-5}$. If we had not been able to compute this p-value based on Eq. (1) and had to rely on Monte-Carlo or permutation tests, we would need at least 10^6, better even more than 10^7 simulations to get a rough estimation of this small probability. And the above mentioned performance measures already need a little bit of computation time for a single biomarker.

6 Conclusions

In this paper, we have presented an approach how to judge biomarker pilot studies with small sample sizes and large numbers of possible biomarker candidates. For binary classification problems we can use the AUC as a measure of performance for the single biomarkers, leading to closed form solutions of the required calculations and therefore to fast computation. Other performance measures could also be applied for the price of high computational costs due to the need of simulations instead of closed form solutions. Efficient algorithms or new solutions will be a topic of further research.

Another question concerns the correlation between the biomarker candidates. The computation of the p-values in the context of AUC values and for the Monte Carlo test assumes independent biomarker candidates. This is definitely an unrealistic assumption because at least subsets of the biomarker candidates – no matter whether they are associated with the disease or not – will show high correlations because these measured values interact in a highly complicated biological system and cannot function independently. In a certain way, this would be taken into account by a permutation test because the correlation among the biomarker candidates is not changed, only the distribution of the classes is rearranged. This aspect will need further investigations.

References

1. De Angelis, G., Rittenhouse, H., Mikolajczyk, S., Blair, S., Semjonow, A.: Twenty years of PSA: from prostate antigen to tumor marker. Rev. Urol. **9**(3), 113–123 (2007)
2. Lichtinghagen, R., Pietsch, D., Bantel, H., Manns, M., Brand, K., Bahr, M.: The enhanced liver fibrosis (ELF) score: normal values, influence factors and proposed cut-off values. J. Hepatol. **59**(2), 236–242 (2013)
3. Ambroise, C., McLachlan, G.J.: Selection bias in gene extraction on the basis of microarray gene-expression data. Proc. Natl. Acad. Sci. **99**(10), 6562–6566 (2002)
4. Varma, S., Simon, R.: Bias in error estimation when using cross-validation for model selection. BMC Bioinform. **7**(91), 1 (2006). doi:10.1186/1471-2105-7-91
5. Omar, M., Klawonn, F., Brand, S., Stiesch, M., Krettek, C., Eberhard, J.: Transcriptome-wide high-density microarray analysis reveals differential gene transcription in periprosthetic tissue from hips with low-grade infection versus aseptic loosening. J. Arthroplasty (2016, to appear). doi:10.1016/j.arth.2016.06.036
6. Hand, D.: Measuring classifier performance: a coherent alternative to the area under the ROC curve. Mach. Learn. **77**, 103–123 (2009)
7. Flach, P., Hernández-Orallo, J., Ferri, C.: A coherent interpretation of AUC as a measure of aggregated classification performance. In: Proceedings of the 28th International Conference on Machine Learning (ICML 2011), pp. 657–664 (2011)

8. Mason, S.J., Graham, N.E.: Areas beneath the relative operating characteristics (ROC) and relative operating levels (ROL) curves: Statistical significance and interpretation. Q. J. Royal Meteorol. Soc. **128**(584), 2145–2166 (2002)
9. Holm, S.: A simple sequentially rejective multiple test procedure. Scand. J. Stat. **6**, 65–70 (1979)
10. Szafranski, S., Wos-Oxley, M., Vilchez-Vargas, R., Jáuregui, R., Plumeier, I., Klawonn, F., Tomasch, J., Meisinger, C., Kühnisch, J., Sztajer, H., Pieper, D., Wagner-Döbler, I.: High-resolution taxonomic profiling of the subgingival microbiome for biomarker discovery and periodontitis diagnosis. Appl. Environ. Microbiol. **81**, 1047–1058 (2015)
11. Demler, O., Pencina, M., D'Agostino, R.S.: Impact of correlation on predictive ability of biomarkers. Stat. Med. **32**, 4196–421 (2013)
12. Montvida, O., Klawonn, F.: Relative cost curves: An alternative to AUC and an extension to 3-class problems. Kybernetika **50**, 647–660 (2014)
13. Hand, D., Till, R.: A simple generalisation of the area under the ROC curve for multiple class classification problems. Mach. Learn. **45**, 171–186 (2001)
14. Li, J., Fine, J.: ROC analysis with multiple classes and multiple tests: methodology and its application in microarray studies. Biostatistics **9**, 566–576 (2008)
15. Li, J., Fine, J.: Nonparametric and semiparametric estimation of the three way receiver operating characteristic surface. J. Stat. Plan. Infer. **139**, 4133–4142 (2009)
16. Hernández-Orallo, J.: Pattern Recogn. ROC curves for regression **46**(12), 3395–3411 (2013)
17. Novoselova, N., Della Beffa, C., Wang, J., Li, J., Pessler, F., Klawonn, F.: HUM calculator and HUM package for R: easy-to-use software tools for multicategory receiver operating characteristic analysis. Bioinformatics **30**, 1635–1636 (2014)

Stability Evaluation of Event Detection Techniques for Twitter

Andreas Weiler[1], Joeran Beel[2(✉)], Bela Gipp[1], and Michael Grossniklaus[1]

[1] Department of Computer and Information Science,
University of Konstanz, 78457 Konstanz, Germany
{andreas.weiler,bela.gipp,michael.grossniklaus}@uni-konstanz.de
[2] Digital Contents and Media Sciences Research Division,
National Institute of Informatics (NII), Tokyo 101-8430, Japan
beel@nii.ac.jp

Abstract. Twitter continues to gain popularity as a source of up-to-date news and information. As a result, numerous event detection techniques have been proposed to cope with the steadily increasing rate and volume of social media data streams. Although most of these works conduct some evaluation of the proposed technique, comparing their effectiveness is a challenging task. In this paper, we examine the challenges to reproducing evaluation results for event detection techniques. We apply several event detection techniques and vary four parameters, namely time window (15 vs. 30 vs. 60 mins), stopwords (include vs. exclude), retweets (include vs. exclude), and the number of terms that define an event (1...5 terms). Our experiments use real-world Twitter streaming data and show that varying these parameters alone significantly influences the outcomes of the event detection techniques, sometimes in unforeseen ways. We conclude that even minor variations in event detection techniques may lead to major difficulties in reproducing experiments.

1 Introduction

The continuous success of Twitter and its freely available data stream have fostered many research efforts specialized on social media data. In this area of research, event detection is one of the most popular topics. In general, all event-detection approaches have in common that they attempt to detect patterns that differ from the normal behavior of the data stream. However, there are different types of techniques that can be used for this task. For example, Weng *et al.* [29] and Cordeiro [9] use techniques that are based on wavelet transformation to detect the events. Other works, such as Alvanaki *et al.* [2] or Mathioudakis and Koudas [16], use statistical models to detect significant abnormalities.

A major challenge in event-detection research is reproducibility. Reproducibility describes the case in which the outcome of two experiments allows drawing the same conclusions [4]. For instance, if an experiment shows that Algorithm A has faster run-times than Algorithm B, the conclusion might be

© Springer International Publishing AG 2016
H. Boström et al. (Eds.): IDA 2016, LNCS 9897, pp. 368–380, 2016.
DOI: 10.1007/978-3-319-46349-0_32

that Algorithm A outperforms Algorithm B. This research would be considered reproducible if a similar experiment also leads to results that support the conclusion that Algorithm A outperforms Algorithm B.

Reproducibility is affected by three factors, namely the similarity of scenarios, algorithms, and evaluation techniques [4]. If two experiments use the same algorithms, in the same scenario and apply the same evaluation techniques, then one would expect the outcome of the experiments to be the same. However, algorithms, scenarios and evaluation techniques typically differ somewhat between two experiments. If these differences are sufficiently small, one would nevertheless expect the outcome of the experiments to be at least similar and to support the same conclusions.

Our previous research in the field of recommender systems showed that minor differences in the experimental setup can at times lead to significant differences in the outcomes of two experiments. In one experiment to assess the effectiveness of a recommendation approach, removing stopwords increased recommendation effectiveness by 50 % [6]. In another experiment, effectiveness was almost the same [5]. Similarly, Lu et al. [14] found that sometimes terms from an article's abstract performed better than terms from the article's body, but in other cases they observed the opposite. Zarrinkalam and Kahani [30] found that terms from the title and abstract were most effective in some cases, while in other experiments terms from the title, abstract, and citation context were most effective. Bethard and Jurafsky [7] reported that using citation counts in the recommendation process strongly increased the effectiveness of their recommendation approach, while He et al. [12] reported that citation counts slightly increased the effectiveness of their approach. In all these examples, the changes in the algorithms, scenarios, and evaluation methods were minor. Nevertheless, even minor changes led to significantly different outcomes of the experiments, meaning that many research results in the recommender system community must be considered as not reproducible.

In the research community that studies event detection in social media data streams, reproducibility has received little attention to date. Based on our previous research and experience in the area of recommender systems for scientific publications, we believe that research on event detection techniques must place more emphasis on the issue of reproducibility. Currently, many evaluations of event detection appear to be non-reproducible. Weiler [24] lists the evaluation methods of a collection of 42 research works on event detection. Half of these evaluations is based on case or user studies. Reproducing these studies can already be challenging due to the inherent human element. Also problematic is the use of different data sets, which makes it hard or even impossible to reproduce the results of an experiment. Often the data sets used are heavily pre-filtered for users and/or regions or obtained by applying keyword filters. To address this issue, some works attempt to create and provide labelled reference data sets to evaluate event detection techniques. For example, McCreadie et al. [18] created a set of approximately 16 million tweets together with a list of 49 reference topics for a two-week period. However, since the corpus focuses on ad-hoc retrieval tasks and no description is given of how the topics were created, this reference data set is ill-suited for the evaluation of event detection techniques. Further reference data sets are proposed by Becker *et al.* [3], Petrović *et al.* [23], and McMinn *et al.* [19]. All of these corpora suffer from the shortcoming that the contained tweets need to be crawled. In the case of Twitter, crawling is a challenging task. With limited requests to the API it is almost impossible to retrieve all the tweets in a reasonable time frame. Also it is possible that a certain number of tweets are no longer available and therefore the final crawled corpora is not complete, which again limits the reproducibility of experimental results.

Based on a literature review of existing research, it can be observed that the terms "reproducibility" or "stability" are never mentioned as evaluation measures. Therefore, our research objective is to study the stability of event detection techniques as a necessary pre-condition for the reproducibility of event detection research. In the long run, the effect of all three factors (changes in algorithms, scenarios, and evaluation methods) need to be researched. However, for now, we focus on the first factor, *i.e.*, the effect of minor variations in event-detection algorithms. The research question of this paper is therefore: "How do minor changes in event detection techniques affect the reproducibility of experimental results?"

2 Methodology

To assess the reproducibility of experiments conducted with state-of-the-art event detection techniques, we study the stability of the obtained results w.r.t. slight variations in the parameter settings of these techniques. The studied event detection techniques all consist of a pre-processing, event detection, and event construction phase. For the evaluations presented in this paper, we varied parameters that affect the pre-processing and event detection. For the pre-processing phase, we conducted two experiments that respectively omitted the operators to suppress retweets and stopwords. In the event detection phase, we varied the size of the time-based window that is processed by the techniques. Based on these parameter variations, we studied the following configurations.

- 1 h windows with stopwords vs. without stopwords (pre-processing)
- 1 h windows with retweets vs. without retweets (pre-processing)
- 15 min vs. 30 min vs. 1 h windows (event detection)

For each of these configurations, we study the stability of the task-based and run-time performance results. In terms of task-based performance, we compare the results of a technique in one configuration to the results of the same technique in a different configuration. As all techniques report events as a set of five terms, we measure on how many terms in the two result lists overlap. In terms of run-time performance, we analyze how the different configurations influence the throughput (tweets/sec) of a technique. The rationale behind these experiments is that in order to be reproducible, small changes in the parameters should not drastically change the detected events. In other words, the more diverse the detected events were, the less stable the algorithms are, and hence, the less likely it would be to reproduce the results obtained in one experiment.

2.1 Experimental Setup

The data sets used in our evaluation consist of 10 % of the public live stream of Twitter for three different days. Using the Twitter Streaming API[1] with the so-called "Gardenhose" access level, we collected data for the randomly chosen days of 15th April 2013, 13th March 2013, and 8th July 2014. On average, the data sets contain a total of 20 million English tweets per day and an average of 850,000 tweets per hour.

All experiments were conducted on server-grade hardware with 1 Intel Xeon E5 processor at 3.5 GHz with 6 cores and 64 GB of main memory, running Oracle Java 1.8.0_40 (64-bit). Regardless of the available physical memory, the $-Xmx$ flag of the Java Virtual Machine (JVM) was used to limit the maximum memory to 24 GB.

2.2 Event Detection Techniques

The studied techniques were all realized as query plans (*cf.* Fig. 1) in the *Niagarino* data stream management system [27]. The operators with a dashed frame are the components that are modified in our experiments. The implementations and parameters of the first three techniques *EDCoW* [29], *WATIS* [9], and *Shifty* [25] have already been described in our previous work on evaluating event detection techniques [27].

In this paper, we additionally study the *LLH* and *enBlogue* (*ENB*) event detection techniques. *LLH* is a reimplementation of Weiler *et al.* [28]. In a first step, the technique aggregates and groups the distinct terms by their counts. Then the log-likelihood ratio operator collects n values per term as input signal. For the calculation of the log-likelihood ratio at least two windows need to be analyzed by the operator. After the analysis of two windows the log-likelihood ratio between all terms in the current window is calculated against the past. Events are reported by selecting the top N terms with the highest log-likelihood ratio together with the corresponding top four most co-occurring terms. Since these are the terms with the highest abnormal behavior in their current frequency with respect to their historical frequency, we define them as events. Note that in contrast to the original technique that detected events for pre-defined geographical areas, we adjusted the approach to calculate the log-likelihood measure for the frequency of all distinct terms in the current time window against their frequency in the past time windows.

ENB is a reimplementation of Alvanaki *et al.* [2], which uses non-overlapping windows to compute statistics about tags and tag pairs. An event consists of a pair of tags and at least one of the two tags needs to be a so-called seed tag. Seed tags are determined by calculating a popularity score. Tags with a popularity

---

[1] https://dev.twitter.com (April 28, 2016).

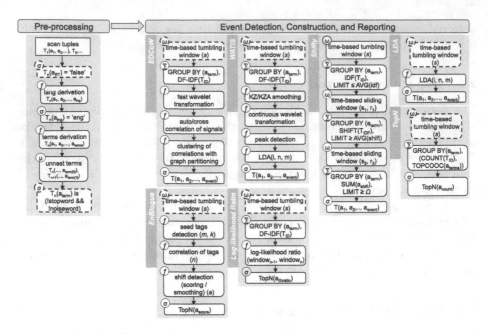

Fig. 1. Query plans of the studied event detection techniques and baselines.

score within a pre-defined range of the top percentage terms (threshold k) are then chosen as seeds. Also a minimum of m tweets need to contain the tag. The correlation of two tags is calculated by a local and global impact factor, which is based on the corresponding sets of tweets that are currently present in the window. If two tags are strongly connected, they are considered to be related. A minimum of n correlations needs to exist. An event is considered as emergent, if its behavior deviates from the expected. In order to detect this condition, the shifting behavior of correlations between terms is calculated by a scoring and smoothing function, which uses the fading parameter a to smooth out past values. Since we require all event detection techniques to output an event as a set of five terms, the three most co-occurring terms of both tags of the pair computed by *ENB* are added to the event. Finally, the technique reports the top N events, which are selected by ranking all events based on the calculated score of shift.

Apart from these event detection techniques, we also implemented two baseline techniques. The *TopN* technique just reports the most frequent N terms per window including their most frequent co-occurrence terms. The *LDA* technique reports topics created by the well-known Latent Dirichlet allocation modeling [8] and is realized by using the Mallet toolkit [17].

3   Results

In this section, we present the results of our experiments as averaged results over all three data sets. First, the impact of changes to the pre-processing phase is studied. Second, we demonstrate the impact of changes to the event detection phase, in particular when varying the size of the time windows.

3.1   Impact of Pre-processing Variations

We evaluate the impact of changes to the pre-processing phase by starting from the parameter settings used in our previous evaluations [26,27]. In the first experiment, we remove the pre-processing operator that suppresses retweets in the input (first operator with a dashed frame in Fig. 1). The results shown in Fig. 2 demonstrate that the inclusion of retweets has a strong impact on the events detected by the studied event detection techniques. In contrast, the influence of this change on the baseline techniques is less pronounced. In the second experiment, we omitted the pre-processing operator that removes stopwords from the input (second operator with a dashed frame in Fig. 1). Figure 3 indicates that the results of the event detection techniques are more stable w.r.t. this second change, with the statistical methods *LLH* and *ENB* proving the most stable. We can also observe that the baseline techniques are more strongly influenced by the inclusion of stopwords than the event detection techniques.

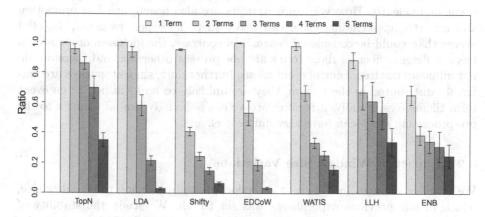

Fig. 2. Impact of including retweets during pre-processing, represented as the ratio of events contained in the results with and without retweets. Each bar presents the ratio of events that share the corresponding number of terms.

Additionally, we measured the throughput (see Fig. 4) for these four different configurations. In the first experiment, the throughput of all techniques decreased by about 30 % to 40 % if retweets are included. In the second experiment, the inclusion of stopwords decreases the throughput by about 10 %, with the exception of *LDA*, where it decreases by almost 30 %.

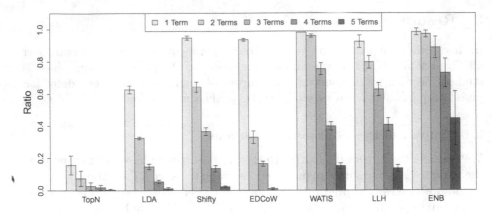

Fig. 3. Impact of including stopwords during pre-processing, represented as the ratio of events contained in the results with and without stopwords. Each bar present the ratio of events that share the corresponding number of terms.

The results observed in these first experiments are as expected. All studied event detection techniques use some form of relative term frequency as a measure for term importance or popularity. In this setting, the inclusion of retweets increases the frequency of terms that are also present if retweets are suppressed. In some cases, this repetition of terms will help to identify an already identified event more clearly. However, since retweets are also heavily used in promotion and advertising, including them can also lead to false positives, *i.e.*, detected events that would be considered "spam". In contrast, the inclusion of stopwords has a different effect as these terms are not present otherwise and therefore do not influence the frequency of event terms. Furthermore, since stopwords are uniformly distributed in the stream, they are unlikely to be identified as an event term themselves. Finally, it is noteworthy that seemingly similar changes to the pre-processing stage can have very different effects.

3.2   Impact of Window Size Variations

We evaluate changes in the event detection phase by varying the window size, which in our previous experiments was set to 1 h. We study the stability of the results by comparing three different configurations with 15, 30, and 60 min, respectively (operators with a dashed frame on the right side of Fig. 1). For techniques that report the top N events as results, we adjust the value of N in accordance to the window size: for 15 min windows the top 5, for 30 min the top 10 and for 60 min the top 20 events are reported. Since the number of events reported per time window can differs substantially depending on the length of the time window, we also adjusted further parameters (*cf.* Table 1). Note that *Shifty* is designed to be independent of the input window size and therefore we have to explicitly stop and restart the processing after 15, 30, or 60 min in order to obtain comparable results.

Table 1. Parameter settings for *Shifty*, *WATIS*, and *EDCoW*.

Technique	Parameters
Shifty15	$s_{input} = 1$ min, $s_1 = 2$ min, $r_1 = 1$ min, $s_2 = 4$ min, $r_2 = 1$ min, $\Omega = 23$
Shifty30	$s_{input} = 1$ min, $s_1 = 2$ min, $r_1 = 1$ min, $s_2 = 4$ min, $r_2 = 1$ min, $\Omega = 22$
Shifty60	$s_{input} = 1$ min, $s_1 = 2$ min, $r_1 = 1$ min, $s_2 = 4$ min, $r_2 = 1$ min, $\Omega = 24$
WATIS15	$s = 25$ s, $N = 3$ intervals, $i_{kza} = 5$, $i_{lda} = 500$
WATIS30	$s = 49$ s, $N = 3$ intervals, $i_{kza} = 5$, $i_{lda} = 500$
WATIS60	$s = 87$ s, $N = 5$ intervals, $i_{kza} = 5$, $i_{lda} = 500$
EDCoW15	$s = 4$ s, $N = 32$ intervals, $\gamma = 2.0$, $\epsilon = 0.1$
EDCoW30	$s = 4$ s, $N = 32$ intervals, $\gamma = 1.5$, $\epsilon = 0.1$
EDCoW60	$s = 4$ s, $N = 32$ intervals, $\gamma = 0.9$, $\epsilon = 0.1$

Figure 5 summarizes the results for this experiment. We can observe that results of the techniques that report a fixed number of N events are more stable than the threshold-based techniques. We can also see that the results of the baseline techniques are very stable in comparison to the results of the event detection techniques. This outcome is explained by the fact that both baseline techniques simply report the most frequent terms, which are bound to be similar in the context of Twitter and independent of a given time frame. Finally, we can observe that *Shifty* is more stable than both *EDCoW* and *WATIS*, which is noteworthy because we introduced artificial interruptions into *Shifty*'s processing to obtain comparable results. By breaking up larger windows into smaller ones, it is possible that *Shifty* misses events that occur across the boundaries of the smaller windows, but would be included entirely in the larger window. Since this effect will increase result instability, *Shifty*'s high stability is a promising result.

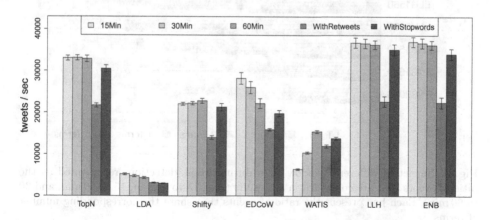

Fig. 4. Impact of all variations for the throughput in tweets/sec.

376   A. Weiler et al.

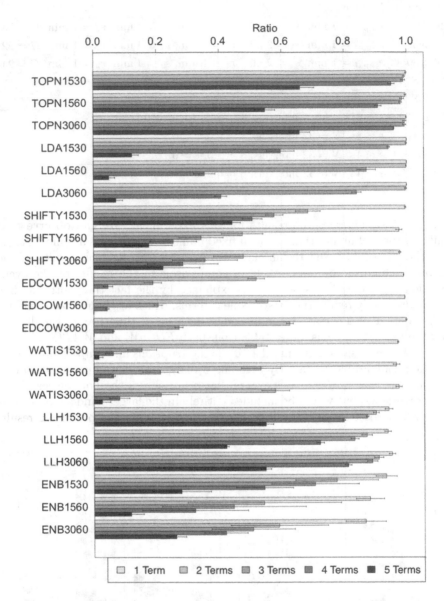

Fig. 5. Impact of different window sizes during event detection, represented as the ratio of events that are contained in all results (*e.g.*, 15 to 30 min, 15 to 60 min, and 30 to 60 min). Each bar present the ratio of events that share the corresponding number of terms.

Again, we also measured the throughput (see Fig. 4) achieved by the different techniques in each configuration. In all our experiments, the throughput of the baselines techniques, as well as the one of *Shifty*, *LLH*, and *ENB* remained stable across the window sizes that we tested. This is due to the fact that the three event detection techniques apply various filtering steps early on and thereby keep the number of terms to analyze within a certain lower bound. The first exception to this observation is *WATIS*. The throughput of *WATIS* when using 30 min windows is twice as high as when using 15 min windows. In the case of 1 h windows, the throughput of *WATIS* is almost three times higher as when using 15 min windows. This is attributable to the processing time of *WATIS* strongly correlating with the number of terms entering the analysis phase, which itself depends on the window size. In the case of 30 min windows, almost twice as many terms are processed as in the case of 15 min. For 1 h windows, the number of terms is three times higher than for 15 min windows. The second exception is *EDCoW*, which exhibits the opposite behavior of *WATIS*, *i.e.*, throughput decreases for longer windows w.r.t. shorter ones. The two most important factors contributing to the run-time of *EDCoW* are the computation of the auto-correlation and the graph partitioning (*cf.* Fig. 1). In the case of the auto-correlation computation, longer windows produce longer signals, which require more time to be processed than shorter signals. The complexity of the graph partitioning also increases with longer windows, since a 15 min window consists on average of about 12,000 edges, while the graphs for 30 min and 1 h windows contain an average of about 25,000 and 70,000 edges, respectively.

4   Conclusions and Future Work

In this paper, we addressed the evaluation of event detection techniques w.r.t. their result stability in an effort to study the reproducibility of experiments in this research area. Our results show that minor modifications in the different phases of the techniques can have a strong impact on the stability of their results. However, we must take into account that by changing the size of the windows, the existing terms in the time frame can vary considerably. Therefore, it is to be expected that the ratio for 3 to 5 terms is very low. Also, the event detection techniques *WATIS* and *EDCoW* are originally designed to analyze even longer time frames, such as days, weeks, and months.

As immediate future work, we plan to take advantage of our platform-based approach to extend our evaluations and study further techniques. As extensions of our evaluations we plan to include further parameter settings and to research the interdependencies of the parameters. By reviewing the surveys of related work (*e.g.*, Nurwidyantoro and Winarko [20], Madani *et al.* [15], or Farzindar and Khreich [10]), we found several candidates for this venture. On the one hand, techniques such as *TwitterMonitor* [16] and *Twevent* [13] are interesting because the techniques they use are closely related to our own techniques. On the other hand, clustering and hashing techniques, such as *ET* [21] or the work of Petrović *et al.* [22] would also be interesting to compare. Since the source code of most of these works is not provided by their authors, it is a challenging task to correctly implement these techniques. Notable exceptions to this lack of reproducibility are *SocialSensor* [1] and *MABED* [11], which are both freely available as source code. In this context, it would be interesting to define measures, which can be used to rank the degree of reproducibility of existing and future research work in the area of event detection. For this purpose, we created a survey [24] about the techniques and evaluations of 42 related works. With this list we can, for example, rank the works based on the availability of source code, pseudo code, or at least a very precise description of the algorithm. Furthermore, the research work could be ranked according to how many parameters the event detection technique needs and how easily the evaluation can be reproduced.

Acknowledgement. The research presented in this paper is funded in part by the Deutsche Forschungsgemeinschaft (DFG), Grant No. GR 4497/4: "Adaptive and Scalable Event Detection Techniques for Twitter Data Streams" and by a fellowship within the FITweltweit programme of the German Academic Exchange Service (DAAD). We would also like to thank the students Christina Papavasileiou, Harry Schilling, and Wai-Lok Cheung for their contributions to the implementations of *WATIS*, *EDCoW*, and *enBlogue*.

References

1. Aiello, L.M., Petkos, G., Martin, C., Corney, D., Papadopoulos, S., Skraba, R., Göker, A., Kompatsiaris, I.: Sensing trending topics in Twitter. IEEE Trans. Multimedia **15**(6), 1268–1282 (2013)
2. Alvanaki, F., Michel, S., Ramamritham, K., Weikum, G.: See what's enBlogue: real-time emergent topic identification in social media. In: Proceedings of International Conference on Extending Database Technology (EDBT), pp. 336–347 (2012)
3. Becker, H., Naaman, M., Gravano, L.: Beyond trending topics: real-world event identification on Twitter. In: Proceedings of International Conference on Weblogs and Social Media (ICWSM), pp. 438–441 (2011)
4. Beel, J., Breitinger, C., Langer, S., Lommatzsch, A., Gipp, B.: Towards reproducibility in recommender-systems research. User Model. User-Adap. Inter. **26**(1), 69–101 (2016)
5. Beel, J., Langer, S.: A Comparison of offline evaluations, online evaluations, and user studies in the context of research-paper recommender systems. In: Kapidakis, S., Mazurek, C., Werla, M. (eds.) TPDL 2015. LNCS, vol. 9316, pp. 153–168. Springer, Heidelberg (2015). doi:10.1007/978-3-319-24592-8_12
6. Beel, J., Langer, S., Genzmehr, M., Nürnberger, A.: Introducing Docear's research paper recommender system. In: Proceedings of Joint Conference on Digital Libraries (JCDL), pp. 459–460 (2013)
7. Bethard, S., Jurafsky, D.: Who should i cite: learning literature search models from citation behavior. In: Proceedings of International Conference on Information and Knowledge Management (CIKM), pp. 609–618 (2010)
8. Blei, D.M., Ng, A.Y., Jordan, M.I.: Latent Dirichlet allocation. J. Mach. Learn. Res. **3**, 993–1022 (2003)
9. Cordeiro, M.: Twitter event detection: combining wavelet analysis and topic inference summarization. In: Proceedings of Doctoral Symposium on Informatics Engineering (DSIE) (2012)

10. Farzindar, A., Khreich, W.: A survey of techniques for event detection in Twitter. Comput. Intell. **31**(1), 132–164 (2015)
11. Guille, A., Favre, C.: Mention-anomaly-based event detection and tracking in Twitter. In: Proceedings of International Conference on Advances in Social Networks Analysis and Mining (ASONAM), pp. 375–382 (2014)
12. He, Q., Pei, J., Kifer, D., Mitra, P., Giles, L.: Context-aware citation recommendation. In: Proceedings of International Conference on World Wide Web (WWW), pp. 421–430 (2010)
13. Li, C., Sun, A., Datta, A.: Twevent: segment-based event detection from tweets. In: Proceedings of International Conference on Information and Knowledge Management (CIKM), pp. 155–164 (2012)
14. Lu, Y., He, J., Shan, D., Yan, H.: Recommending citations with translation model. In: Proceedings of International Conference on Information and Knowledge Management (CIKM), pp. 2017–2020 (2011)
15. Madani, A., Boussaid, O., Zegour, D.E.: What's happening: a survey of tweets event detection. In: Proceedings of International Conference on Communications, Computation, Networks and Technologies (INNOV), pp. 16–22 (2014)
16. Mathioudakis, M., Koudas, N.: TwitterMonitor: trend detection over the Twitter stream. In: Proceedings of International Conference on Management of Data (SIGMOD), pp. 1155–1158 (2010)
17. McCallum, A.K.: MALLET: A Machine Learning for Language Toolkit (2002). http://mallet.cs.umass.edu
18. McCreadie, R., Soboroff, I., Lin, J., Macdonald, C., Ounis, I., McCullough, D.: On building a reusable Twitter corpus. In: Proceedings of International Conference on Research and Development in Information Retrieval (SIGIR), pp. 1113–1114 (2012)
19. McMinn, A.J., Moshfeghi, Y., Jose, J.M.: Building a large-scale corpus for evaluating event detection on Twitter. In: Proceedings of International Conference on Information and Knowledge Management (CIKM), pp. 409–418 (2013)
20. Nurwidyantoro, A., Winarko, E.: Event detection in social media: a survey. In: Proceedings of International Conference on ICT for Smart Society (ICISS), pp. 1–5 (2013)
21. Parikh, R., Karlapalem, K.: ET: Events from Tweets. In: Proceedings of International Conference Companion on World Wide Web (WWW), pp. 613–620 (2013)
22. Petrović, S., Osborne, M., Lavrenko, V.: Streaming first story detection with application to Twitter. In: Proceedings of Conference on the North American Chapter of the Association for Computational Linguistics (HLT), pp. 181–189 (2010)
23. Petrović, S., Osborne, M., Lavrenko, V.: Using paraphrases for improving first story detection in news and Twitter. In: Proceedings of Conference of the North American Chapter of the Association for Computational Linguistics: Human Language Technologies (NAACL HLT), pp. 338–346 (2012)
24. Weiler, A.: Design and evaluation of event detection techniques for social media data streams. Ph.D. thesis, University of Konstanz, Konstanz (2016)
25. Weiler, A., Grossniklaus, M., Scholl, M.H.: Event identification and tracking in social media streaming data. In: Proceedings of EDBT Workshop on Multimodal Social Data Management (MSDM), pp. 282–287 (2014)
26. Weiler, A., Grossniklaus, M., Scholl, M.H.: Evaluation measures for event detection techniques on Twitter data streams. In: Maneth, S. (ed.) BICOD 2015. LNCS, vol. 9147, pp. 108–119. Springer, Heidelberg (2015). doi:10.1007/978-3-319-20424-6_11

27. Weiler, A., Grossniklaus, M., Scholl, M.H.: Run-time and task-based performance of event detection techniques for Twitter. In: Zdravkovic, J., Kirikova, M.,
Johannesson, P. (eds.) CAiSE 2015. LNCS, vol. 9097, pp. 35–49. Springer,
Heidelberg (2015). doi:10.1007/978-3-319-19069-3_3
28. Weiler, A., Scholl, M.H., Wanner, F., Rohrdantz, C.: Event identification for local
areas using social media streaming data. In: Proceedings of SIGMOD Workshop
on Databases and Social Networks (DBSocial), pp. 1–6 (2013)
29. Weng, J., Lee, B.S.: Event detection in Twitter. In: Proceedings of International
Conference on Weblogs and Social Media (ICWSM), pp. 401–408 (2011)
30. Zarrinkalam, F., Kahani, M.: SemCiR - a citation recommendation system based
on a novel semantic distance measure. Program: Electron. Libr. Inf. Syst. **47**(1),
92–112 (2013)

IDA 2016 Industrial Challenge: Using Machine Learning for Predicting Failures

Camila Ferreira Costa[✉] and Mario A. Nascimento

Department of Computing Science, University of Alberta, Edmonton, Canada
{camila.costa,mario.nascimento}@ualberta.ca

Abstract. This paper presents solutions to the IDA 2016 Industrial Challenge which consists of using machine learning in order to predict whether a specific component of the Air Pressure System of a vehicle faces imminent failure. This problem is modelled as a classification problem, since the goal is to determine if an unobserved instance represents a failure or not. We evaluate various state-of-the-art classification algorithms and investigate how to deal with the imbalanced dataset and with the high amount of missing data. Our experiments showed that the best classifier was cost-wise 92.56 % better than a baseline solution where a random classification is performed.

1 Introduction

This paper discusses the solutions investigated for the IDA 2016 Industrial Challenge. The challenge was to deliver a prediction model that is able to judge whether a vehicle faces imminent failure of a specific component of the Air Pressure System (APS). The dataset provided consists of data collected from heavy Scania trucks. Each instance of the training set is classified as positive or negative. The positive class represents failures for the specific component considered, while negative instances represent trucks with failures for components not related to the APS. Based on this scenario, the goal is to develop a prediction model that minimizes a total misclassification cost where there is a much greater cost for predicting false negatives (FNs) than false positives (FPs), namely 500 and 10 units, respectively. (For more details please refer to https://ida2016.blogs. dsv.su.se/?page_id=1387.)

This problem can be modelled as a classification problem, thus we evaluate well-known classification algorithms such as Logistic Regression [4], k-nearest neighbors [3], Support Vector Machine (SVM) [2], Decision Trees [10] and Random Forests [7]. Moreover, the dataset provided has highly imbalanced classes as well as non-trivial amounts of missing data in both training and test sets, thus we also investigate techniques for dealing with both of these problems.

Our experiments showed that the best classifier was cost-wise 92.56 % better than a straightforward solution where a random classification is performed.

© Springer International Publishing AG 2016
H. Boström et al. (Eds.): IDA 2016, LNCS 9897, pp. 381–386, 2016.
DOI: 10.1007/978-3-319-46349-0_33

2   Related Work

2.1   Classification Algorithms

Logistic Regression (LR) [4] is commonly used for binary classification where the prediction can take on two values, 0 (negative) or 1 (positive). LR calculates the probability of an instance belonging to the positive class. If this probability is higher than some chosen cutoff value, which is generally 0.5, the instance is classified as positive. A disadvantage of this algorithm is that it is only able to find a linear decision boundary and therefore LR will not perform well on datasets where the two classes cannot be separated by a straight line.

The k-nearest neighbor (k-NN) [3] classifier is among the simplest classification algorithms and, unlike LR, it performs a non-linear classification. It takes as input a positive number k and an unseen instance and finds the k nearest neighbours of the new instance, according to a given distance measure. It then classifies the new instance as the most common class among the classes of its k nearest neighbours. A drawback of this strategy is that a frequent class may dominate the prediction of the new instance. A possible solution to this problem is to assign to each of the k-NN a weight inversely proportional to the distance from it to the new example. Thus, a greater weight is given to closer neighbours.

The original Support Vector Machine (SVM) algorithm [2] is also a binary linear classifier, but differently from LR, it is not probabilistic. More specifically, it constructs a hyperplane that separates two classes. The best hyperplane is chosen such that the margin between the two classes is maximized, i.e. the best hyperplane is the one that has the largest distance to the nearest data point of any class. SVMs can also efficiently perform a non-linear classification using the "kernel trick" [1,2], which implicitly maps their inputs into high-dimensional feature spaces. In fact, SVMs are the most suitable option when there is the need to deal with highly dimensional space. However, they are not very efficient when many training examples are used besides being memory-intensive. Moreover, the choice of the kernel and its parameters are not trivial.

A Decision Tree (DT) [10] is a tree composed of internal nodes and leaves. Each internal node represents a test on a feature. Each branch represents the outcome of the test and are labeled with the possible values of the feature. The leaf nodes are labeled with a class. The construction of the tree is done recursively. At each step the feature that best splits the set of items is chosen according to a given metric. This process is repeated on each derived subset and is completed when all the instances in a node belong to the same class or the path from the root to a leave contains all features. An advantage of this technique is that it is easy to interpret. However, decision trees are very prone to overfitting. Moreover, it builds decision boundaries parallel to the axes and will work best if the class labels roughly lie in hyper-rectangular regions.

A random forest (RF) [7] is an ensemble of decision trees, however, unlike decision trees, it does suffer from overfitting. In this method, each decision tree is constructed by using a random subset of the training data. The ensemble of simple trees then vote for the most popular class.

2.2   Dealing with Missing Data

The easiest way to deal with missing data is to consider only test cases with complete information, i.e. the ones that do not have any missing values are kept. This is not feasible in the case of this challenge because (1) most of the training dataset would have to be discarded and (2) the testing dataset has missing data that have to be dealt with anyway.

Another approach is to impute missing values. In this case, the missing values are filled with estimated ones based on information available in the data set and they are then treated as if they were observed. The mean imputation is the most simple and frequently used method. However, its use is not recommended, since it can distort the distribution of the features and underestimate the standard deviation [6].

Expectation Maximization (EM) [5] and Multiple Imputation [11] are considered state-of-the-art missing data techniques [6]. The EM algorithm is an iterative procedure that produces maximum likelihood estimates and is composed of two steps. The first E-step computes the expected value of the sum of the variables with missing data based on initial values of the parameters of the imputation model: mean and variance-covariance matrix. In the M-step the parameters are estimated based on the information calculated in the E-step of the same iteration. The algorithm iterates until the estimates do not change substantially.

Finally, instead of filling in a single value for each missing value, a Multiple Imputation procedure replaces each missing value with a set of plausible values, which are predicted using existing values from other variables. Multiple data sets are produced in order to account for the uncertainty of the missing values. These data sets are then analyzed by using the desired method, producing multiple analysis results, which are combined to produce one overall result.

3   Solutions and Implementation

The following were the two main challenges that we had to deal with when solving the problem of classifying new instances as positive or negative. First, the dataset provided is highly imbalanced. From the 60000 instances in the training set, only 1000 belong to the positive class. This can lead the prediction models to not classify positive instances properly. Moreover, we also have to take into consideration the high penalty for misclassifying such instances. Second, the missing data rate in the dataset is very high. If we considered only the instances that have no missing data, less than 1 % of the original dataset would remain.

We evaluated all the classification algorithms presented in Sect. 2.1 and resorted to the implementations provided by the Scikit-learn library [9]. With this library we are able to apply class-specific weights in the loss function of the SVM and LR algorithms, which allows us to cope with the imbalanced dataset and the different penalties for misclassifying positives and negatives instances. The weight of each class was set to be inversely proportional to the fraction of cases of the corresponding class.

In the k-NN and RF classifiers the likelihood of an instance belonging to each class is calculated. Generally, an instance would be classified as positive, for example, if the probability of belonging to this class were greater than 50 %. By applying this threshold, these two algorithms are not able to classify positive instances properly, which greatly increases the misclassification cost. In order to deal with this problem, we have increased this threshold, meaning that a instance is only classified as negative if we are very confident of that. We explain how the values for these thresholds were chosen in Sect. 4. This same strategy was used for DTs, however, we do not show the results obtained using this algorithm since in general, and as evidenced in our experiments, it is outperformed by RFs.

In order to deal with missing data, we have considered two methods: mean imputation and a more sophisticated method named Soft-Impute [8]. Soft-Impute is an efficient algorithm for large-scale matrix completion. It works like EM, replacing missing values with the current guess, and then solving the optimization problem, a nuclear norm regularized least squares problem, on the complete matrix using a soft-thresholded Singular Value Decomposition (SVD).

To simulate how well the classifiers would perform in the test set, we use k-fold Cross-Validation. We chose k to be equal to 10, since it is a commonly used value. It is worth mentioning that the missing data techniques were applied separately in each fold. The cost in each of the 10 test sets were summed up to obtain the total misclassification cost.

4   Results

Table 1 shows the results obtained using the various classifiers studied in this paper. Since the results obtained using the Soft-Impute algorithm were consistently better than the ones using mean imputation, we only report the results obtained using the former. In order to evaluate how well the classifiers performed, we compared their performances with that of a straightforward baseline where the classes are randomly assigned.

For the SVM algorithm we have considered two different kernels: *gaussian* and *poly*. The best results were obtained using the *gaussian* kernel. Even though this classifier was significantly better than the baseline solution, it was outperformed by all other ones, even by the relative simple LR.

The k-NN algorithm was the second best classifier. Instead of using the traditional algorithm, we considered the probability of an instance of belonging to the negative class for classifying it. More specifically, we find the 60 NN of each instance and if at least one neighbour belongs to the positive class, we classify the instance as positive. This value was chosen considering the distribution of the classes in the training set. That way we only classify an instance as negative if its probability of belonging to this class is very high, thus reducing the risk of false negatives, which were much more costly than false positives.

RFs presented the best results. The result shown in Table 1 was obtained using 50 estimators (trees) and a cutoff of 95 %, both chosen empirically. As in the k-NN classifier, this threshold determines whether an instance belongs to

Table 1. Classifiers and respective performances. (FPs and FNs denote false positives and false negatives, each costing 10 and 500 units, respectively.)

Classifier	Misclassification cost	Gain wrt random	% Misclassification
Random	545000	N/A	FP: 50 % FN: 50 %
SVM	74310	86.36 %	FP: 1.15 % FN: 13.5 %
LR	61470	88.72 %	FP: 2.36 % FN: 9.5 %
k-NN	49850	90.84 %	FP: 2.94 % FN: 6.5 %
RF	40570	92.56 %	FP: 3.74 % FN: 3.7 %

the negative class. The total cost of this algorithm was 92.56 % better than the misclassification cost of the baseline solution. It is noteworthy that the number of false negatives, a much more expensive misclassification than false positives in this challenge's scenario, was also significantly smaller using RFs than any of the other approaches, making it the overall winner.

5   Conclusion

We have studied several classification algorithms for solving the problem of determining whether a vehicle faces imminent failure of a specific component of the APS system. We had the challenge of dealing with highly unbalanced data and a large amount of missing values. A combination of the Soft-Impute alogrithm, which deals with missing data, and the RF classifier that tries to avoid false negatives by using an empirically chosen cutoff, led to a model that yielded 92.56 % better cost than a solution that assigns a class to a new instance randomly.

Acknowledgements. We acknowledge partial financial support by NSERC Canada, as well as preliminary discussions on this challenge with Philippe Gaudreau.

References

1. Aizerman, M.A., Braverman, E.A., Rozonoer, L.: Theoretical foundations of the potential function method in pattern recognition learning. In: Automation and Remote Control, pp. 821–837 (1964)
2. Boser, B.E., Guyon, I.M., Vapnik, V.N.: A training algorithm for optimal margin classifiers. In: Proceedings of the 5th COLT, pp. 144–152 (1992)
3. Cover, T.M., Hart, P.E.: Nearest neighbor pattern classification. IEEE Trans. Inf. Theor. **13**(1), 21–27 (1967)
4. Cox, D.R.: The regression analysis of binary sequences. J. R. Stat. Soc. Ser. B **20**(2), 215–242 (1958)
5. Dempster, A.P., Laird, N.M., Rubin, D.B.: Maximum likelihood from incomplete data via the EM algorithm. J. R. Stat. Soc. Ser. B **39**(1), 1–38 (1977)
6. Graham, J.W.: Missing data analysis: making it work in the real world. Annu. Rev. Psychol. **60**, 549–576 (2009)

7. Ho, T.K.: Random decision forests. In: Proceedings of the 3rd IJDAR, pp. 278–282 (1995)
8. Mazumder, R., Hastie, T., Tibshirani, R.: Spectral regularization algorithms for learning large incomplete matrices. J. Mach. Learn. Res. **11**, 2287–2322 (2010)
9. Pedregosa, F., et al.: Scikit-learn: machine learning in python. J. Mach. Learn. Res. **12**, 2825–2830 (2011)
10. Quinlan, J.R.: Induction of decision trees. Mach. Learn. **1**(1), 81–106 (1986)
11. Rubin, D.B.: Multiple Imputation for Nonresponse in Surveys. Wiley, New York (1987)

An Optimized k-NN Approach
for Classification on Imbalanced Datasets
with Missing Data

Ezgi Can Ozan[1(✉)], Ekaterina Riabchenko[1], Serkan Kiranyaz[2],
and Moncef Gabbouj[1]

[1] Tampere University of Technology, Tampere, Finland
{ezgi.ozan,ekaterina.riabchenko,moncef.gabbouj}@tut.fi
[2] Electrical Engineering Department, College of Engineering,
Qatar University, Doha, Qatar
mkiranyaz@qu.edu.qa

Abstract. In this paper, we describe our solution for the machine learning prediction challenge in IDA 2016. For the given problem of 2-class classification on an imbalanced dataset with missing data, we first develop an imputation method based on k-NN to estimate the missing values. Then we define a tailored representation for the given problem as an optimization scheme, which consists of learned distance and voting weights for k-NN classification. The proposed solution performs better in terms of the given challenge metric compared to the traditional classification methods such as SVM, AdaBoost or Random Forests.

Keywords: k-NN classifier · Missing data · Imbalanced datasets

1 Introduction

Missing or unknown data is a common problem in real life machine learning applications. It may not always be possible to access all sensor readings at a given time, which may occur because of various reasons. While the researchers focus on minimizing the moments of failure for the data sources, it is also an important field of research to predict the missing values with the information at hand.

In IDA 2016 machine learning prediction challenge, a dataset obtained from the Scania trucks is presented, with missing readings from various sensors. The dataset consists of two classes. The first class corresponds to trucks with failures for a specific component of the Air Pressure System (APS) and the second class corresponds to trucks with failures for components not related to the APS. As expected, in real life the number of samples belonging to the second class will be much greater than the first class. This is also the case in the given dataset, thus the distribution of samples among the classes is strongly imbalanced. In this paper we explain our solution to the problem of classification on an imbalanced dataset with missing data. We propose a k-NN based approach, with an additional optimization scheme which is developed to improve the overall performance.

© Springer International Publishing AG 2016
H. Boström et al. (Eds.): IDA 2016, LNCS 9897, pp. 387–392, 2016.
DOI: 10.1007/978-3-319-46349-0_34

2   Problem Definition

In this section, we define the given problem and present some background information about the terms used in the proposed method. The given problem can be defined as a 2-class prediction problem with missing data on imbalanced dataset. To measure the performance of the prediction task, a cost function defined by IDA2016 is given, as shown in (1), where FP stands for a false positive and FN for a false negative prediction.

$$\mathbf{Cost} = 10 \times FP + 500 \times FN, \tag{1}$$

The cost function is designed to give a higher cost for a missed break down (FN) compared to a false prediction (FP). Considering the nature of the problem this makes sense, as missing a break down for an expensive truck is much more costly than calling the truck for maintenance unnecessarily.

The second aspect of the given problem is the missing data. In the training set, several attributes of many given samples are marked as NA, which states that those attributes are missing. In the literature the missing data problem is studied in three subsections depending on how the data got missing [1]. These three subsections are:

- Missing-At-Random (MAR)
- Missing-Completely-At-Random (MCAR)
- Not-Missing-At-Random (NMAR).

For, MAR missingness is independent of the missing variables but the pattern of data missingness is traceable, i.e. missingness only depends on the observed input. For MCAR, the probability that a variable is missing does not depend on anything, as it is completely at random. Finally for NMAR, the missing data depends on the missing variable itself, meaning there is another reason for the data to be lost, which is fully related to the variable itself, so it is not possible to estimate the missing value. Since no information about the reasons of the occurrence of missing data is provided, we assume the missing data is MCAR.

3   Proposed Method

As the missing data problem is also present in the provided test set, the first problem to solve is to estimate the values of the missing attributes. Here we propose a machine learning based imputation method called k-NN imputation.

3.1   k-NN Imputation of Missing Attributes

In this method, we propose to use the mean of the nearest k samples to estimate a missing value for a given sample. This brings the problem of finding the nearest k samples in a dataset for a given incomplete sample. Considering that the dataset includes many incomplete samples, a new distance metric should be defined, to sort the reference samples according to their distance to the given sample. In our solution, we use a slightly modified version of the Heterogeneous Euclidean Overlap Metric (HEOM) [2]. The distance between the samples x and y is calculated as given in (2) and (3).

$$D(x,y) = \sum_{i}^{L} D_i(x_i, y_i),$$ (2)

$$D_i(x_i, y_i) = \begin{cases} 1, & x_i \text{ or } y_i \text{ is missing}; \\ d_i(x_i, y_i), & \text{otherwise}. \end{cases}$$ (3)

where $d_i(x_i, y_i)$ is the distance metric which corresponding to the attribute i, and L is the total number of attributes $(d_i(x_i, y_i) \leq 1)$. Using the HEOM as defined above, we first obtain the nearest k neighbours (we select K = 50) for a given sample, in which the given attribute is not missing and calculate the mean in order to estimate the missing value.

In the given set, since there are both numerical attributes and histograms, our approach to each one is different. We use the square or the Euclidean distance for the numerical values and the cosine distance for histograms. We perform normalization for each attribute and histogram, i.e., the sum of all the bins of a histogram is equalized to 1 and each attribute is linearly mapped to [0, 1].

$$d_i(x_i, y_i) = \begin{cases} 1 - \frac{\langle x_i, y_i \rangle}{\|x_i\| \|y_i\|}, & x_i \text{ is a histogram}; \\ \|x_i - y_i\|^2, & x_i \text{ is a numerical attribute}. \end{cases}$$ (4)

3.2 Optimized k-NN Classification

In this challenge, we face a highly imbalanced classification problem, as the provided training set consists of 59000 negative samples and only 1000 positive samples. In order to handle such a great imbalance, also taking the desired cost calculation into account, we propose a k-NN based classification approach. The Nearest Neighbour classification is one of the most popular methods in data mining [3]. This idea is very simple and easy to envision. The main advantage of using a NN based approach is that it does not require a training stage, since it is an instance based learning approach. However, it is still possible to adapt different cost functions and integrate further optimizations in order to achieve a better performance.

In k-NN approaches, a given input is classified according to the class of its nearest neighbours among a stored set of reference samples. One can also suggest that, the reference samples which are closer to the test sample should have a higher impact on the classification decision. In [4] Dudani proposes a weighted voting scheme and shows that the weighted k-NN outperforms the majority voting based k-NN. This weighting is linear, and the weights of the neighbours are linearly mapped between [0, 1]. Using this scheme, the probability of a given sample x belongs to the class c can be defined as given below:

$$P(c) = \frac{\sum_{k \in C} \left(\frac{d^K - d^k}{d^K - d^1}\right)}{\sum_k \left(\frac{d^K - d^k}{d^K - d^1}\right)}, \tag{5}$$

where d^k is the distance between the given sample x and the kth nearest neighbour y^k, i.e., $d^k = d(x, y^k)$ and C is the set of samples which belong to class c. The nominator is the sum of weighted votes from the neighbors which belong to the class c and it is divided by the denominator, which is the sum of all weighted votes. To simplify the equation further we can get rid of the denominators in both summations assuming that $d^K \neq d^1$.

$$P(c) = \frac{\sum_{k \in C} d^K - d^k}{\sum_k d^K - d^k}, \tag{6}$$

As there are 59 times more negative samples than positive ones, one can expect that it is highly probable to find a negative sample within the neighbourhood of a positive one. In order to handle this situation, we propose also to weight the votes of samples according to their class labels. Samples belonging to the positive class should have higher weight as they are outnumbered by the negative samples. This is formulated as:

$$P(c) = \frac{\sum_{k \in C} \left((1 + c^k \theta)(d^K - d^k)\right)}{\sum_k \left((1 + c^k \theta)(d^K - d^k)\right)}, \tag{7}$$

where $c^k \in \{0, 1\}$ is the label of the neighbour y^k. θ is the voting weight for the positive class.

Finally, the contribution of each attribute can also be weighted. In the proposed approach, we define a weighted distance metric and aim to learn the optimal parameters for the corresponding weights. This is formulated as:

$$P(c) = \frac{\sum_{k \in C}^{K} \left((1 + c^k \theta) \sum_i^L w_i^2 (d_i^K - d_i^k)\right)}{\sum_k^K \left((1 + c^k \theta) \sum_i^L w_i^2 (d_i^K - d_i^k)\right)}, \tag{8}$$

where w_i^2 is the square of the weight for the ith distance. We use the square term to make sure that the weight is greater than zero. The decision of a k-NN classifier for a given sample x will be the class $\dot{c} = \text{argmax}_c P(c)$. In other words if $P(\dot{c}) > 0.5$ the sample x is classified as $\dot{c}$. Given that a sample x belongs to class $\dot{c}$, the probability that it is misclassified is $1 - P(\dot{c})$. Using this, we can calculate the total cost of classification given in (1) as given below:

$$Cost = \sum_n^N (10 + 490c^n)(1 - \varphi(P(c^n))), \tag{9}$$

where c^n is the class label corresponding to the nth sample, N is the total number of samples and φ is the decision function, i.e., $\varphi(x) = \begin{cases} 1 & x > 0.5 \\ 0 & o.w \end{cases}$. We can search for the optimal parameters w_i^2 and θ to minimize this cost, using the stochastic gradient descent approach. In order to make the function fully derivable, we select φ as a sigmoid, as given below, where β is used to adjust the steepness level of the sigmoid. (we select $\beta = 25$)

$$\varphi(x) = \frac{1}{1 + e^{-(x-0.5)\beta}}, \tag{10}$$

4 Experiments

We evaluate the performance of the proposed method on 5 different tests and perform 5-fold cross-validation. We compare our results with traditional classifiers such as SVM, Random Forest and AdaBoost as baselines [5]. We set the class weights 1 and 60. The cost value obtained for each test is presented in Table 1.

Table 1. Classification cost per subset

Method	Subset 1	Subset 2	Subset 3	Subset 4	Subset 5	Average
SVM	**7900**	11640	10640	14060	**11940**	11236
RF	35100	43960	43100	26180	32350	36264
Adaboost	43700	50190	54320	45260	44780	47650
Ours	9120	**9850**	**9380**	**13640**	11970	**10792**

5 Conclusion

In this paper, we propose a solution for the class prediction problem on imbalanced datasets with missing data. We first apply a k-NN based imputation algorithm to estimate the missing attributes. Then we define a tailored cost function for the given problem and minimize it using a stochastic gradient descent based approach, in order to create a k-NN classifier with a learned distance metric. Our results are shown to outperform the baseline classification methods.

References

1. García-Laencina, P.J., Sancho-Gómez, J.-L., Figueiras-Vidal, A.R.: Pattern classification with missing data: a review. Neural Comput. Appl. 19(2), 263–282 (2009)
2. Batista, G., Monard, M.C.: A study of k-nearest neighbour as an imputation method. Hybrid Intell. Syst. 87(48), 251–260 (2002)

3. Wu, X., Kumar, V., Ross, Q.J., Ghosh, J., Yang, Q., Motoda, H., McLachlan, G.J., Ng, A., Liu, B., Yu, P.S., Zhou, Z.-H., Steinbach, M., Hand, D.J., Steinberg, D.: Top 10 algorithms in data mining. Knowl. Inf. Syst. **14**(1), 1–37 (2008)
4. Dudani, S.A.: The distance-weighted k-nearest-neighbor rule. IEEE Trans. Syst. Man Cybern. **SMC-6**(4), 325–327 (1976)
5. Pedregosa, F., Grisel, O., Weiss, R., Passos, A., Brucher, M.: Scikit-learn: machine learning in python. J. Mach. Learn. Res. **12**(1), 2825–2830 (2011)

Combining Boosted Trees with Metafeature Engineering for Predictive Maintenance

Vítor Cerqueira[(⊠)], Fábio Pinto, Claudio Sá, and Carlos Soares

INESC TEC, Universidade do Porto,
Rua Dr. Roberto Frias, s/n, 4200-465 Porto, Portugal
{vmac,claudio.r.sa}@inesctec.pt, fhpinto@inescporto.pt, csoares@fe.up.pt

Abstract. We describe a data mining workflow for predictive maintenance of the Air Pressure System in heavy trucks. Our approach is composed by four steps: (i) a filter that excludes a subset of features and examples based on the number of missing values (ii) a metafeatures engineering procedure used to create a meta-level features set with the goal of increasing the information on the original data; (iii) a biased sampling method to deal with the class imbalance problem; and (iv) boosted trees to learn the target concept. Results show that the metafeatures engineering and the biased sampling method are critical for improving the performance of the classifier.

Keywords: Predictive maintenance · Anomaly detection · Boosting · Metalearning

1 Introduction

This paper describes a data mining workflow for predictive maintenance of heavy trucks. This type of vehicles are typically operated in a daily basis and used for large trips. They are an important tool in several industrial sectors such as transportation or construction. In this context, it is of fundamental importance that all components comprising these vehicles are regularly maintained. A well done maintenance is key to avoid undesired breakdowns, which can be costly to the company operating these vehicles.

One of those components is the Air Pressure System (APS). The APS generates pressurised air that is used for different tasks in a truck, such as braking and gear changing, making it a core component for maintenance purposes.

The data collected describes several components from heavy Scania trucks in everyday usage. Moreover, in order to guarantee the quality of the predictive model, the data has been sampled from all available data by experts.

In the Data Mining terminology this problem is presented as a binary classification problem, where the positive class of the target concept consists of failures for a specific component of the APS. The negative class consists of trucks with failures not related to the APS. The exploratory analysis of the data enabled to outlined two important conditions: (1) high quantity of missing values and

© Springer International Publishing AG 2016
H. Boström et al. (Eds.): IDA 2016, LNCS 9897, pp. 393–397, 2016.
DOI: 10.1007/978-3-319-46349-0_35

(2) high imbalance in the class distribution. These characteristics of the data raises challenges from the data scientist perspective.

For dealing with (1), we use a filter that excludes the features and examples that present an higher percentage of missing values. This not only reduced the size of the data but also enabled to remove some of the noisy features that were part of the original data.

For dealing with (2), we use SMOTE [1], an over-sampling technique that creates synthetic examples of the minority class. SMOTE enables to balance the class distribution of the data which leads to a better generalization of the classifier.

On top of these issues, the data is completely anonymized for proprietary reasons. This is particularly cumbersome in the feature engineering step. We choose to generate new features by taking a metafeature engineering approach that does not require knowledge about the domain. Most of these metafeatures were generated using unsupervised techniques for anomaly detection problems.

As for the modeling step, we use an ensemble of boosted trees. Ensemble learning has shown to be a good solution in variety of data mining tasks and we also verified a great improvement in the performance of our system by using ensemble approaches. The workflow is summarised in Fig. 1.

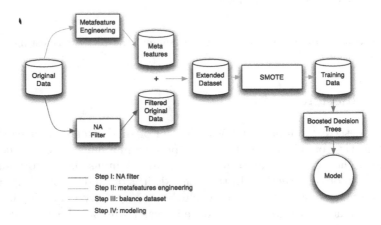

Fig. 1. Schema of our workflow. It starts by dealing with the missing values (Step I). In Step II the metafeatures are computed, which are then added to the original, after applying the missing values filter. Then, in Step III, we use SMOTE to balance the dataset and finally in Step IV we generate our ensemble of boosted decision trees.

The paper is structured as follows. In Sect. 2, we present our approach to deal with the missing values in the data. Section 3 explains the metafeature engineering step that we followed. In Sect. 4 we detail the modeling approach that we used to learn our final model and finally, in Sect. 5, we conclude the paper by discussing the results obtained and the lessons learned from this challenge.

2   Dealing with Missing Values

Dealing with missing values has been widely studied in the literature [2]. The usual techniques, such as listwise deletion, pairwise deletion, indicator variable, and mean substitution could have been an option. However, for simplicity and in this particular case, we decided to remove features with the greater amounts of missing values. It is rather an ad-hoc approach, but there is also no clear consensus in the literature on which approach is better [2].

Some features presented a great number of missing values, up to 80 % in the most extreme case. As an indicator, 8 out of the 170 independent variables, had more than 50 % of missing values.

While paying special attention to the performance of the models, we tested how much features we could remove without affecting the accuracy. We also realized that after removing the features with the most missing values, there was quite a number of duplicates in the data. This seems to indicate that the removed features have little effect on the target.

3   Metafeature Engineering

From the statistical point of view, anomalies are associated with observations with high deviation from the typical behaviour, i.e., outliers. Since the positive class of the data is characterized by rare events in the domain (malfunctions in the APS), we can regard this problem as an anomaly detection one.

In this context, we perform an outlier analysis to the data in order to assess how far each observation is from the norm. The results from this outlier analysis are then embedded as attributes in our data. By doing this meta-level analysis we aim at increasing the information related to the outlyingness of each observation. Therefore, we generated metafeatures using three different outlier detection techniques that we present in the following subsections.

3.1   Boxplot Analysis

The Box plot is a typical way of describing the distribution of some data through some summary statistics (e.g. median, Inter-Quartile Range (IQR)).

Since all the attributes of our data are numeric, we can perform this analysis to each predictor separately. For each attribute we compare each value with the typical value of that same attribute. Then, as explained in [3], if the difference between these two values is high that might be an indicator that something is not right and the respective observation might be an outlier. Furthermore, the size of this difference can be regarded as a measure of outlyingness.

One drawback of this analysis is that it is a uni-variate approach. By analyzing only one attribute at a time, we lose potentially useful global information. To overcome this issue we approach the problem using the Local Outlier Factor (LOF) method, which measures the degree of outlyingness of observations as a whole.

3.2   LOF

LOF [4] is a method for quantifying the outlyingness of an observation by comparing it to its local neighbourhood through density estimation. Essentially, an observation with a very low density has greater probability of being an outlier.

3.3   Clustering-Based Outlier Ranking

The third method we used to analyze the outliers in the dataset is based on an hierarchical Agglomerative Clustering algorithm [5].

Hierarchical Agglomerative Clustering starts with Z groups (Z being the number of observations), each initially containing one object, and then at each step it merges the two most similar groups until there is only one single group, containing all data.

The rationale for this method is that the last observation that are merged might still be significantly different from the group they are merged into. By definition outliers are different cases and will typically not fit well into a cluster, unless that cluster is comprised by other outliers itself. Yet again, since these are not ordinary data points, we do not expect them to form large groups.

4   Modeling

In this section we detail the modeling approach that we used for this challenge.

The model was generated using the XGBoost library [6]. The parameter tuning of the learning algorithm was done using 10-fold cross validation. We payed particular attention to parameter setting in order to avoid overfitting. Given the experimental results that we gathered, overfitting can be a pitfall in this challenge.

The performance of the modeling algorithms should be measured according to a cost sensitive metric defined by the challenge organizes. The intuition for this is because the two error types (false negative - FN and false positive - FP) do not have the same meaning. Sending a vehicle for an unnecessary maintenance (FP) is clearly less costly than facing an unexpected breakdown (FN). For this reason the evaluation metric is computed as follows: $\text{Cost} = \text{FP} \times 10 + \text{FN} \times 500$.

5   Conclusions

Table 1 presents the results for three workflows that we tested. We concluded from the several experiments we carried out that XGBoost seems to be a good option for this problem, particularly when the metafeatures are available for learning. However, we did encounter some issues regarding the tuning of the algorithm in terms of overfitting. For this particular data, the algorithm showed high sensitivity regarding its parameters.

Overall, we think that SMOTE and the metafeature engineering are the most important steps in our proposal.

Table 1. Results estimated using 10-fold cross validation for four methods. A Random Forest with and without metafeatures and the XGBoost algorithm with and without metafeatures. The XGBoost with metafeatures shows the minimum average cost and with the lowest deviance.

Algorithm	Average cost	SD cost
RF without metafeatures	4721	882
RF with metafeatures	4440	900
XGBoost without metafeatures	4030	910
XGBoost with metafeatures	3750	810

Acknowledgements. This research has received funding from the ECSEL Joint Undertaking, the framework programme for research and innovation horizon 2020 (2014–2020) under grant agreement n 662189-MANTIS-2014-1. It was also financed by the ERDF European Regional Development Fund through the Operational Programme for Competitiveness and Internationalisation - COMPETE 2020 Programme within project POCI-01-0145-FEDER-006961, and by National Funds through the FCT Fundação para a Ciência e a Tecnologia (Portuguese Foundation for Science and Technology) as part of project UID/EEA/50014/2013.

References

1. Chawla, N.V., Bowyer, K.W., Hall, L.O., Kegelmeyer, W.P.: Smote: synthetic minority over-sampling technique. J. Artif. Intell. Res. **16**, 321–357 (2002)
2. Acock, A.C.: Working with missing values. J. Marriage Fam **67**(4), 1012–1028 (2005)
3. Torgo, L.: Data Mining with R: Learning with Case Studies, 1st edn. Chapman & Hall/CRC, Boca Raton (2010)
4. Breunig, M.M., Kriegel, H.P., Ng, R.T., Sander, J.: LOF: identifying density-based local outliers. In: ACM SIGMOD International Conference on Management of Data, pp. 93–104 (2000)
5. Torgo, L.: Resource-bounded fraud detection. In: Neves, J., Santos, M.F., Machado, J.M. (eds.) EPIA 2007. LNCS, vol. 4874, pp. 449–460. Springer, Heidelberg (2007). doi:10.1007/978-3-540-77002-2_38
6. Chen, T., Guestrin, C.: XGBoost: a scalable tree boosting system. arXiv:1603.02754 (2016)

Prediction of Failures in the Air Pressure System of Scania Trucks Using a Random Forest and Feature Engineering

Christopher Gondek, Daniel Hafner$^{(\boxtimes)}$, and Oliver R. Sampson

University of Konstanz, Konstanz, Germany
{christopher.gondek,daniel.hafner,oliver.sampson}@uni-konstanz.de
https://www.uni-konstanz.de

Abstract. This paper demonstrates an approach in data analysis to minimize overall maintenance costs for the air pressure system of Scania trucks. Feature creation on histograms was used. Randomly chosen subsets of attributes were then evaluated to generate an order and a final subset of features. Finally, a RANDOM FOREST was applied and fine-tuned. The results clearly show that data analysis in the field is beneficial and improves upon the naive approaches of checking every truck or no truck until failure.

Keywords: Data mining · Feature extraction · Dimension reduction · Random forest

1 Introduction

Given a high dimensional dataset by this year's Industrial Challenge, we chose a combination of feature engineering and feature reduction whilst constantly evaluating the results using a RANDOM FOREST. [1] Our work is structured closely to the KDD process, the model we used throughout the challenge.

2 Project Understanding

The goal of the task, as presented by the Industrial Challenge for IDA 2016, was to minimize maintenance costs of the air pressure system (APS) of Scania trucks. Therefore, failures should be predicted before they occur. Falsely predicting a failure has a cost of 10, missing a failure a cost of 500. This leads to the need of cost minimization.

3 Data Understanding

The data given to us contains a training set and a test set. The training set contains 60,000 rows, of which 1,000 belong to the positive class and 171 columns,

© Springer International Publishing AG 2016
H. Boström et al. (Eds.): IDA 2016, LNCS 9897, pp. 398–402, 2016.
DOI: 10.1007/978-3-319-46349-0_36

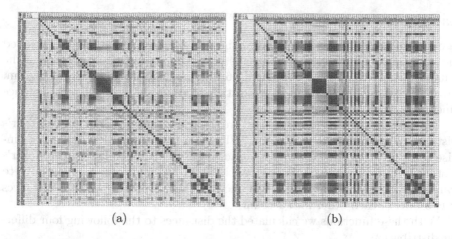

Fig. 1. Correlation matrices of the attributes for the (a) positive class and the (b) negative. A blue color means a positive correlation, red a negative one. The saturation indicates the strength. (Color figure online)

of which one is the class column. All attributes are numeric. 70 of these attributes belong to 7 histograms with ten bins each. Based on visual inspection we guessed, that the sum across each histogram indicates the age of the APS. Also, most failures could be predicted by using one or two features. It appeared that the hard part is to correctly predict failures for records that are actually very close to the non-failure class. Some visual inspection methods we used were:

- Box plots to get an overview of the variance of the values.
- Correlation matrices for identifying features that correlate. (see Fig. 1)
- Scatter plots to see how the classes are spread.
- Radar charts to recognize outliers.

4   Data Preperation

4.1   Data Cleaning

The dataset contains up to 82 % missing values per attribute. Furthermore, many of the attributes contain outliers. Therefore, we chose to replace the missing values by the median.

4.2   Normalization

After evaluating several classifiers including NAIVE BAYES, MULTILAYER PERCEPTRON, and SUPPORT VECTOR MACHINES, we determined that a RANDOM FOREST would perform best. Hence, a normalization was not necessary.

4.3 Feature Engineering

In our first models we only considered the sum of each histogram and excluded the single bins. This lead to good results but disregarded the distribution of the values. Therefore, we then calculated 16 different features for each histogram. All of these are distances to other distributions using two different distance functions.

The two distance functions we used are the χ^2-distance and the Earth Mover's Distance. The χ^2-distance, proposed by Pearson in the early 1900's [2], is a bin-wise comparison of the observed value to the expected one. The Earth Mover's Distance, introduced by Rubner et al. in 1998 [3], finds the cheapest way to transform one histogram into another one. For this purpose it takes the distances of two bins to each other into account.

With these functions we calculated the distances to the following four different distributions:

1. Mean distribution of the positive examples. It is calculated by filtering the data points with the positive class and computing the mean value for each bin. Based on these, the distance from the computed to the measured histogram is calculated.
2. Mean distribution of the negative examples. The calculation was done in a similar manner as the first one but considers the negative class instead of the positive.
3. Normal distribution with the parameters $\mu = 5$, $\sigma = 1.5$.
4. Mirrored normal distribution. It was achieved by mirroring the normal distribution along the x-axis and shifting it on the y-axis into the positive.

All the above mentioned distances are highly correlated to the sum of the bins. To resolve this dependency the histograms are normalized by their sum. With these additional histograms the same distances as before are calculated.

4.4 Feature Selection

The features given and calculated combined resulted in 282 dimensions, excluding the class column. As stated in Sect. 3, many of them were correlated and hence probably not needed. Consequently, feature selection was introduced. It was done in two steps: Ranking the features by their expressiveness and testing the performance of the feature sets varying in size.

Feature Expressiveness. Our approach to rank the dimensions according to their expressiveness was the following:

1. Take 200 random features out of the 282.
2. Learn a RANDOM FOREST and predict the class.
3. Store the precision of the results together with the features used.
4. Repeat steps 1–3 2,000 times.
5. Calculate the mean precision of each dimension and rank them in decreasing order.

Testing the Feature Sets. Using the ranked features from Sect. 4.4, we could compute the costs of the prediction model with a different number of dimensions.

This was done by training a RANDOM FOREST and predicting the class using a 10-fold cross-validation and calculating the average costs starting with the feature set containing only the most expressive feature, i.e., am_0. Afterwards, the set got expanded by the second best feature and the prediction was repeated. This was done until all dimensions were included.

The analysis showed that the average costs per data record as determined by the given cost function of only one dimension are about 3.1 and decrease rapidly. Between 10 to all 282 features, the costs fluctuates between 0.85 and 0.6. This led us to the conclusion that we do not need all dimensions and can reduce them.

5   Modeling

The RANDOM FOREST algorithm always tries to minimize the prediction error. It assumes that all wrong predicted classes are equally expensive. But that is not the case for the IDA Challenge. In fact, the cost of a false negative is 50 times higher than a false positive. We tried to overcome this problem by correcting the predicted class based on the confidence of our classifier. For that we slightly adjusted the procedure described in Sect. 4.4. For every feature subset we set a threshold for the prediction confidence and changed it in steps of one percent. Whenever the confidence was below or equal to the threshold, the predicted class was set to "pos."

An analysis of the results showed that in most cases the best threshold was 95 %. Using that, we got the costs that are shown in Fig. 2. With at least 10 dimensions the costs fluctuates between about 0.75 and 0.57. To get the best prediction possible, we used the global minimum with 210 features.

6   Evaluation

The naive approach to solve the challenge would be to label all records as negative which has a mean cost of $500 \cdot 1000/60000 = 8.33$ or to label all records as positive which results in a mean cost of 9.83. Therefore, our approach with average costs of around 0.6 is able to reduce the mean cost by the factor of 13.9.

To get reproducible results with the stratified sampling in the 10-fold cross-validation a fixed seed was used. Since this may lead to an adaption of the subsets of features to said seed, we did repeat the cross-validation with several others. Overall, the costs per truck stayed approximately the same. Therefore, we assumed that overfitting is not a major problem and the expected costs are in the neighborhood of around 0.6.

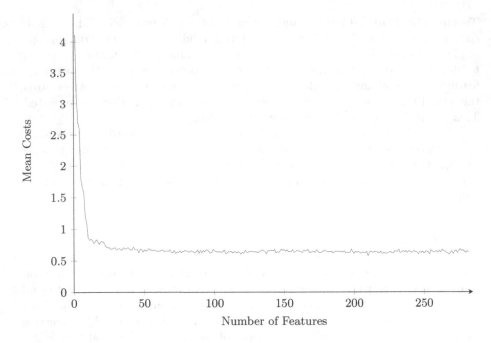

Fig. 2. Curve of the mean costs evaluated using a confidence threshold of 95 % and 10-fold cross-validation with a fixed seed.

7 Conclusion

An early detection of a failure in an Air Pressure System in trucks can save the company a lot money. The prediction of a fault can be performed even if the meaning of the measured values is unknown or only histograms are available. We demonstrated how meaningful features of histograms can be computed to improve the prediction. Also, we showed how the forecasts can be adapted to a cost function using a threshold on the confidence of a Random Forest. Finally, a significantly lowered main cost compared to the naive approaches was achieved.

References

1. Breiman, L.: Random forests. Mach. Learn. **45**(1), 5–32 (2001)
2. Pearson, K.: On the criterion that a given system of derivations from the probable in the case of a correlated system of variables is such that it can be reasonably supposed to have arisen from random sampling. Lond. Edinb. Dublin Philos. Mag. J. Sci. **50**(5), 157–175 (1900)
3. Rubner, Y., Tomasi, C., Guibas, L.J.: A metric for distributions with applications to image databases. In: Proceedings of the Sixth International Conference on Computer Vision, p. 59. IEEE Computer Society, Washington, DC (1998)

Author Index

Ah-Pine, Julien 320
Artières, Thierry 284
Avvenuti, Marco 86

Bacciu, Clara 86
Barhoumi, Walid 183
Battistelli, Delphine 192
Beel, Joeran 368
Ben Amor, Nahla 38
Ben Ishak, Mouna 38
Berthold, Michael R. 215
Bissyandé, Tegawendé F. 1
Boekhout, Hanjo D. 295
Bradley, Elizabeth 343
Braune, Christian 61
Brofos, James 146
Brunner-La Rocca, Hans-Peter 14
Buscaldi, Davide 237

Caroli, Frederico 332
Cats, Oded 98
Cerqueira, Vítor 393
Charnois, Thierry 192, 237
Contardo, Gabriella 284
Costa, Camila Ferreira 381
Costa, Gianni 110

da Silva, Josenildo C. 261
Dankel, Marco 61
Davarzani, Nasser 14
Del Vigna, Fabio 86
Deluca, Paolo 86
Denoyer, Ludovic 284
Dhahbi, Sami 183

Eberhard, Jörg 356

Freitas, André 332

Gabbouj, Moncef 387
Gábor, Kata 237
Gama, João 123
Garland, Joshua 343

Gipp, Bela 368
Gondek, Christopher 398
Grossniklaus, Michael 368

Hafner, Daniel 398
Handschuh, Siegfried 332
Hassan, Syed Murtaza 98
Holat, Pierre 192
Höppner, Frank 73

James, Ryan G. 343
Jaulent, Marie-Christine 192
Jones, Tyler R. 343

Kamioka, Eduardo Haruo 332
Karel, Joël 14
Khiari, Jihed 98
Kiranyaz, Serkan 387
Klawonn, Frank 356
Klein, Jacques 1
Klusch, Matthias 261
Koch, Ina 356
Kruse, Rudolf 61

Labiod, Lazhar 273
Le Traon, Yves 1
Leray, Philippe 38
Li, Daoyuan 1
Lodi, Stefano 261

Marchetti, Andrea 86
Métivier, Jean-Philippe 192
Moreira-Matias, Luis 98

Nadif, Mohamed 273
Nascimento, Mario A. 381
Neme, Antonio 226
Neme, Omar 226
Noordegraaf-Eelens, Liesbeth 26

O'Donoghue, Jim 134
Omar, Mohamed 356
Ortale, Riccardo 110
Ozan, Ezgi Can 387

Peeters, Ralf   14
Petrocchi, Marinella   86
Pinto, Fábio   393
Plaat, Aske   50
Popelínský, Luboš   308
Post, Martijn J.   158

Riabchenko, Ekaterina   387
Roantree, Mark   134
Ryšavý, Petr   204

Sá, Claudio   393
Saleiro, Pedro   171
Sampson, Oliver R.   215, 398
Shu, Rui   146
Smirnov, Evgueni   14
Soares, Carlos   171, 393
Sobek, Tobias   73
Soekhoe, Deepak   50
Sousa, Ricardo   123

Takes, Frank W.   295
Teernstra, Livia   26
Tellier, Isabelle   237
Tesconi, Maurizio   86
Tomeh, Nadi   192

Vaculík, Karel   308
van der Putten, Peter   26, 50, 158
van Engelen, Jesper E.   295
van Rijn, Jan N.   158
Verbeek, Fons   26

Wang, Junxi   356
Wang, Xinyu   320
Weiler, Andreas   368
Weston, David J.   249
White, James W.C.   343

Zagrouba, Ezzeddine   183
Zargayouna, Haïfa   237
Železný, Filip   204
Zhang, Frank   146